Lecture Notes in Mathematics

Edited by J.-M. Morel, F. Takens and B. Teissier

Editorial Policy
for the publication of monographs

1. Lecture Notes aim to report new developments in all areas of mathematics and their applications- quickly, informally and at a high level. Mathematical texts analysing new developments in modelling and numerical simulation are welcome.

   Monograph manuscripts should be reasonably self-contained and rounded off. Thus they may, and often will, present not only results of the author but also related work by other people. They may be based on specialised lecture courses. Furthermore, the manuscripts should provide sufficient motivation, examples and applications. This clearly distinguishes Lecture Notes from journal articles or technical reports which normally are very concise. Articles intended for a journal but too long to be accepted by most journals, usually do not have this "lecture notes" character. For similar reasons it is unusual for doctoral theses to be accepted for the Lecture Notes series, though habilitation theses may be appropriate.

2. Manuscripts should be submitted (preferably in duplicate) either to Springer's mathematics editorial in Heidelberg, or to one of the series editors (with a copy to Springer). In general, manuscripts will be sent out to 2 external referees for evaluation. If a decision cannot yet be reached on the basis of the first 2 reports, further referees may be contacted: The author will be informed of this. A final decision to publish can be made only on the basis of the complete manuscript, however a refereeing process leading to a preliminary decision can be based on a pre-final or incomplete manuscript. The strict minimum amount of material that will be considered should include a detailed outline describing the planned contents of each chapter, a bibliography and several sample chapters.

   Authors should be aware that incomplete or insufficiently close to final manuscripts almost always result in longer refereeing times and nevertheless unclear referees' recommendations, making further refereeing of a final draft necessary.

   Authors should also be aware that parallel submission of their manuscript to another publisher while under consideration for LNM will in general lead to immediate rejection.

3. Manuscripts should in general be submitted in English. Final manuscripts should contain at least 100 pages of mathematical text and should always include

   - a table of contents;
   - an informative introduction, with adequate motivation and perhaps some historical remarks: it should be accessible to a reader not intimately familiar with the topic treated;
   - a subject index: as a rule this is genuinely helpful for the reader.

Continued on inside back-cover

Lecture Notes in Mathematics 1865

Editors:
J.-M. Morel, Cachan
F. Takens, Groningen
B. Teissier, Paris

David Applebaum   B.V. Rajarama Bhat
Johan Kustermans   J. Martin Lindsay

Quantum Independent Increment Processes I

From Classical Probability to Quantum Stochastic Calculus

Editors:

Michael Schürmann
Uwe Franz

 Springer

Editors and Authors

David Applebaum
Probability and Statistics Department
University of Sheffield
Hicks Building
Hounsfield Road
Sheffield, S3 7RH, UK

E-mail: *D.Applebaum@sheffield.ac.uk*

B.V. Rajarama Bhat
Indian Statistical Institute
8th Mile, Mysore Road
Bangalore 560 059, India

E-mail: *bhat@isibang.ac.in*

Uwe Franz
Inst. Mathematik und Informatik
Universität Greifswald
Friedrich-Ludwig-Jahn-Str. 15a
17487 Greifswald, Germany

E-mail: *franz@uni-greifswald.de*

Johan Kustermans
KU Leuven
Department Wiskunde
Celestijnenlaan 200B
3001 Heverlee, Belgium

E-mail: *johan.kustermans@wis.kuleuven.ac.be*

J. Martin Lindsay
School of Mathematical Sciences
University of Nottingham
University Park
Nottingham, NG7 2RD, UK

E-mail: *martin.lindsay@nottingham.ac.uk*

Michael Schürmann
Inst. Mathematik und Informatik
Universität Greifswald
Friedrich-Ludwig-Jahn-Str. 15a
17487 Greifswald, Germany

E-mail: *schurman@uni-greifswald.de*

Library of Congress Control Number: 2004117328

Mathematics Subject Classification (2000): 60G51, 81S25, 46L60, 58B32, 47A20, 16W30

ISSN 0075-8434
ISBN 3-540-24406-9 Springer-Verlag Berlin Heidelberg New York

DOI 10.1007/b105131

Springer-Verlag is a part of Springer Science + Business Media

springeronline.com

Typesetting: Camera-ready TeX output by the authors

41/3142/du - 543210 - Printed on acid-free paper

Dedicated to the memory of Paul-André Meyer

Preface

This volume is the first of two volumes containing the lectures given at the School "Quantum Independent Increment Processes: Structure and Applications to Physics". This school was held at the Alfried Krupp Wissenschafts}-kolleg in Greifswald during the period March 9 – 22, 2003. We thank the lecturers for all the hard work they accomplished. Their lectures give an introduction to current research in their domains that is accessible to Ph. D. students. We hope that the two volumes will help to bring researchers from the areas of classical and quantum probability, operator algebras and mathematical physics together and contribute to developing the subject of quantum independent increment processes.

We are greatly indebted to the Volkswagen Foundation for their financial support, without which the school would not have been possible.

Special thanks go to Mrs. Zeidler who helped with the preparation and organisation of the school and who took care of the logistics.

Finally, we would like to thank the students for coming to Greifswald and helping to make the school a success.

Greifswald,
February 2005

Michael Schürmann
Uwe Franz

Contents

Contents of Volume II
Structure of Quantum Lévy Processes,
Classical Probability and Physics

Lévy Processes on Quantum Groups and Dual Groups

List of Contributors

David Applebaum
Probability and Statistics Dept.
University of Sheffield
Hicks Building
Hounsfield Road
Sheffield, S3 7RH, UK
D.Applebaum@sheffield.ac.uk

Ole E. Barndorff-Nielsen
Dept. of Mathematical Sciences
University of Aarhus
Ny Munkegade
DK-8000 Århus, Denmark
oebn@imf.au.dk

B. V. Rajarama Bhat
Indian Statistical Institute
Bangalore, India
bhat@isibang.ac.in

Uwe Franz
Universität Greifswald
Friedrich-Ludwig-Jahnstrasse 15 A
D-17487 Greifswald, Germany
franz@uni-greifswald.de

Rolf Gohm
Universität Greifswald
Friedrich-Ludwig-Jahnstrasse 15 A
D-17487 Greifswald, Germany
gohm@uni-greifswald.de

Burkhard Kümmerer
Fachbereich Mathematik
Technische Universität Darmstadt
Schloßgartenstraße 7
64289 Darmstadt, Germany
kuemmerer@mathematik.
tu-darmstadt.de

Johan Kustermans
KU Leuven
Departement Wiskunde
Celestijnenlaan 200B
3001 Heverlee, Belgium
johan.kustermans@wis.kuleuven.
ac.be

J. Martin Lindsay
School of Mathematical Sciences
University of Nottingham
University Park
Nottingham, NG7 2RD, UK
martin.lindsay@nottingham.ac.
uk

Steen Thorbjørnsen
Dept. of Mathematics & Computer
Science
University of Southern Denmark
Campusvej 55
DK-5230 Odense, Denmark
steenth@imada.sdu.dk

Introduction

Random variables and stochastic processes are used to describe the behaviour of systems in a vast range of areas including statistics, finance, actuarial mathematics and computer science, as well as engineering, biology and physics. Due to an unavoidable lack of information about the state of the system concerned at a given moment in time, it is often impossible to predict these fluctuations with certainty — think of meteorology, for example. The unpredictable behaviour may be due to more fundamental reasons, as is the case in quantum mechanics. Here Heisenberg uncertainty limits the accuracy of simultaneous predictions of so-called complementary observables such as the position and momentum of a particle.

If the random fluctuations do not depend on time or position, then they should be described by stochastic processes which are homogeneous in space and time. In Euclidean space this leads to the important class of stochastic processes called Lévy processes, which have independent and stationary increments ([Lév65]). These processes have been attracting increasing interest over the last decade or so (see [Sko91], [Ber96], [Sat99], [BNMR01] and [App04]).

In quantum mechanics complete knowledge of the state is still insufficient to predict with certainty the outcomes of all possible measurements. Therefore its statistical interpretation has to be an essential part of the theory. Quantum probability starts from von Neumann's formulation of quantum mechanics ([vN96]) and studies quantum theory from a probabilistic point of view. Two key papers in the field are [AFL82] and [HP84].

A typical situation where quantum noise plays a role is in the description of a 'small' quantum system interacting with its 'large' environment. The state of the environment, also called heat bath or reservoir, cannot be measured or controlled completely. However it is reasonable to assume, at least as a first approximation, that it is homogeneous in time and space, and that the influence of the system on the heat bath can be neglected.

In concrete models the heat bath is generally described by a Fock space. The Hilbert space for the joint 'system plus heat bath' is then the tensor product of the Hilbert space representing the system with this Fock space. The separate time evolutions of the heat bath and system are coupled through their interaction to yield a unitary evolution of the system plus heat bath which is a cocycle with respect to the free evolution of the heat bath. Thus, through interaction (in other words, considered as an open system), the evolution of the system becomes non-unitary. In the Heisenberg picture this is given by a quantum dynamical semigroup, that is a one-parameter semigroup of completely positive maps (rather than *-automorphisms) on the system observables, see *Quantum Markov processes and applications to physics*, by Burkhard Kümmerer, in volume two of these notes. In the physics literature the dual Schrödinger picture is usually preferred; this is adopted in the influential monograph [Dav76].

Fock spaces arose in quantum field theory and in representation theory as continuous tensor products. The close connection between independent increment processes on the one hand, and current representations and Fock space on the other, was realised in the late sixties and early seventies ([Ara70], [PS72] and [Gui72], see also the survey article [Str00]). The development of a quantum stochastic calculus was a natural sequel to this discovery. This calculus involves the integration of operator 'processes', that is time-indexed families of operators adapted to a Fock-space filtration, with respect to the so-called creation, preservation and annihilation processes. It is modelled on the Itô integral, but in fact may be based on the nonadapted stochastic calculus of Hitsuda and Skorohod, see part three of this volume, *Quantum stochastic analysis — an introduction*, by Martin Lindsay. The relationship between classical and quantum stochastic calculus is also the subject of the final lecture of part one, *Lévy processes in Euclidean spaces and groups*, by David Applebaum.

Part four of this volume, *Dilations, cocycles and product systems* by Rajarama Bhat, concerns the relation between the unitary evolution of the *closed* system plus heat bath and the quantum dynamical semigroup which is the evolution of the *open* system itself. It addresses the question of which unitary evolutions correspond to a given quantum dynamical semigroup.

Formally, quantum groups arise from groups in a similar way to how quantum probability arises from classical probability, and to how C^*-algebra theory is now commonly viewed as noncommutative topology. Namely, one casts the axioms for a group (or probability space, or topological space) in terms of the appropriate class of functions on the group (respectively, probability, or topological space). This yields a commutative algebra with extra structure, and the quantum object is then defined by dropping the commutativity axiom. This procedure has been successfully applied to differential geometry ([Con94]).

For example taking the algebra of representative functions on a group (i.e. those functions which can be written as matrix elements of a finite-dimensional representation of the group), one obtains the axioms of a commutative Hopf algebra ([Swe69]). Dropping commutativity, one arrives at one definition of a Hopf algebra. At least in finite dimension, the Hopf algebra axioms give a satisfactory definition of a (finite) quantum group.

Similarly the essentially bounded measurable functions on a probability space, with functions equal almost everywhere identified, form a commutative von Neumann algebra on which the expectation functional yields a state which is faithful and normal. Conversely, every commutative von Neumann algebra with faithful normal state is isomorphic to such an algebra of (measure equivalence classes of) random variables on a probability space with state given by the expectation functional.

Thus the axioms depend on the choice of functions. For example *all* functions on a group form a Hopf algebra only if the group is finite. The guiding principle for finding the 'right' set of axioms is that it should yield a rich theory which incorporates a good measure of the classical theory. In the case of quantum probability there is a straightforward choice. A unital *-algebra with a state is called an algebraic noncommutative probability space, and simply a noncommutative probability space when the algebra is a von Neumann algebra and the state is normal. In the latter case the state is often, but not always, assumed to be faithful. In fact recent progress in the understanding of noncommutative stochastic independence has benefitted from a loosening of the axioms to allow noninvolutive algebras, see *Lévy processes on quantum groups and dual groups*, by Uwe Franz in volume two of these notes.

In what is now known as topological quantum group theory, the search for the 'right' foundations has a long history. Only recently have Kustermans and Vaes obtained a relatively simple set of axioms that is both rich enough to contain all the examples one would want to consider as quantum groups whilst still having a satisfactory duality theory, see part two of this volume, *Locally compact quantum groups*, by Johan Kustermans.

References

[AFL82] L. Accardi, A. Frigerio, and J.T. Lewis. Quantum stochastic processes. *Publ. RIMS*, 18:97–133, 1982.

[App04] D. Applebaum. *Lévy Processes and Stochastic Calculus*. Cambridge University Press, Cambridge, 2004.

[Ara70] H. Araki. Factorizable representations of current algebra. *Publ. RIMS Kyoto University*, 5:361–422, 1970.

[Ber96] J. Bertoin. *Lévy Processes*. Cambridge University Press, Cambridge, 1996.

[BNMR01] O. E. Barndorff-Nielsen, T. Mikosch, and S. I. Resnick, editors. *Lévy Processes*. Birkhäuser Boston Inc., Boston, MA, 2001. Theory and applications.

[Con94] A. Connes. *Noncommutative Geometry*. Academic Press, San Diego, 1994.

[Dav76] E. B. Davies. *Quantum Theory of open Systems*. Academic Press, London, 1976.

[Gui72] A. Guichardet. *Symmetric Hilbert spaces and Related Topics*, volume 261 of *Lecture Notes in Math*. Springer-Verlag, Berlin, 1972.

[HP84] R. L. Hudson and K. R. Parthasarathy. Quantum Ito's formula and stochastic evolutions. *Comm. Math. Phys.*, 93(3):301–323, 1984.

[Lév65] Paul Lévy. *Processus Stochastiques et Mouvement Brownien*. Gauthier-Villars & Cie, Paris, 1965.

[PS72] K.R. Parthasarathy and K. Schmidt. *Positive Definite Kernels, Continuous Tensor Products, and Central Limit Theorems of Probability Theory*, volume 272 of *Lecture Notes in Math.* Springer-Verlag, Berlin, 1972.

[Sat99] Ken-iti Sato. *Lévy processes and Infinitely Divisible Distributions.* Cambridge University Press, Cambridge, 1999. Translated from the 1990 Japanese original, Revised by the author.

[Sko91] A. V. Skorohod. *Random Processes with Independent Increments.* Kluwer Academic Publishers Group, Dordrecht, 1991. Translated from the second Russian edition by P. V. Malyshev.

[Str00] R. F. Streater. Classical and quantum probability. *J. Math. Phys.*, 41(6):3556–3603, 2000.

[Swe69] M. E. Sweedler. *Hopf Algebras.* Benjamin, New York, 1969.

[vN96] J. von Neumann. *Mathematical foundations of quantum mechanics.* Princeton Landmarks in Mathematics. Princeton University Press, Princeton, 1996. Translated from the German, with preface by R.T. Beyer.

Lévy Processes in Euclidean Spaces and Groups

David Applebaum

Probability and Statistics Department
University of Sheffield
Hicks Building
Hounsfield Road
Sheffield, S3 7RH, UK
D.Applebaum@sheffield.ac.uk

1 Introduction

"Probability theory has always generated its problems by its contact with other areas. There are very few problems that are generated by its own internal structure. This is partly because, once stripped of everything else, a probability space is essentially the unit interval with Lebesgue measure."

S.R.S. Varadhan, AMS Bulletin January (2003)

One of the most beautiful and fruitful ideas in probability theory is that of *infinite divisibility*. For a random variable to be infinitely divisible, we require that it can be decomposed as the sum of n independent, identically distributed random variables, for any natural number n. Many distributions of importance for both pure and applied probability have been shown to be infinitely divisible and some of the best known in a very long list are the normal law, the Poisson and compound Poisson laws, the t-distribution, the χ^2 distribution, the log-normal distribution, the stable laws, the normal inverse Gaussian and the hyperbolic distributions. The basic ideas of infinite divisibility chrystallised during the heroic age of classical probability in the 1920s and 1930s - the key result is the beautiful Lévy-Khintchine formula which gives the general form of the characteristic function for an infinitely divisible probability distribution. Another important discovery from this era is that such distributions are precisely those which arise as limit laws for row sums of asymptotically negligible triangular arrays of independent random variables. Gnedenko and Kolmogorov [40] is a classic text for these results - for a more modern viewpoint, see Jacod and Shiryaev [51].

When we pass from single random variables to stochastic processes, the analogue of infinite divisibility is the requirement that the process has stationary and independent increments. Such processes were first investigated systematically by Paul Lévy (see e.g. Chapter 5 of [56]) and now bear his name in honour of his groundbreaking contributions.

Many important stochastic processes are Lévy processes - these include Brownian motion, Poisson and compound Poisson processes, stable processes and subordinators. Note that any infinitely divisible probability distribution can be embedded as the law of $X(1)$ in some Lévy process $(X(t), t \geq 0)$. A key structural result, which gives great insight into sample path behaviour, is the Lévy-Itô decomposition which asserts that any Lévy process can be decomposed as the sum of four terms - a deterministic (drift) which increases linearly with time, a diffusion term which is controlled by Brownian motion, a compensated sum of small jumps and a (finite) sum of large jumps. In particular, this shows that Lévy processes are a natural subclass of semimartingales with jumps (see e.g. [66], [51]).

Lévy processes are also Markov (in fact Feller) processes and their infinitesimal generators are represented as integral perturbations of a second order elliptic differential operator, in a structure which mirrors the Lévy-Khintchine form. Alternatively, the generator is represented as a pseudo-differential operator with a symbol determined by the Lévy-Khintchine formula. This latter structure is paradigmatic of a wide class of Feller processes, wherein the symbol has the same form but an additional spatial dependence. This is a major theme of Niels Jacob's books ([48, 49, 50]).

The last decade has seen Lévy processes come to the forefront of activity in probability theory and there have been several major developments from both theoretical and applied perspectives. These include fluctuation theory ([19]), codification of the genealogical structure of continuous branching processes ([55]), investigations of turbulence via Burger's equation ([20]), the study of stochastic differential equations with jumps and associated stochastic flows [54], construction of Euclidean random fields [2], properties of linearly viscoelastic materials [23], new examples of times series [24] and a host of applications to option pricing in "incomplete" financial markets (see e.g. [75], chapter 5 of [13], and references therein). In addition, two important monographs have appeared which are devoted to the subject ([19], [74]) and a third is to appear shortly ([13]). Since 1998, conferences to review and discuss new developments have taken place on an annual basis - the proceedings of the first of these are collected in [15].

The first four sections of these notes aim to give an overview of the key structural properties of Lévy processes taking values in Euclidean space, and of the associated stochastic calculus. They are based very closely on parts of

Chapters 1 to 4 of [13], but except in a few vital instances, the detailed proofs have been omitted. Section 1 introduces the concepts of infinite divisibility and Lévy process and presents the vital Lévy-Khintchine formula, section 2 introduces important concepts such as martingale and stopping times and concludes with the celebrated Lévy-Itô decomposition. In section 3, we describe the representations of the generator of the process while section 4 gives an account of stochastic integration, Itô's formula and stochastic differential equations.

The notions of infinite divisibility and Lévy process are sufficiently robust to allow extensive generalisation from the basic theory in Euclidean space. To see the Lévy-Khintchine formula in a Hilbert space setting, consult [64], while the Banach space version is in [59]. Section 5 herein describes group-valued Lévy processes and this can seen as the classical theory which underlines the notion of quantum group valued Lévy process, which is described in the lecture notes of Uwe Franz. We remark that the Lévy process concept also generalises to Riemannian manifolds [7], to hypergroups [21] and indeed to quantum hypergroups [36]. Another interesting generalisation, in the spirit of quantum probability, is the study of infinitely divisible completely positive mappings from a group to the algebra of bounded operators on a Hilbert space (see [32]).

Probability theory on groups describes the joyful interplay of the concepts of chance and symmetry. For Lévy processes, there are three different subcategories of topological groups where a good theory can be developed - the locally compact abelian groups (LCA groups), Lie groups and general locally compact groups. In the LCA case, the ability to define a Fourier transform means that many features of the theory are similar to the Euclidean space case - an account of the Lévy-Khintchine formula can again be found in [64]. The most extensively studied case is that of a Lie group. Mathematically, one of the joys of working on this topic is the interplay of the different techniques from semigroup theory, non-commutative harmonic analysis and stochastic calculus. The first work on this area was an outstanding paper by G.A.Hunt [45] which gave a Lévy-Khintchine style characterisation of the generator. Stochastic calculus techniques were introduced in [33], and more recently, [8]. Many new and interesting results can be found in the forthcoming monograph by Liao [57]. On more general locally compact groups, projective limit techniques arising from the solution of Hilbert's 5th problem enable us to gain insight into the structure of the generator, and this is described in H.Heyer's classic book [43]. More recent progress in this area can be found in [22] and [12].

Section 6 of these notes paves the way for Martin Lindsay's contribution to this volume, by indicating two mechanisms whereby classical processes may be embedded into the quantum formalism. The first approach employs group representations to demonstrate how group-valued Lévy processes induce operator-valued stochastic differential equations whose form is generic for quantum stochastics. Secondly, we give an account of how Lévy processes may be represented by suitable combinations of creation, conservation and annihilation operators, acting in a suitable Fock space. This beautiful interplay of ideas, which evolved in the 1960s and 1970s out of work on factorisable representations of current groups, reveals the probabilistic origins of quantum stochastic calculus.

Notation: If T is a topological space, $\mathcal{B}(T)$ is the Borel σ-algebra of all Borel sets in T. $B_b(T)$ is the Banach space (with respect to the supremum norm) of all (real valued) bounded Borel measurable functions on T. $C_b(T)$ is the Banach sub-space of all bounded continuous functions on T.

If T is locally compact, $C_0(T)$ is the Banach subspace of all continuous functions on T which vanish at infinity. The linear space $C_c(T)$ of continuous functions with compact support is norm-dense in $C_0(T)$.

Acknowledgements: I would like to thank the editors,Uwe Franz and Michael Schürmann, for giving me the opportunity to contribute to this volume, and also for the invitation to Greifswald to deliver the lecture course on which these notes are based, and the superb hospitality which was extended to me there. I would also like to thank the participants in the school for a number of observations which have improved the accuracy of these notes. Particular thanks are due to Uwe Franz who read through the whole article with great attention to detail and to Robin Hudson, who made a number of helpful comments about Lecture 6.

Thanks are also due to Cambridge University Press for granting me permission to include material herein which is taken from [13].

2 Lecture 1: Infinite Divisibility and Lévy Processes in Euclidean Space

2.1 Some Basic Ideas of Probability

Let $(\Omega, \mathcal{F}, P)$ be a probability space, so that Ω is a set, $\mathcal{F}$ is a σ-algebra of subsets of Ω and P is a probability measure defined on $(\Omega, \mathcal{F})$. Random variables are measurable functions $X : \Omega \to \mathbb{R}^d$. The law of X is p_X, where for each $A \in \mathcal{B}(\mathbb{R}^d)$, $p_X(A) = P(X \in A)$. $(X_n, n \in \mathbb{N})$ are *independent* if for all $i_1, i_2, \ldots i_r \in \mathbb{N}, A_{i_1}, A_{i_2}, \ldots, A_{i_r} \in \mathcal{B}(\mathbb{R}^d)$,

$$P(X_{i_1} \in A_1, X_{i_2} \in A_2, \ldots, X_{i_r} \in A_r)$$
$$= P(X_{i_1} \in A_1)P(X_{i_2} \in A_2) \cdots P(X_{i_r} \in A_r).$$

If X and Y are independent, the law of $X + Y$ is given by *convolution*

$$p_{X+Y} = p_X * p_Y, \text{ where } p_X * p_Y(A) = \int_{\mathbb{R}^d} p_X(A - y) p_Y(dy).$$

Equivalently $\int_{\mathbb{R}^d} f(y) p_X * p_Y(dy) = \int_{\mathbb{R}^d} \int_{\mathbb{R}^d} f(x + y) p_X(dx) p_Y(dy)$, for all $f \in B_b(\mathbb{R}^d)$.

Characteristic function of X is $\phi_X : \mathbb{R}^d \to \mathbb{C}$, where $\phi_X(u) = \int_{\mathbb{R}^d} e^{i(u,x)} p_X(dx)$.

Exercise 1.1. If X and Y are independent, show that $\phi_{X+Y}(u) = \phi_X(u)\phi_Y(u)$, for all $u \in \mathbb{R}^d$. (Note - the converse is false, e.g. consider $X + X$, where X is Cauchy distributed).

More generally:-

Theorem 2.1 (Kac's theorem). $X_1, \ldots, X_n$ *are independent if and only if*

$$\mathbb{E}\left(\exp\left(i \sum_{j=1}^{n}(u_j, X_j)\right)\right) = \phi_{X_1}(u_1) \cdots \phi_{X_n}(u_n)$$

for all $u_1, \ldots, u_n \in \mathbb{R}^d$.

The characteristic function of a probability measure μ on $\mathbb{R}^d$ is $\phi_\mu(u) = \int_{\mathbb{R}^d} e^{i(u,x)} \mu(dx)$. Important properties are:-

1. $\phi_\mu(0) = 1$.
2. ϕ_μ is *positive definite* i.e. $\sum_{i,j} c_i \bar{c}_j \phi_\mu(u_i - u_j) \geq 0$, for all $c_i \in \mathbb{C}, u_i \in \mathbb{R}^d, 1 \leq i, j \leq n, n \in \mathbb{N}$. (*Exercise 1.2*)
3. ϕ_μ is uniformly continuous (*Exercise 1.3*) - Hint: Look at $|\phi_\mu(u+h) - \phi_\mu(u)|$ and use dominated convergence)).

Conversely *Bochner's theorem* states that if $\phi : \mathbb{R}^d \to \mathbb{C}$ satisfies (1), (2) and is continuous at $u = 0$, then it is the characteristic function of some probability measure μ on $\mathbb{R}^d$.
(For a nice functional analytic proof based on spectral theory of self-adjoint operators - see Reed and Simon [67], p.330).

$\psi : \mathbb{R}^d \to \mathbb{C}$ is *conditionally positive definite* if for all $n \in \mathbb{N}$ and $c_1, \ldots, c_n \in \mathbb{C}$ for which $\sum_{j=1}^{n} c_j = 0$ we have

$$\sum_{j,k=1}^{n} c_j \bar{c}_k \psi(u_j - u_k) \geq 0,$$

for all $u_1, \ldots, u_n \in \mathbb{R}^d$. $\psi : \mathbb{R}^d \to \mathbb{C}$ will be said to be *hermitian* if $\overline{\psi(u)} = \psi(-u)$, for all $u \in \mathbb{R}^d$.

Theorem 2.2 (Schoenberg correspondence). $\psi : \mathbb{R}^d \to \mathbb{C}$ *is hermitian and conditionally positive definite if and only if* $e^{t\psi}$ *is positive definite for each* $t > 0$.

Proof. We only give the easy part here. For the full story see Berg and Forst [18], p.41 or Parthasarathy and Schmidt [63] pp. 1-4.
Suppose that $e^{t\psi}$ is positive definite for all $t > 0$. Fix $n \in \mathbb{N}$ and choose $c_1, \ldots, c_n$ and $u_1, \ldots, u_n$ as above. We then find that for each $t > 0$,

$$\frac{1}{t} \sum_{j,k=1}^{n} c_j \bar{c}_k (e^{t\psi(u_j - u_k)} - 1) \geq 0,$$

and so

$$\sum_{j,k=1}^{n} c_j \bar{c}_k \psi(u_j - u_k) = \lim_{t \to 0} \frac{1}{t} \sum_{j,k=1}^{n} c_j \bar{c}_k (e^{t\psi(u_j - u_k)} - 1) \geq 0.$$

$\square$

To see the need to be hermitian, define $\tilde{\psi} = \psi + ix$, where ψ is hermitian and conditionally positive definite and $x \in \mathbb{R}, x \neq 0$. $\tilde{\psi}$ is clearly conditionally positive definite, but not hermitian and it is then easily verified that $e^{t\tilde{\psi}}$ cannot be positive definite for any $t > 0$.

Note the analyst's convention of using $-\psi$ which they call "negative-definite".

2.2 Infinite Divisibility

Let μ be a probability measure on $\mathbb{R}^d$. Define $\mu^{*^n} = \mu * \cdots * \mu$ (n times). We say that μ has a *convolution nth root*, if there exists a probability measure $\mu^{\frac{1}{n}}$ for which $(\mu^{\frac{1}{n}})^{*^n} = \mu$.

μ is *infinitely divisible* if it has a convolution nth root for all $n \in \mathbb{N}$. In this case $\mu^{\frac{1}{n}}$ is unique.

Theorem 2.3. μ *is infinitely divisible iff for all $n \in \mathbb{N}$, there exists a probability measure μ_n with characteristic function ϕ_n such that*

$$\phi_\mu(u) = (\phi_n(u))^n,$$

for all $u \in \mathbb{R}^d$. Moreover $\mu_n = \mu^{\frac{1}{n}}$.

Proof. If μ is infinitely divisible, take $\phi_n = \phi_{\mu^{\frac{1}{n}}}$. Conversely, for each $n \in \mathbb{N}$, by Fubini's theorem,

$$\phi_\mu(u) = \int_{\mathbb{R}^d} \cdots \int_{\mathbb{R}^d} e^{i(u, y_1 + \cdots + y_n)} \mu_n(dy_1) \cdots \mu_n(dy_n)$$

$$= \int_{\mathbb{R}^d} e^{i(u,y)} \mu_n^{*^n}(dy)$$

But $\phi_\mu(u) = \int_{\mathbb{R}^d} e^{i(u,y)} \mu(dy)$ and ϕ determines μ uniquely. Hence $\mu = \mu_n^{*^n}$. $\square$

- If μ and ν are each infinitely divisible, then so is $\mu * \nu$.

- If $(\mu_n, n \in \mathbb{N})$ are infinitely divisible and $\mu_n \overset{w}{\Rightarrow} \mu$, then μ is infinitely divisible.

[Note: *Weak convergence.* $\mu_n \overset{w}{\Rightarrow} \mu$ means

$$\lim_{n \to \infty} \int_{\mathbb{R}^d} f(x) \mu_n(dx) = \int_{\mathbb{R}^d} f(x) \mu(dx),$$

for each $f \in C_b(\mathbb{R}^d)$. For the even weaker topology of *vague convergence*, replace $C_b(\mathbb{R}^d)$ by $C_0(\mathbb{R}^d)$.]

A random variable X is *infinitely divisible* if its law p_X is infinitely divisible, e.g. $X \stackrel{d}{=} Y_1^{(n)} + \cdots + Y_n^{(n)}$, where $Y_1^{(n)}, \ldots, Y_n^{(n)}$ are i.i.d., for each $n \in \mathbb{N}$.

Examples of Infinite Divisibility

In the following, we will demonstrate infinite divisibility of a random variable X by finding i.i.d $Y_1^{(n)}, \ldots, Y_n^{(n)}$ such that $X \stackrel{d}{=} Y_1^{(n)} + \cdots + Y_n^{(n)}$, for each $n \in \mathbb{N}$.

Example 1 - Gaussian Random Variables

Let $X = (X_1, \ldots, X_d)$ be a random vector.
We say that it is $(non - degenerate)\,Gaussian$ if there exists a vector $m \in \mathbb{R}^d$ and a strictly positive-definite symmetric $d \times d$ matrix A such that X has a pdf (probability density function) of the form:-

$$f(x) = \frac{1}{(2\pi)^{\frac{d}{2}} \sqrt{\det(A)}} \exp\left(-\frac{1}{2}(x - m, A^{-1}(x - m))\right), \qquad (2.1)$$

for all $x \in \mathbb{R}^d$.
In this case we will write $X \sim N(m, A)$. The vector m is the mean of X , so $m = \mathbb{E}(X)$ and A is the covariance matrix so that $A = \mathbb{E}((X - m)(X - m)^T)$. A standard calculation yields

$$\phi_X(u) = e^{i(m,u) - \frac{1}{2}(u, Au)}, \qquad (2.2)$$

and hence

$$(\phi_X(u))^{\frac{1}{n}} = e^{i(\frac{m}{n}, u) - \frac{1}{2}(u, \frac{1}{n}Au)},$$

so we see that X is infinitely divisible with each $Y_j^{(n)} \sim N(\frac{m}{n}, \frac{1}{n}A)$ for each $1 \leq j \leq n$.
We say that X is a *standard normal* whenever $X \sim N(0, \sigma^2 I)$ for some $\sigma > 0$.

We say that X is *degenerate Gaussian* if (2.2) holds with $\det(A) = 0$, and these random variables are also infinitely divisible.

Example 2 - Poisson Random Variables

In this case, we take $d = 1$ and consider a random variable X taking values in the set $n \in \mathbb{N} \cup \{0\}$. We say that X is *Poisson* if there exists $c > 0$ for which

$$P(X = n) = \frac{c^n}{n!} e^{-c}.$$

In this case we will write $X \sim \pi(c)$. We have $e(X) = \text{Var}(X) = c$. It is easy to verify (*Exercise 1.4*) that

$$\phi_X(u) = \exp[c(e^{iu} - 1)],$$

from which we deduce that X is infinitely divisible with each $Y_j^{(n)} \sim \pi(\frac{c}{n})$, for $1 \leq j \leq n, n \in \mathbb{N}$.

Example 3 - Compound Poisson Random Variables

Let $(Z(n), n \in \mathbb{N})$ be a sequence of i.i.d. random variables taking values in $\mathbb{R}^d$ with common law μ_Z and let $N \sim \pi(c)$ be a Poisson random variable which is independent of all the $Z(n)$'s. The *compound Poisson random variable* X is defined as follows:-

$$X = Z(1) + \cdots + Z(N).$$

Proposition 2.4. *For each* $u \in \mathbb{R}^d$,

$$\phi_X(u) = \exp\left[\int (e^{i(u,y)} - 1)c\mu_Z(dy)\right].$$

Proof. Let ϕ_Z be the common characteristic function of the Z_n's. By conditioning and using independence we find,

$$\phi_X(u) = \sum_{n=0}^{\infty} e(e^{i(u,Z(1)+\cdots+Z(N))}|N = n)P(N = n)$$

$$= \sum_{n=0}^{\infty} e(e^{i(u,Z(1))+\cdots+Z(n))})e^{-c}\frac{c^n}{n!}$$

$$= e^{-c} \sum_{n=0}^{\infty} \frac{[c\phi_Z(u)]^n}{n!}$$

$$= \exp[c(\phi_Z(u) - 1)],$$

and the result follows on writing $\phi_Z(u) = \int e^{i(u,y)}\mu_Z(dy)$. $\square$

If X is compound Poisson as above, we write $X \sim \pi(c, \mu_Z)$. It is clearly infinitely divisible with each $Y_j^{(n)} \sim \pi(\frac{c}{n}, \mu_Z)$, for $1 \leq j \leq n$.

The Lévy-Khintchine Formula

de Finetti (1920's) suggested that the most general infinitely divisible random variable could be written $X = Y + W$, where Y and W are independent, $Y \sim N(m, A), W \sim \pi(c, \mu_Z)$. Then $\phi_X(u) = e^{\eta(u)}$, where

$$\eta(u) = i(m, u) - \frac{1}{2}(u, Au) + \int_{\mathbb{R}^d} (e^{i(u,y)} - 1)c\mu_Z(dy). \tag{2.3}$$

This is WRONG! $\nu(\cdot) = c\mu_Z(\cdot)$ is a finite measure here. Lévy and Khintchine showed that ν can be σ-finite, provided it is what is now called a *Lévy measure* on $\mathbb{R}^d - \{0\} = \{x \in \mathbb{R}^d, x \neq 0\}$, i.e.

$$\int (|y|^2 \wedge 1)\nu(dy) < \infty. \tag{2.4}$$

Since $|y|^2 \wedge \epsilon \leq |y|^2 \wedge 1$ whenever $0 < \epsilon \leq 1$, it follows from (2.4) that

$$\nu((-\epsilon, \epsilon)^c) < \infty \quad \text{for all } \epsilon > 0.$$

Exercise 1.5. Show that every Lévy measure on $\mathbb{R}^d - \{0\}$ is σ-finite.

Exercise 1.6. Deduce that ν is a Lévy measure if and only if

$$\int \frac{|y|^2}{1 + |y|^2}\nu(dy) < \infty. \tag{2.5}$$

[Hint: Verify the inequalities

$$\frac{|y|^2}{1 + |y|^2} \leq |y|^2 \wedge 1 \leq 2\frac{|y|^2}{1 + |y|^2},$$

for each $y \in \mathbb{R}^d$.]
Here is the fundamental result of this lecture:-

Theorem 2.5 (Lévy-Khintchine). *A Borel probability measure μ on $\mathbb{R}^d$ is infinitely divisible if there exists a vector $b \in \mathbb{R}^d$, a non-negative symmetric $d \times d$ matrix A and a Lévy measure ν on $\mathbb{R}^d - \{0\}$ such that for all $u \in \mathbb{R}^d$,*

$$\phi_\mu(u) = \exp\left[i(b, u) - \frac{1}{2}(u, Au) + \int_{\mathbb{R}^d - \{0\}} (e^{i(u,y)} - 1 - i(u, y)\chi_{\hat{B}}(y))\nu(dy)\right]. \tag{2.6}$$

where $\hat{B} = B_1(0) = \{y \in \mathbb{R}^d; |y| < 1\}$.
Conversely, any mapping of the form (2.6) is the characteristic function of an infinitely divisible probability measure on $\mathbb{R}^d$.

The triple (b, A, ν) is called the *characteristics* of the infinitely divisible random variable X. Define $\eta = \log \phi_\mu$, where we take the principal part of the logarithm. η is called the *Lévy symbol* by me, the *characteristic exponent* by others.
We're not going to prove this result here. To understand it, it is instructive to let $(U_n, n \in \mathbb{N})$ be a sequence of Borel sets in $B_1(0)$ with $U_n \downarrow \{e\}$. Observe that

$$\eta(u) = \lim_{n \to \infty} \eta_n(u) \quad \text{where each}$$

$$\eta_n(u) = i\left[\left(m - \int_{U_n^c \cap \hat{B}} y\nu(dy), u\right)\right] - \frac{1}{2}(u, Au) + \int_{U_n^c} (e^{i(u,y)} - 1)\nu(dy),$$

so η is in some sense (to be made more precise later) the limit of a sequence of sums of Gaussians and independent compound Poissons. Interesting phenomena appear in the limit as we'll see below. First, we classify Lévy symbols analytically:-

Theorem 2.6. η *is a Lévy symbol if and only if it is a continuous, hermitian conditionally positive definite function for which* $\eta(0) = 0$.

Proof. Suppose that η is a Lévy symbol - then so is $t\eta$, for each $t > 0$. Then there exists a probability measure $\mu(t)$, for each $t \geq 0$ such that $\phi_{\mu(t)}(u) = e^{t\eta(u)}$ for each $u \in \mathbb{R}^d$. η is continuous and $\eta(0) = 0$. Since ϕ_μ is positive definite then η is hermitian and conditionally positive definite by the Schoenberg correspondence.

Conversely, suppose that η is continuous, hermitian and conditionally positive definite with $\eta(0) = 0$. By the Schoenberg correspondence and Bochner's theorem, there exists a probability measure μ for which $\phi_\mu(u) = e^{\eta(u)}$ for each $u \in \mathbb{R}^d$. Since, for each $n \in \mathbb{N}$, $\frac{\eta}{n}$ is another continuous, hermitian conditionally positive definite function which vanishes at the origin, we see that μ is infinitely divisible and the result follows. $\qquad\square$

Stable Laws

This is one of the most important subclasses of infinitely divisible laws.

We consider the general central limit problem in dimension $d = 1$, so let $(Y_n, n \in \mathbb{N})$ be a sequence of real valued random variables and consider the rescaled partial sums

$$S_n = \frac{Y_1 + Y_2 + \cdots + Y_n - b_n}{\sigma_n},$$

where $(b_n, n \in \mathbb{N})$ is an arbitrary sequence of real numbers and $(\sigma_n, n \in \mathbb{N})$ an arbitrary sequence of positive numbers. We are interested in the case where there exists a random variable X for which

$$\lim_{n \to \infty} P(S_n \leq x) = P(X \leq x), \tag{2.7}$$

for all $x \in \mathbb{R}$ i.e. $(S_n, n \in \mathbb{N})$ converges in distribution to X. If each $b_n = nm$ and $\sigma_n = \sqrt{n}\sigma$ for fixed $m \in R, \sigma > 0$ then $X \sim N(m, \sigma^2)$ by the usual Laplace - de-Moivre central limit theorem.

More generally a random variable is said to be *stable* if it arises as a limit as in (2.7). It is not difficult (see e.g. Gnedenko and Kolmogorov [40]) to show that (2.7) is equivalent to the following:-

There exist real valued sequences $(c_n, n \in \mathbb{N})$ and $(d_n, n \in \mathbb{N})$ with each $c_n > 0$ such that

$$X_1 + X_2 + \cdots + X_n \overset{d}{=} c_n X + d_n \tag{2.8}$$

where $X_1, \ldots, X_n$ are independent copies of X. X is said to be *strictly stable* if each $d_n = 0$.

To see that (2.8) $\Rightarrow$ (2.7) take each $Y_j = X_j, b_n = d_n$ and $\sigma_n = c_n$. In fact it can be shown (see Feller [35], p.166) that the only possible choice of c_n in

(2.8) is $c_n = \sigma n^{\frac{1}{\alpha}}$, where $0 < \alpha \leq 2$ and $\sigma > 0$. The parameter α plays a key role in the investigation of stable random variables and is called the *index of stability*.

Note that (2.8) can also be expressed in the equivalent form

$$\phi_X(u)^n = e^{iud_n}\phi_X(c_n u),$$

for each $u \in \mathbb{R}$.

It follows immediately from (2.8) that all stable random variables are infinitely divisible and the characteristics in the Lévy-Khintchine formula are given by the following result.

Theorem 2.7. *If X is a stable real-valued random variable, then its characteristics must take one of the two following forms.*

1. *When $\alpha = 2$, $\nu = 0$ (so $X \sim N(b, A)$).*
2. *When $\alpha \neq 2$, $A = 0$ and $\nu(dx) = \dfrac{c_1}{x^{1+\alpha}}\chi_{(0,\infty)}(x)dx + \dfrac{c_2}{|x|^{1+\alpha}}\chi_{(-\infty,0)}(x)dx$,*
   where $c_1 \geq 0, c_2 \geq 0$ and $c_1 + c_2 > 0$.

A proof is given in Sato [74], p.80.

A careful transformation of the integrals in the Lévy-Khintchine formula gives a different form for the characteristic function which is often more convenient (see Sato [74], p.86).

Theorem 2.8. *A real-valued random variable X is stable if and only if there exists $\sigma > 0, -1 \leq \beta \leq 1$ and $\mu \in \mathbb{R}$ such that for all $u \in \mathbb{R}$,*

1.
$$\phi_X(u) = \exp\left[i\mu u - \frac{1}{2}\sigma^2 u^2\right] \quad when \ \ \alpha = 2.$$

2.
$$\phi_X(u) = \exp\left[i\mu u - \sigma^\alpha|u|^\alpha(1 - i\beta sgn(u)\tan(\frac{\pi\alpha}{2}))\right] \quad when \ \ \alpha \neq 1, 2.$$

3.
$$\phi_X(u) = \exp\left[i\mu u - \sigma|u|(1 + i\beta\frac{2}{\pi}sgn(u)\log(|u|))\right] \quad when \ \ \alpha = 1.$$

It can be shown that $\mathbb{E}(X^2) < \infty$ if and only if $\alpha = 2$ (i.e. X is Gaussian) and $\mathbb{E}(|X|) < \infty$ if and only if $1 < \alpha \leq 2$.

All stable random variables have densities f_X, which can in general be expressed in series form (see Feller [35], Chapter 17, section 6). In three important cases, there are closed forms.

1. **The Normal Distribution**

$$\alpha = 2, \quad X \sim N(\mu, \sigma^2).$$

2. The Cauchy Distribution

$$\alpha = 1, \beta = 0 \quad f_X(x) = \frac{\sigma}{\pi[(x-\mu)^2 + \sigma^2]}.$$

3. The Lévy Distribution

$$\alpha = \frac{1}{2}, \beta = 1 \quad f_X(x) = \left(\frac{\sigma}{2\pi}\right)^{\frac{1}{2}} \frac{1}{(x-\mu)^{\frac{3}{2}}} \exp\left(-\frac{\sigma}{2(x-\mu)}\right), \quad \text{for } x > \mu.$$

Exercise 1.7. (The Cauchy Distribution)
Prove directly that

$$\int_{-\infty}^{\infty} e^{iux} \frac{\sigma}{\pi[(x-\mu)^2 + \sigma^2]} dx = e^{i\mu u - \sigma|u|}.$$

[Hint. One approach is to use the calculus of residues. Alternatively, by integrating from $-\infty$ to 0 and then 0 to ∞, separately, deduce that $\int_{-\infty}^{\infty} e^{-itx} e^{-|x|} dx = \frac{2}{1+t^2}$.
Now use Fourier inversion.]

Note that if a stable random variable is symmetric then Theorem 2.8 yields

$$\phi_X(u) = \exp(-\rho^\alpha |u|^\alpha) \quad \text{for all } 0 < \alpha \le 2, \tag{2.9}$$

where $\rho = \sigma$, when $0 < \alpha < 2$, and $\rho = \frac{\sigma}{\sqrt{2}}$, when $\alpha = 2$, and we will write $X \sim S\alpha S$ in this case.
Although it does not have a closed form density, the symmetric stable distribution with $\alpha = \frac{3}{2}$ is of considerable practical importance. It is called the *Holtsmark distribution* and its three-dimensional generalisation has been used to model the gravitational field of stars (see Feller [35], p.173).

One of the reasons why stable laws are so important in applications is the nice decay properties of the tails. The case $\alpha = 2$ is special in that we have exponential decay, indeed for a standard normal X there is the elementary estimate

$$P(X > y) \sim \frac{e^{-\frac{1}{2}y^2}}{\sqrt{2\pi}y} \quad \text{as } y \to \infty,$$

(see Feller [34], Chapter 7, section 1).
When $\alpha \ne 2$ we have the slower polynomial decay as expressed in the following,

$$\lim_{y \to \infty} y^\alpha P(X > y) = C_\alpha \frac{1+\beta}{2} \sigma^\alpha,$$

$$\lim_{y \to \infty} y^\alpha P(X < -y) = C_\alpha \frac{1-\beta}{2} \sigma^\alpha,$$

where $C_\alpha > 1$ (see Samorodnitsky and Taqqu [73], p.16-18 for a proof and an explicit expression for the constant C_α). The relatively slow decay of the tails for non-Gaussian stable laws makes them ideally suited for modelling a wide range of interesting phenomena, some of which exhibit "long-range dependence".

The generalisation of stability to random vectors is straightforward - just replace $X_1, \ldots, X_n$, X and each d_n in (2.8) by vectors and the formula in Theorem 2.7 extends directly. Note however that when $\alpha \neq 2$ in the random vector version of Theorem 2.7, the Lévy measure takes the form

$$\nu(dx) = \frac{c}{|x|^{d+\alpha}} dx$$

where $c > 0$.

The corresponding extension of Theorem 2.8 can be found in e.g. Sato [74], p.83.

We can generalise the definition of stable random variables if we weaken the conditions on the random variables $(Y(n), n \in \mathbb{N})$ in the general central limit problem by requiring these to be independent, but no longer necessarily identically distributed. In this case the limiting random variables are called *self-decomposable* (or of class L) and they are also infinitely divisible. Alternatively a random variable X is self-decomposable if and only if for each $0 < a < 1$, there exists a random variable Y_a which is independent of X such that

$$X \stackrel{d}{=} aX + Y_a \Leftrightarrow \phi_X(u) = \phi_X(au)\phi_{Y_a}(u),$$

for all $u \in \mathbb{R}^d$. Self-decomposable distributions are discussed in Sato [74] p.90-99, where it is shown that an infinitely divisible law on $\mathbb{R}$ is self-decomposable if and only if the Lévy measure is of the form:

$$\nu(dx) = \frac{k(x)}{|x|} dx,$$

where k is decreasing on $(0, \infty)$ and increasing on $(-\infty, 0)$. There has recently been increasing interest in these distributions both from a theoretical and applied perspective.

Other examples of infinitely divisible distributions:-

- gamma distribution (χ^2 is a special case).

- lognormal distribution

- Student t distribution

- Hyperbolic distributions (important in finance !)

- Riemann zeta distribution $\phi_u(v) = \dfrac{\zeta(u + iv)}{\zeta(u + i0)}$ where ζ is the Riemann zeta function.

- Relativistic distribution - minus the relativistic free energy

$$\eta_{m,c}(p) = -[\sqrt{m^2c^4 + c^2|p|^2} - mc^2].$$

(p is momentum, c is velocity of light, m is rest mass).

3 Lévy Processes

Let $X = (X(t), t \geq 0)$ be a stochastic process defined on a probability space $(\Omega, \mathcal{F}, P)$. We say that it has *independent increments* if for each $n \in \mathbb{N}$ and each $0 \leq t_1 < t_2 < \cdots < t_{n+1} < \infty$, the random variables $(X(t_{j+1}) - X(t_j), 1 \leq j \leq n)$ are independent and it has *stationary increments* if each $X(t_{j+1}) - X(t_j) \overset{d}{=} X(t_{j+1} - t_j) - X(0)$.

We say that X is a *Lévy process* if

(L1)   Each $X(0) = 0$ (a.s),

(L2) X has independent and stationary increments,

(L3) X is *stochastically continuous* i.e. for all $a > 0$ and for all $s \geq 0$,

$$\lim_{t \to s} P(|X(t) - X(s)| > a) = 0.$$

Note that in the presence of (L1) and (L2), (L3) is equivalent to the condition

$$\lim_{t \downarrow 0} P(|X(t)| > a) = 0$$

for all $a > 0$ (indeed, this follows easily from the fact that $P(|Y| > a) = P(Y \in B_a(0)^c)$, for any $\mathbb{R}^d$-valued random variable Y). Recall that the *sample paths* of a process are the maps $t \to X(t)(\omega)$ from $\mathbb{R}^+$ to $\mathbb{R}^d$, for each $\omega \in \Omega$.

We are now going to explore the relationship between Lévy processes and infinite divisibility.

Proposition 3.1. *If X is a Lévy process, then $X(t)$ is infinitely divisible for each $t \geq 0$.*

Proof. For each $n \in \mathbb{N}$, we can write

$$X(t) = Y_1^{(n)}(t) + \cdots + Y_n^{(n)}(t)$$

where each $Y_k^{(n)}(t) = X(\frac{kt}{n}) - X(\frac{(k-1)t}{n})$. The $Y_k^{(n)}(t)$'s are i.i.d. by (L2). $\square$

By Proposition 3.1, we can write $\phi_{X(t)}(u) = e^{\eta(t,u)}$ for each $t \geq 0, u \in \mathbb{R}^d$, where each $\eta(t,.)$ is a Lévy symbol.

Exercise 1.6 Show that if $X = (X(t), t \geq 0)$ is stochastically continuous, then the map $t \to \phi_{X(t)}(u)$ is continuous for each $u \in \mathbb{R}^d$.

Theorem 3.2. *If X is a Lévy process, then*

$$\phi_{X(t)}(u) = e^{t\eta(u)},$$

for each $u \in \mathbb{R}^d, t \geq 0$, where η is the Lévy symbol of $X(1)$.

Proof. Suppose that X is a Lévy process and for each $u \in \mathbb{R}^d, t \geq 0$, define $\phi_u(t) = \phi_{X(t)}(u)$ then by (L2) we have for all $s \geq 0$,

$$
\begin{aligned}
\phi_u(t+s) &= \mathbb{E}(e^{i(u,X(t+s))}) \\
&= \mathbb{E}(e^{i(u,X(t+s)-X(s))}e^{i(u,X(s))}) \\
&= \mathbb{E}(e^{i(u,X(t+s)-X(s))})\mathbb{E}(e^{i(u,X(s))}) \\
&= \phi_u(t)\phi_u(s)\ldots\text{(i)}
\end{aligned}
$$

Now $\phi_u(0) = 1\ldots$ (ii) by (L1), and the map $t \to \phi_u(t)$ is continuous. However the unique continuous solution to (i) and (ii) is given by $\phi_u(t) = e^{t\alpha(u)}$, where $\alpha : \mathbb{R}^d \to \mathbb{C}$. Now by Proposition 3.1, $X(1)$ is infinitely divisible, hence α is a Lévy symbol and the result follows.   □

We now have the Lévy-Khinchine formula for a Lévy process $X = (X(t), t \geq 0)$:-

$$
\mathbb{E}(e^{i(u,X(t))}) = \tag{3.1}
$$

$$
\exp\left(t\left[i(b,u) - \frac{1}{2}(u,Au) + \int_{\mathbb{R}^d-\{0\}} (e^{i(u,y)} - 1 - i(u,y)\chi_{\hat{B}}(y))\nu(dy)\right]\right),
$$

for each $t \geq 0, u \in \mathbb{R}^d$, where (b, A, ν) are the characteristics of $X(1)$. We will define the Lévy symbol and the characteristics of a Lévy process X to be those of the random variable $X(1)$. We will sometimes write the former as η_X when we want to emphasise that it belongs to the process X.

Exercise 1.8. Let X be a Lévy process with characteristics (b, A, ν) show that $-X = (-X(t), t \geq 0)$ is also a Lévy process and has characteristics $(-b, A, \tilde{\nu})$ where $\tilde{\nu}(A) = \nu(-A)$ for each $A \in \mathcal{B}(\mathbb{R}^d)$. Show also that for each $c \in \mathbb{R}$, the process $(X(t) + tc, t \geq 0)$ is a Lévy process and find its characteristics.

Exercise 1.9. Show that if X and Y are stochastically continuous processes then so is their sum $X + Y = (X(t) + Y(t), t \geq 0)$. [Hint: Use the elementary inequality

$$
P(|A+B| > c) \leq P\left(|A| > \frac{c}{2}\right) + P\left(|B| > \frac{c}{2}\right),
$$

where A and B are random vectors and $c > 0$].

Exercise 1.10. Show that the sum of two independent Lévy processes is again a Lévy process (Hint: Use Kac's theorem to establish independent increments).

Let p_t be the law of $X(t)$, for each $t \geq 0$. By (L2), we have for all $s, t \geq 0$ that:

$$
p_{t+s} = p_t * p_s.
$$

By (L3), we have $p_t \xrightarrow{w} \delta_0$ as $t \to 0$, i.e. $\lim_{t\to 0} \int f(x)p_t(dx) = f(0)$. $(p_t, t \geq 0)$ is a *weakly continuous convolution semigroup* of probability measures on $\mathbb{R}^d$. Conversely, given any such semigroup, we can always construct a Lévy process on path space via Kolmogorov's construction.

3.1 Examples of Lévy Processes

Example 1, Brownian Motion and Gaussian Processes

A *(standard) Brownian motion* in $\mathbb{R}^d$ is a Lévy process $B = (B(t), t \geq 0)$ for which

(B1) $B(t) \sim N(0, tI)$ for each $t \geq 0$,
(B2) B has continuous sample paths.

It follows immediately from (B1) that if B is a standard Brownian motion, then its characteristic function is given by

$$\phi_{B(t)}(u) = \exp\left\{-\frac{1}{2}t|u|^2\right\},$$

for each $u \in \mathbb{R}^d, t \geq 0$.

We introduce the marginal processes $B_i = (B_i(t), t \geq 0)$ where each $B_i(t)$ is the ith component of $B(t)$, then it is not difficult to verify that the B_i's are mutually independent Brownian motions in $\mathbb{R}$. We will call these *one-dimensional Brownian motions* in the sequel.

Brownian motion has been the most intensively studied Lévy process. In the early years of the twentieth century, it was introduced as a model for the physical phenomenon of Brownian motion by Einstein and Smoluchowski and as a description of the dynamical evolution of stock prices by Bachelier. The theory was placed on a rigorous mathematical basis by Norbert Wiener in the 1920's. The first part of Nelson [61] contains a nice historical account of these developments from the physical point of view.

We could try to use the Kolmogorov existence theorem to construct Brownian motion from the following prescription on cylinder sets of the form $I^H_{t_1,\ldots,t_n} = \{\omega \in \Omega; \omega(t_1) \in [a_1, b_1], \ldots, \omega(t_n) \in [a_n, b_n]\}$ where $H = [a_1, b_1] \times \cdots [a_n, b_n]$ and we have taken Ω to be the set of all mappings from $\mathbb{R}^+$ to $\mathbb{R}^d$:

$$P(I^H_{t_1,\ldots,t_n}) = \int_H \frac{1}{(2\pi)^{\frac{n}{2}}\sqrt{t_1(t_2 - t_1)\ldots(t_n - t_{n-1})}}$$
$$\exp\left(-\frac{1}{2}\left(\frac{x_1^2}{t_1} + \frac{(x_2 - x_1)^2}{t_2 - t_1} + \cdots + \frac{(x_n - x_{n-1})^2}{t_n - t_{n-1}}\right)\right) dx_1 \cdots dx_n.$$

However there is then no guarantee that the paths are continuous. The literature contains a number of ingenious methods for constructing Brownian motion. One of the most delightful of these (originally due to Paley and Wiener) obtains this, in the case $d = 1$, as a random Fourier series

$$B(t) = \frac{\sqrt{2}}{\pi} \sum_{n=0}^{\infty} \frac{\sin(\pi t(n + \frac{1}{2}))}{n + \frac{1}{2}} \xi(n),$$

for each $t \geq 0$, where $(\xi(n), n \in \mathbb{N})$ is a sequence of i.i.d. $N(0, 1)$ random variables (see Chapter 1 of Knight [52]) for a modern account). A nice construction of Brownian motion from a wavelet point of view can be found in Steele [78], pp. 35-9.

We list a number of useful properties of Brownian motion in the case $d = 1$. This is far from exhaustive and for further examples as well as details of the proofs, the reader is advised to consult the literature such as Sato [74], pp.22-28, Revuz and Yor [68], Rogers and Williams [69].

- Brownian motion is locally Hölder continuous with exponent α for every $0 < \alpha < \frac{1}{2}$ i.e. for every $T > 0, \omega \in \Omega$ there exists $K = K(T, \omega)$ such that

$$|B(t)(\omega) - B(s)(\omega)| \le K|t - s|^\alpha,$$

  for all $0 \le s < t \le T$.
- The sample paths $t \to B(t)(\omega)$ are almost surely nowhere differentiable.
- For any sequence, $(t_n, n \in \mathbb{N})$ in $\mathbb{R}^+$ with $t_n \uparrow \infty$,

$$\liminf_{n \to \infty} B(t_n) = -\infty \text{ a.s.} \quad \limsup_{n \to \infty} B(t_n) = \infty \text{ a.s.}$$

- The law of the iterated logarithm:-

$$P\left(\limsup_{t \downarrow 0} \frac{B(t)}{(2t \log(\log(\frac{1}{t})))^{\frac{1}{2}}} = 1\right) = 1.$$

Let A be a non-negative symmetric $d \times d$ matrix and let σ be a square root of A so that σ is a $d \times m$ matrix for which $\sigma\sigma^T = A$. Now let $b \in \mathbb{R}^d$ and let B be a Brownian motion in R^m. We construct a process $C = (C(t), t \ge 0)$ in $\mathbb{R}^d$ by

$$C(t) = bt + \sigma B(t), \tag{3.2}$$

then C is a Lévy process with each $C(t) \sim N(tb, tA)$. It is not difficult to verify that C is also a Gaussian process, i.e. all its finite dimensional distributions are Gaussian. It is sometimes called *Brownian motion with drift*. The Lévy symbol of C is

$$\eta_C(u) = i(b, u) - \frac{1}{2}(u, Au).$$

We will see in the next section that a Lévy process has continuous sample paths if and only if it is of the form (3.2).

Example 2 - The Poisson Process

The Poisson process of intensity $\lambda > 0$ is a Lévy process N taking values in $\mathbb{N} \cup \{0\}$ wherein each $N(t) \sim \pi(\lambda t)$ so we have

$$P(N(t) = n) = \frac{(\lambda t)^n}{n!} e^{-\lambda t},$$

for each $n = 0, 1, 2, \ldots$. The Poisson process is widely used in applications and there is a wealth of literature concerning it and its generalisations. We define non-negative random variables $(T_n, \mathbb{N} \cup \{0\})$ (usually called waiting times) by $T_0 = 0$ and for $n \in \mathbb{N}$,

$$T_n = \inf\{t \geq 0; N(t) = n\},$$

then it is well known that the T_n's are gamma distributed. Moreover, the inter-arrival times $T_n - T_{n-1}$ for $n \in \mathbb{N}$ are i.i.d. and each has exponential distribution with mean $\frac{1}{\lambda}$. The sample paths of N are clearly piecewise constant, on finite intervals, with "jump" discontinuities of size 1 at each of the random times $(T_n, n \in \mathbb{N})$.

For later work it is useful to introduce the *compensated Poisson process* $\tilde{N} = (\tilde{N}(t), t \geq 0)$ where each $\tilde{N}(t) = N(t) - \lambda t$. Note that $\mathbb{E}(\tilde{N}(t)) = 0$ and $\mathbb{E}(\tilde{N}(t)^2) = \lambda t$ for each $t \geq 0$.

Example 3 - The Compound Poisson Process

Let $(Z(n), n \in \mathbb{N})$ be a sequence of i.i.d. random variables taking values in $\mathbb{R}^d$ with common law μ_Z and let N be a Poisson process of intensity λ which is independent of all the $Z(n)$'s. The *compound Poisson process* Y is defined as follows:-

$$Y(t) = Z(1) + \ldots + Z(N(t)), \tag{3.3}$$

for each $t \geq 0$, so each $Y(t) \sim \pi(\lambda t, \mu_Z)$.

By Proposition 2.4 we see that Y has Lévy symbol

$$\eta_Y(u) = \left[\int (e^{i(u,y)} - 1) \lambda \mu_Z(dy) \right].$$

Again the sample paths of Y are piecewise constant, on finite intervals, with "jump discontinuities" at the random times $(T(n), n \in \mathbb{N})$, however this time the size of the jumps is itself random, and the jump at $T(n)$ can be any value in the range of the random variable $Z(n)$.

Example 4 - Interlacing Processes

Let C be a Gaussian Lévy process as in Example 1 and Y be a compound Poisson process as in Example 3, which is independent of C. Define a new process X by

$$X(t) = C(t) + Y(t),$$

for all $t \geq 0$, then it is not difficult to verify that X is a Lévy process with Lévy symbol of the form (2.3). Using the notation of Examples 2 and 3, we see that the paths of X have jumps of random size occurring at random times. In fact we have,

$$\begin{aligned}
X(t) &= C(t) \quad \text{for } 0 \leq t < T_1, \\
&= C(T_1) + Z_1 \quad \text{when } t = T_1, \\
&= X(T_1) + C(t) - C(T_1) \quad \text{for } T_1 < t < T_2, \\
&= X(T_2-) + Z_2 \quad \text{when } t = T_2,
\end{aligned}$$

and so on recursively. We call this procedure an *interlacing* as a continuous path process is "interlaced" with random jumps. From the remarks after Theorem 2.5, it seems reasonable that the most general Lévy process might arise as the limit of a sequence of such interlacings, and we will investigate this further in the next section.

Example 5 - Stable Lévy Processes

A *stable Lévy process* is a Lévy process X in which the Lévy symbol is given by theorem 2.7. So, in particular, each $X(t)$ is a stable random variable. Of particular interest is the rotationally invariant case whose Lévy symbol is given by

$$\eta(u) = -\sigma^{\alpha}|u|^{\alpha},$$

where α is the index of stability ($0 < \alpha \leq 2$). One of the reasons why these are important in applications is that they display self-similarity. In general, a stochastic process $Y = (Y(t), t \geq 0)$ is *self-similar with Hurst index $H > 0$* if the two processes $(Y(at), t \geq 0)$ and $(a^{H}Y(t), t \geq 0)$ have the same finite-dimensional distributions for all $a \geq 0$. By examining characteristic functions, it is easily verified that a rotationally invariant stable Lévy process is self-similar with Hurst index $H = \frac{1}{\alpha}$, so that e.g. Brownian motion is self-similar with $H = \frac{1}{2}$. A nice general account of self-similar processes can be found in Embrechts and Maejima [31]. In particular, it is shown therein that a Lévy process X is self-similar if and only if each $X(t)$ is strictly stable.

Just as with Gaussian processes, we can extend the notion of stability beyond the class of stable Lévy processes. In general then, we say that a stochastic process $X = (X(t), t \geq 0)$ is *stable* if all its finite-dimensional distributions are stable. For a comprehensive introduction to such processes, see Samorodnitsky and Taqqu [73], Chapter 3.

3.2 Subordinators

A *subordinator* is a one-dimensional Lévy process which is increasing a.s. Such processes can be thought of as a random model of time evolution, since if $T = (T(t), t \geq 0)$ is a subordinator we have

$$T(t) \geq 0 \text{ for each } t > 0 \text{ a.s.} \quad \text{and} \quad T(t_1) \leq T(t_2) \text{ whenever } t_1 \leq t_2 \text{ a.s.}$$

Now since for $X(t) \sim N(0, At)$ we have $P(X(t) \geq 0) = P(X(t) \leq 0) = \frac{1}{2}$, it is clear that such a process cannot be a subordinator. More generally we have

Theorem 3.3. *If T is a subordinator then its Lévy symbol takes the form*

$$\eta(u) = ibu + \int_{(0,\infty)} (e^{iuy} - 1)\lambda(dy), \tag{3.4}$$

where $b \geq 0$, and the Lévy measure λ satisfies the additional requirements

$$\lambda(-\infty, 0) = 0 \quad \text{and} \quad \int_{(0,\infty)} (y \wedge 1)\lambda(dy) < \infty.$$

Conversely, any mapping from $\mathbb{R}^d \to \mathbb{C}$ of the form (3.4) is the Lévy symbol of a subordinator.

A proof of this can be found in Rogers and Williams [69], pp.78-9.

We call the pair (b, λ), the *characteristics* of the subordinator T.

Exercise 1.11. Show that the additional constraint on Lévy measures of subordinators is equivalent to the requirement $\int_{(0,\infty)} \frac{y}{1+y}\lambda(dy) < \infty$.

Now for each $t \geq 0$, the map $u \to \mathbb{E}(e^{iuT(t)})$ can clearly be analytically continued to the region $\{iu, u > 0\}$ and we then obtain the following expression for the Laplace transform of the distribution

$$\mathbb{E}(e^{-uT(t)}) = e^{-t\psi(u)},$$

where $\psi(u) = -\eta(iu) = bu + \int_{(0,\infty)} (1 - e^{-uy})\lambda(dy)$ (3.5)

for each $t, u \geq 0$. We note that this is much more useful for both theoretical and practical application than the characteristic function.

The function ψ is usually called the *Laplace exponent* of the subordinator.

Examples

(1) The Poisson Case

Poisson processes are clearly subordinators. More generally a compound Poisson process will be a subordinator if and only if the $Z(n)$'s in equation (3.3) are all $\mathbb{R}^+$ valued.

(2) α-Stable Subordinators

Using straightforward calculus, we find that for $0 < \alpha < 1$, $u \geq 0$,

$$u^\alpha = \frac{\alpha}{\Gamma(1 - \alpha)} \int_0^\infty (1 - e^{-ux}) \frac{dx}{x^{1+\alpha}}.$$

Hence by (3.5), Theorem 3.3 and Theorem 2.7, we see that for each $0 < \alpha < 1$ there exists an α-stable subordinator T with Laplace exponent

$$\psi(u) = u^\alpha.$$

and the characteristics of T are $(0, \lambda)$ where $\lambda(dx) = \frac{\alpha}{\Gamma(1-\alpha)} \frac{dx}{x^{1+\alpha}}$.
Note that when we analytically continue this to obtain the Lévy symbol we obtain the form given in Theorem 2.8(2) with $\mu = 0, \beta = 1$ and $\sigma^\alpha = \cos\left(\frac{\alpha\pi}{2}\right)$.

(3) The Lévy Subordinator

The $\frac{1}{2}$-stable subordinator has a density given by the Lévy distribution (with $\mu = 0$ and $\sigma = \frac{t^2}{2}$)

$$f_{T(t)}(s) = \left(\frac{t}{2\sqrt{\pi}}\right) s^{-\frac{3}{2}} e^{-\frac{t^2}{4s}},$$

for $s \geq 0$. The Lévy subordinator has a nice probabilistic interpretation as a first hitting time for one-dimensional standard Brownian motion $(B(t), t \geq 0)$, more precisely

$$T(t) = \inf\{s > 0; B(s) = \frac{t}{\sqrt{2}}\}. \tag{3.6}$$

Exercise 1.12. Show directly that for each $t \geq 0$,

$$\mathbb{E}(e^{-uT(t)}) = \int_0^\infty e^{-us} f_{T(t)}(s) ds = e^{-tu^{\frac{1}{2}}},$$

where $(T(t), t \geq 0)$ is the Lévy subordinator.

Hint: Write $g_t(u) = \mathbb{E}(e^{-uT(t)})$. Differentiate with respect to u and make the substitution $x = \frac{t^2}{4us}$ to obtain the differential equation $g_t'(u) = -\frac{t}{2\sqrt{u}} g_t(u)$. Via the substitution $y = \frac{t}{2\sqrt{s}}$ we see that $g_t(0) = 1$ and the result follows (see also Sato [74] p.12).

(4) Inverse Gaussian Subordinators

We generalise the Lévy subordinator by replacing Brownian motion by the Gaussian process $C = (C(t), t \geq 0)$ where each $C(t) = B(t) + \mu t$ and $\mu \in \mathbb{R}$. The *inverse Gaussian subordinator* is defined by

$$T(t) = \inf\{s > 0; C(s) = \delta t\}$$

where $\delta > 0$ and is so-called since $t \to T(t)$ is the generalised inverse of a Gaussian process.

Using martingale methods, we can show that for each $t, u > 0$,

$$\mathbb{E}(e^{-uT(t)}) = e^{-t\delta(\sqrt{2u+\mu^2}-\mu)}, \tag{3.7}$$

In fact each $T(t)$ has a density:-

$$f_{T(t)}(s) = \frac{\delta t}{\sqrt{2\pi}} e^{\delta t \mu} s^{-\frac{3}{2}} \exp\left\{-\frac{1}{2}(t^2\delta^2 s^{-1} + \mu^2 s)\right\}, \tag{3.8}$$

for each $s, t \geq 0$.

In general any random variable with density $f_{T(1)}$ is called an *inverse Gaussian* and denoted as $\text{IG}(\delta, \mu)$.

(5) Gamma Subordinators

Let $(T(t), t \geq 0)$ be a *gamma process* with parameters $a, b > 0$ so that each $T(t)$ has density

$$f_{T(t)}(x) = \frac{b^{at}}{\Gamma(at)} x^{at-1} e^{-bx},$$

for $x \geq 0$; then it is easy to verify that for each $u \geq 0$,

$$\int_0^\infty e^{-ux} f_{T(t)}(x) dx = \left(1 + \frac{u}{b}\right)^{-at} = \exp\left(-ta \log\left(1 + \frac{u}{b}\right)\right).$$

From here it is a straightforward exercise in calculus to show that

$$\int_0^\infty e^{-ux} f_{T(t)}(x)dx = \int_0^\infty (1 - e^{-ux})ax^{-1}e^{-bx}dx,$$

From this we see that $(T(t), t \geq 0)$ is a subordinator with $b = 0$ and $\lambda(dx) = ax^{-1}e^{-bx}dx$. Moreover $\psi(u) = a \log \left(1 + \frac{u}{b}\right)$ is the associated Bernstein function (see below).

Before we go further into the probabilistic properties of subordinators we'll make a quick diversion into analysis.

Let $f \in C^\infty((0, \infty))$ with $f \geq 0$. We say it is *completely monotone* if $(-1)^n f^{(n)} \geq 0$ for all $n \in \mathbb{N}$, and a *Bernstein function* if $(-1)^n f^{(n)} \leq 0$ for all $n \in \mathbb{N}$. We then have the following

Theorem 3.4. *1. f is a Bernstein function if and only if the mapping $x \to e^{-tf(x)}$ is completely monotone for all $t \geq 0$.*

2. f is a Bernstein function if and only if it has the representation

$$f(x) = a + bx + \int_0^\infty (1 - e^{-yx})\lambda(dy),$$

for all $x > 0$ where $a, b \geq 0$ and $\int_0^\infty (y \wedge 1)\lambda(dy) < \infty$.

3. g is completely monotone if and only if there exists a measure μ on $[0, \infty)$ for which

$$g(x) = \int_0^\infty e^{-xy}\mu(dy).$$

A proof of this result can be found in Berg and Forst [18], pp.61-72. To interpret this theorem, first consider the case $a = 0$. In this case, if we compare the statement of Theorem 3.4 with equation (3.5), we see that there is a one to one correspondence between Bernstein functions for which $\lim_{x \to 0} f(x) = 0$ and Laplace exponents of subordinators. The Laplace transforms of the laws of subordinators are always completely monotone functions and a subclass of all possible measures μ appearing in Theorem 3.4 (3) is given by all possible laws $p_{T(t)}$ associated to subordinators. A general Bernstein function with $a > 0$ can be given a probabilistic interpretation by means of "killing".

One of the most important probabilistic applications of subordinators is to "time change". Let X be an arbitrary Lévy process and let T be a subordinator defined on the same probability space as X such that X and T are independent. We define a new stochastic process $Z = (Z(t), t \geq 0)$ by the prescription

$$Z(t) = X(T(t)),$$

for each $t \geq 0$ so that for each $\omega \in \Omega, Z(t)(\omega) = X(T(t)(\omega))(\omega)$. The key result is then the following.

Theorem 3.5. *Z is a Lévy process.*

For the proof, see [13], section 1.3.2 or Sato [74], pp.199-200.

Exercise 1.13. Show that for each $A \in \mathcal{B}(\mathbb{R}^d), t \geq 0$,

$$p_{Z(t)}(A) = \int_{(0,\infty)} p_{X(u)}(A) p_{T(t)}(du).$$

We compute the Lévy symbol of the subordinated process Z.

Proposition 3.6.
$$\eta_Z = -\psi_T \circ (-\eta_X).$$

Proof. For each $u \in \mathbb{R}^d, t \geq 0$,

$$\begin{aligned}
\mathbb{E}(e^{i\eta_{Z(t)}(u)}) &= \mathbb{E}(e^{i(u,X(T(t)))}) \\
&= \int \mathbb{E}(e^{i(u,X(s))}) p_{T(t)}(ds) \\
&= \int e^{-s(-\eta_X(u))} p_{T(t)}(ds) \\
&= \mathbb{E}(e^{-\eta_X(u)T(t)}) \\
&= e^{-t\psi_T(-\eta_X(u))}. \qquad \qquad \square
\end{aligned}$$

Note: The penultimate step in the above proof necessitates analytic continuation of the map $u \to \mathbb{E}(e^{iuT(t)})$ to the region $\mathrm{Ran}(\eta_X)$.

Example 1: From Brownian Motion to 2α-stable Processes

Let T be an α-stable subordinator (with $0 < \alpha < 1$) and X be a d-dimensional Brownian motion with covariance $A = 2I$, which is independent of T. Then for each $s \geq 0, u \in \mathbb{R}^d$, $\psi_T(s) = s^\alpha$ and $\eta_X(u) = -|u|^2$, and hence $\eta_Z(u) = -|u|^{2\alpha}$, i.e. Z is a rotationally invariant 2α-stable process.

In particular, if $d = 1$ and T is the Lévy subordinator, then Z is the *Cauchy process*, so each $Z(t)$ has a symmetric Cauchy distribution with parameters $\mu = 0$ and $\sigma = 1$. It is interesting to observe from (3.6) that Z is constructed from two independent standard Brownian motions.

Example 2 : From Brownian Motion to Relativity

Let T be the Lévy subordinator and for each $t \geq 0$ define

$$f_{c,m}(s;t) = e^{-m^2 c^4 s + mc^2 t} f_{T(t)}(s)$$

for each $s \geq 0$ where $m, c > 0$.
It is then an easy exercise to deduce that

$$\int_0^\infty e^{-us} f_{c,m}(s;t) ds = e^{-t[(u+m^2c^4)^{\frac{1}{2}} - mc^2]}.$$

Since the map $u \to -t[(u + m^2 c^4)^{\frac{1}{2}} - mc^2]$ is a Bernstein function which vanishes at the origin, we deduce that there is a subordinator $T_{c,m} = (T_{c,m}(t), t \geq 0)$ where each $T_{c,m}(t)$ has density $f_{c,m}(\cdot; t)$. Now let B be a Brownian motion with covariance $A = 2c^2 I$ which is independent of $T_{c,m}$, then for the subordinated process, we find

$$\eta_Z(p) = -[(c^2 p^2 + m^2 c^4)^{\frac{1}{2}} - mc^2]$$

so that Z is a relativistic process.

Another important example, which has been applied to option pricing by Ole Barndorff-Nielsen, is the *normal inverse Gaussian process* obtained by subordinating Brownian motion with drift by an independent inverse Gaussian subordinator (see [13], section 1.3.2 and references therein).

Question: What is a "quantum subordinator" ?

4 Lecture 2: Semigroups Induced by Lévy Processes

4.1 Conditional Expectation, Filtrations

Recall our probability space $(\Omega, \mathcal{F}, P)$. Let $\mathcal{G}$ be a sub-σ-algebra of $\mathcal{F}$. If $\mathbb{E}(|X|) < \infty$, $\mathbb{E}(X|\mathcal{G})$ is the associated *conditional expectation of X given $\mathcal{G}$.* It is a $\mathcal{G}$-measurable random variable.
Some properties:-

- $\mathbb{E}(\mathbb{E}(X|\mathcal{G})) = \mathbb{E}(X)$.
- $|\mathbb{E}(X|\mathcal{G})| \leq \mathbb{E}(|X||\mathcal{G})$.
- If Y is a $\mathcal{G}$-measurable random variable and $\mathbb{E}(|(X, Y)|) < \infty$, then

$$\mathbb{E}((X, Y)|\mathcal{G}) = (\mathbb{E}(X|\mathcal{G}), Y) \quad \text{a.s.}$$

- If $\mathcal{H}$ is a sub-σ-algebra of $\mathcal{G}$ then

$$\mathbb{E}(\mathbb{E}(X|\mathcal{G})|\mathcal{H}) = \mathbb{E}(X|\mathcal{H}) \quad \text{a.s.}$$

- If X is independent of $\mathcal{G}$ then $\mathbb{E}(X|\mathcal{G}) = \mathbb{E}(X)$ a.s..
- The mapping $\mathbb{E}_{\mathcal{G}} : L^2(\Omega, \mathcal{F}, P) \to L^2(\Omega, \mathcal{G}, P)$ defined by $\mathbb{E}_{\mathcal{G}}(X) = \mathbb{E}(X|\mathcal{G})$ is an orthogonal projection.

A less-well known result which is very useful in proving Markovianity is:-

Lemma 4.1. *Let $\mathcal{G}$ be a sub-σ-algebra of $\mathcal{F}$. If X and Y are $\mathbb{R}^d$-valued random variables such that X is $\mathcal{G}$-measurable and Y is independent of $\mathcal{G}$, then*

$$\mathbb{E}(f(X, Y)|\mathcal{G}) = G_f(X) \quad a.s.$$

for all $f \in B_b(\mathbb{R}^{2d})$, where $G_f(x) = \mathbb{E}(f(x, Y))$, for each $x \in \mathbb{R}^d$.

One way of proving this is using approximation by simple functions - see Sato [74] p.7.

A *filtration* is an increasing family $(\mathcal{F}_t, t \geq 0)$ of sub-σ-algebras of $\mathcal{F}$. A stochastic process $X = (X(t), t \geq 0)$ is *adapted* to the given filtration if each $X(t)$ is $\mathcal{F}_t$-measurable.

e.g. any process is adapted to its *natural filtration*, $\mathcal{F}_t^X = \sigma\{X(s); 0 \leq s \leq t\}$.

4.2 Markov and Feller Processes

An adapted process $X = (X(t), t \geq 0)$ is a *Markov process* if for all $f \in B_b(\mathbb{R}^d), 0 \leq s \leq t < \infty$,

$$\mathbb{E}(f(X(t))|\mathcal{F}_s) = \mathbb{E}(f(X(t))|X(s)) \quad (a.s.). \tag{4.1}$$

(i.e. "past" and "future" are independent, given the present).

Define a family of operators $\{T_{s,t}, 0 \leq s \leq t < \infty\}$ on $B_b(\mathbb{R}^d)$ by the prescription

$$(T_{s,t}f)(x) = \mathbb{E}(f(X(t))|X(s) = x),$$

for each $f \in B_b(\mathbb{R}^d), x \in \mathbb{R}^d$. We recall that I is the identity operator, $If = f$, for each $f \in B_b(\mathbb{R}^d)$.

Theorem 4.2. *(a) $T_{s,t}$ is a linear operator on $B_b(\mathbb{R}^d)$ for each $0 \leq s \leq t < \infty$.*

(b) $T_{s,s} = I$ for each $s \geq 0$.

(c) $T_{r,s}T_{s,t} = T_{r,t}$ whenever $0 \leq r \leq s \leq t < \infty$.

(d) $f \geq 0 \Rightarrow T_{s,t}(f) \geq 0$ for all $0 \leq s \leq t < \infty, f \in B_b(\mathbb{R}^d)$.

(e) $T_{s,t}$ is a contraction, i.e. $\|T_{s,t}\| \leq 1$ for each $0 \leq s \leq t < \infty$.
(f) $T_t(1) = 1$ for all $t \geq 0$.

Proof. (a), (b), (d) and (f) are obvious. (e) is *Exercise* 2.1.
For (c), let $f \in B_b(\mathbb{R}^d), x \in \mathbb{R}^d$, then for each $0 \leq r \leq s \leq t < \infty$, applying conditioning and the Markov property (4.1) yields,

$$\begin{aligned}
(T_{r,t}f)(x) &= \mathbb{E}(f(X(t))|X(r) = x) \\
&= \mathbb{E}(\mathbb{E}(f(X(t))|\mathcal{F}_s)|X(r) = x) \\
&= \mathbb{E}(\mathbb{E}(f(X(t))|X(s))|X(r) = x) \\
&= \mathbb{E}((T_{s,t}f)(X(s))|X(r) = x) \\
&= (T_{r,s}(T_{s,t}f))(x). \qquad \square
\end{aligned}$$

Transition probabilities. These are defined as follows:-

$$p_{s,t}(x, A) = P(X(t) \in A|X(s) = x) = (T_{s,t}\chi_A)(x),$$

for each $x \in \mathbb{R}^d, A \in \mathcal{B}(\mathbb{R}^d)$.

By the properties of conditional probability:

$$(T_{s,t}f)(x) = \int_{\mathbb{R}^d} f(y)p_{s,t}(x, dy). \qquad (4.2)$$

We say that a Markov process is *normal* if for each $A \in \mathcal{B}(\mathbb{R}^d), 0 \leq s \leq t < \infty$, the mappings $x \to p_{s,t}(x, A)$ are measurable.

Theorem 4.3 (The Chapman Kolmogorov Equations). *If X is a normal Markov process, then for each $0 \leq r \leq s \leq t < \infty, x \in \mathbb{R}^d, A \in \mathcal{B}(\mathbb{R}^d)$,*

$$p_{r,t}(x, A) = \int_{\mathbb{R}^d} p_{s,t}(y, A)p_{r,s}(x, dy). \qquad (4.3)$$

Proof. Note that since X is normal, the mappings $y \to p_{s,t}(y, A)$ are integrable. Now applying Theorem 4.2 and (4.2), we obtain

$$\begin{aligned} p_{r,t}(x, A) &= (T_{r,t}\chi_A)(x) \\ &= (T_{r,s}(T_{s,t}\chi_A))(x) \\ &= \int_{\mathbb{R}^d} (T_{s,t}\chi_A)(y)p_{r,s}(x, dy) \\ &= \int_{\mathbb{R}^d} p_{s,t}(y, A)p_{r,s}(x, dy). \end{aligned}$$

$\qquad\qquad\qquad\qquad\qquad\qquad\qquad\qquad\qquad\qquad\qquad\qquad\qquad\qquad\square$

Important fact: Adapted Lévy processes are Markov processes.

To see this, let X be a Lévy process with associated convolution semigroup of laws $(q_t, t \geq 0)$. Use (L2) and Lemma 4.1 to write

$$\begin{aligned} \mathbb{E}(f(X(t))|\mathcal{F}_s) &= \mathbb{E}(f(X(s) + X(t) - X(s))|\mathcal{F}_s) \\ &= \int_{\mathbb{R}^d} f(X(s) + y)q_{t-s}(dy) \end{aligned}$$

It follows that

$$(T_{s,t}f)(x) = (T_{0,t-s}f)(x) = \int_{\mathbb{R}^d} f(x + y)q_{t-s}(dy).$$

$$\text{and} \quad p_{s,t}(x, A) = q_{t-s}(A - x).$$

Writing $T_{0,t} = T_t$, Theorem 4.2 (b) reduces to the semigroup law $T_s T_t = T_{s+t}$.

For a Lévy process $(T_t f)(x) = \mathbb{E}(f(x + X(t)))$, so Lévy processes induce *translation invariant semigroups*.

Exercise 2.2 A Markov process is said to have a *transition density* if for each $x \in \mathbb{R}^d, 0 \leq s \leq t < \infty$ there exists a measurable function $y \to \rho_{s,t}(x, y)$ such that

$$p_{s,t}(x, A) = \int_A \rho_{s,t}(x, y)dy.$$

Deduce that a Lévy process $X = (X(t), t \geq 0)$ has a transition density if and only if q_t has a density f_t for each $t \geq 0$, and hence show that

$$\rho_{s,t}(x, y) = f_{t-s}(y - x),$$

for each $0 \leq s \leq t < \infty, x, y \in \mathbb{R}^d$.

Write down the transition densities for (a) standard Brownian motion, (b) the Cauchy process.

Exercise 2.3 Suppose that the Markov process X has a transition density. Deduce that

$$\rho_{r,t}(x, z) = \int_{\mathbb{R}^d} \rho_{r,s}(x, y)\rho_{s,t}(y, z)dy,$$

for each $0 \leq r \leq s \leq t < \infty, x, y, z \in \mathbb{R}^d$.

In general a Markov process is *(time)-homogeneous* if

$$T_{s,t} = T_{0,t-s},$$

for all $0 \leq s \leq t < \infty$ and using (4.2), it is easily verified that this holds if and only if

$$p_{s,t}(x, A) = p_{0,t-s}(x, A),$$

for each $0 \leq s \leq t < \infty, x \in \mathbb{R}^d, A \in \mathcal{B}(\mathbb{R}^d)$.

A homogeneous Markov process X is said to be a *Feller process* if

1. $T_t : C_0(\mathbb{R}^d) \subseteq C_0(\mathbb{R}^d)$ for all $t \geq 0$.
2. $\lim_{t \downarrow 0} \|T_t f - f\| = 0$ for all $f \in C_0(\mathbb{R}^d)$.

In this case, the semigroup associated to X is called a *Feller semigroup*.

Theorem 4.4. *If X is a Feller process, then its transition probabilities are normal.*

Proof. See Revuz and Yor [68] page 83.   □

The class of all Feller processes is far from empty as the following result shows.

Theorem 4.5. *Every Lévy process is a Feller process.*

Proof. (sketch) Easy use of dominated convergence (*Exercise* 2.4) shows that each $T_t : C_0(\mathbb{R}^d) \subseteq C_0(\mathbb{R}^d)$.

For the second part:-

$$\|T_t f - f\| = \sup_{x \in \mathbb{R}^d} |T_t f(x) - f(x)|$$

$$\leq \int_{B_\delta(0)} \sup_{x \in \mathbb{R}^d} |f(x+y) - f(x)| q_t(dy) + \int_{B_\delta(0)^c} \sup_{x \in \mathbb{R}^d} |f(x+y) - f(x)| q_t(dy)$$

As $t \to 0$, the first term $\to 0$ by uniform continuity of f, the second term $\leq 2\|f\| q_t(B_\delta(0)^c) \to 0$ by stochastic continuity. $\square$.

5 Analytic Diversions

5.1 Semigroups and Generators

Let B be a real Banach space and $L(B)$ be the algebra of all bounded linear operators on B. A *one-parameter semigroup of contractions* on B is a family of bounded, linear operators $(T_t, t \geq 0)$ on B for which

1. $T_{s+t} = T_s T_t$ for all $s, t \geq 0$.
2. $T_0 = I$.
3. $\|T_t\| \leq 1$ for all $t \geq 0$.
4. The map $t \to T_t$ from $\mathbb{R}^+$ to $L(B)$ is strongly continuous at zero, i.e. $\lim_{t \to 0} \|T_t \psi - \psi\| = 0$ for all $\psi \in B$.

From now on we will say that $(T_t, t \geq 0)$ is a *semigroup* whenever it satisfies the above conditions.

Exercise 2.5 If $(T_t, t \geq 0)$ is a semigroup in a Banach space B, show that the map $t \to T_t$ is strongly continuous from $\mathbb{R}^+$ to $L(B)$, i.e. $\lim_{s \to t} \|T_t \psi - T_s \psi\| = 0$ for all $t \geq 0, \psi \in B$.

Exercise 2.6 Let A be a bounded operator in a Banach space B and for each $t \geq 0, \psi \in B$, define

$$T_t \psi = \sum_{n=0}^{\infty} \frac{t^n}{n!} A^n \psi = \text{``} e^{tA} \psi \text{''}.$$

Show that $(T_t, t \geq 0)$ is a strongly continuous semigroup of bounded operators in B. Show further that $(T_t, t \geq 0)$ is norm continuous, in that $\lim_{t \to 0} \|T_t - I\| = 0$.

In general, define

$$D_A = \{\psi \in B; \exists \phi_\psi \in B \text{ such that } \lim_{t \to 0} \left\| \frac{T_t \psi - \psi}{t} - \phi_\psi \right\| = 0\}.$$

It is easy to verify that D_A is a linear space and we may thus define a linear operator A in B with domain D_A, by the prescription $A\psi = \phi_\psi$, so that for each $\psi \in D_A$,

$$A\psi = \lim_{t \to 0} \frac{T_t\psi - \psi}{t}.$$

A is called the *infinitesimal generator*, or sometimes just the *generator* of the semigroup $(T_t, t \geq 0)$. A commonly used notation is "$T_t = e^{tA}$". In the case where $(T_t, t \geq 0)$ is the Feller semigroup associated to a Feller process $X = (X(t), t \geq 0)$, we sometimes call A the *generator of* X.

Some facts about generators:-

- D_A is dense in B.
- $T_t D_A \subseteq D_A$ for each $t \geq 0$.
- $T_t A\psi = A T_t \psi$ for each $t \geq 0, \psi \in D_A$.
- A is closed.
- Define the *resolvent set*, $\rho(A) = \{\lambda \in \mathbb{C}; \lambda I - A \text{ is invertible}\}$, then $(0, \infty) \subseteq \rho(A)$ and for each $\lambda > 0$, *the resolvent*,

$$R_\lambda(A) = (I - A)^{-1} = \int_0^\infty e^{-\lambda t} T_t dt.$$

A good source for material on semigroups is Davies [29].

5.2 The Fourier Transform and Pseudo-differential Operators

Let $f \in L^1(\mathbb{R}^d, \mathbb{C})$, then its *Fourier transform* is the mapping $\hat{f} \in L^1(\mathbb{R}^d, \mathbb{C})$, where

$$\hat{f}(u) = (2\pi)^{-\frac{d}{2}} \int_{\mathbb{R}^d} e^{-i(u,x)} f(x) dx \tag{5.1}$$

for all $u \in \mathbb{R}^d$. If we define $\mathcal{F}(f) = \hat{f}$, then $\mathcal{F}$ is a bounded linear operator on $L^1(\mathbb{R}^d, \mathbb{C})$ which is called the *Fourier transformation*.

We introduce two important families of linear operators in $L^1(\mathbb{R}^d, \mathbb{C})$, *translations* $(\tau_x, x \in \mathbb{R}^d)$ and *phase multiplications* $(e_x, x \in \mathbb{R}^d)$ by

$$(\tau_x f)(y) = f(y - x), \quad (e_x f)(y) = e^{i(x,y)} f(y),$$

for each $f \in L^1(\mathbb{R}^d, \mathbb{C}), x, y \in \mathbb{R}^d$.

It is easy to show that each of τ_x and e_x are isometric isomorphisms of $L^1(\mathbb{R}^d, \mathbb{C})$. Two key, easily verified properties of the Fourier transform are

$$\widehat{\tau_x f} = e_{-x}\hat{f} \quad \text{and} \quad \widehat{e_x f} = \tau_x \hat{f}, \tag{5.2}$$

for each $x \in \mathbb{R}^d$.

Furthermore, if we define the *convolution* $f * g$ of $f, g \in L^1(\mathbb{R}^d, \mathbb{C})$ by

$$(f * g)(x) = (2\pi)^{-\frac{d}{2}} \int_{\mathbb{R}^d} f(x - y)g(y) dy,$$

for each $x \in \mathbb{R}^d$, then we have $\widehat{(f * g)} = \hat{f}\hat{g}$.

Perhaps the most natural context in which to discuss $\mathcal{F}$ is the Schwartz space of rapidly decreasing functions. These are smooth functions which are such that they, and all their derivatives decay to zero at infinity faster than any negative power of $|x|$. To make this precise, we first need some standard notation for partial differential operators. Let $\alpha = (\alpha_1, \ldots, \alpha_d)$ be a *multi-index* so $\alpha \in (\mathbb{N} \cup \{0\})^d$. We define $|\alpha| = \alpha_1 + \cdots + \alpha_d$ and

$$D^\alpha = \frac{1}{i^{|\alpha|}} \frac{\partial^{\alpha_1}}{\partial x_1^{\alpha_1}} \cdots \frac{\partial^{\alpha_d}}{\partial x_d^{\alpha_d}}.$$

Similarly, if $x = (x_1, \ldots, x_d) \in \mathbb{R}^d$, then $x^\alpha = x_1^{\alpha_1} \cdots x_d^{\alpha_d}$.

Now we define *Schwartz space* $S(\mathbb{R}^d, \mathbb{C})$ to be the linear space of all $f \in C^\infty(\mathbb{R}^d, \mathbb{C})$ for which

$$\sup_{x \in \mathbb{R}^d} |x^\beta D^\alpha f(x)| < \infty,$$

for all multi-indices α and β. Note that $C_c^\infty(\mathbb{R}^d, \mathbb{C}) \subset S(\mathbb{R}^d, \mathbb{C})$ and the "Gaussian function" $x \to e^{-|x|^2}$ is in $S(\mathbb{R}^d, \mathbb{C})$. $S(\mathbb{R}^d, \mathbb{C})$ is dense in $C_0(\mathbb{R}^d, \mathbb{C})$ and in $L^p(\mathbb{R}^d, \mathbb{C})$ for all $1 \leq p < \infty$. These statements remain true when $\mathbb{C}$ is replaced by $\mathbb{R}$.

$S(\mathbb{R}^d, \mathbb{C})$ is a Fréchet space with respect to the family of norms $\{\|.\|_N, N \in \mathbb{N} \cup \{0\}\}$ where for each $f \in S(\mathbb{R}^d, \mathbb{C})$,

$$\|f\|_N = \max_{|\alpha| \leq N} \sup_{x \in \mathbb{R}^d} (1 + |x|^2)^N |D^\alpha f(x)|.$$

The dual of $S(\mathbb{R}^d, \mathbb{C})$ with this topology is the space $S'(\mathbb{R}^d, \mathbb{C})$ of *tempered distributions*.

$\mathcal{F}$ is a continuous bijection of $S(\mathbb{R}^d, \mathbb{C})$ into itself with a continuous inverse and we have the important

Theorem 5.1 (Fourier inversion). *If $f \in S(\mathbb{R}^d, \mathbb{C})$ then*

$$f(x) = (2\pi)^{-\frac{d}{2}} \int_{\mathbb{R}^d} \hat{f}(u) e^{i(u,x)} du.$$

In the final part of this section, we show how the Fourier transform allows us to build pseudo-differential operators. We begin by examining the Fourier transform of differential operators. More or less everything flows from the following simple fact:

$$D^\alpha e^{i(u,x)} = u^\alpha e^{i(u,x)},$$

for each $x, u \in \mathbb{R}^d$ and each multi-index α.

Using Fourier inversion and dominated convergence, we then find that

$$(D^\alpha f)(x) = (2\pi)^{-\frac{d}{2}} \int_{\mathbb{R}^d} u^\alpha \hat{f}(u) e^{i(u,x)} du,$$

for all $f \in S(\mathbb{R}^d, \mathbb{C}), x \in \mathbb{R}^d$.

If p is a polynomial in u of the form $p(u) = \sum_{|\alpha| \le k} c_\alpha u^\alpha$ where $k \in \mathbb{N}$ and each $c_\alpha \in \mathbb{C}$, we can form the associated differential operator $P(D) = \sum_{|\alpha| \le k} c_\alpha D^\alpha$ and by linearity

$$(P(D)f)(x) = (2\pi)^{-\frac{d}{2}} \int_{\mathbb{R}^d} p(u)\hat{f}(u)e^{i(u,x)} du.$$

The next step is to employ variable coefficients. If each $c_\alpha \in C^\infty(\mathbb{R}^d)$, for example, we may define $p(x, u) = \sum_{|\alpha| \le k} c_\alpha(x)u^\alpha$ and $P(x, D) = \sum_{|\alpha| \le k} c_\alpha(x)D^\alpha$. We then find that

$$(P(x, D)f)(x) = (2\pi)^{-\frac{d}{2}} \int_{\mathbb{R}^d} p(x, u)\hat{f}(u)e^{i(u,x)} du.$$

The passage from D to $P(x, D)$ has been rather straightforward, but now we will take a leap into the unknown and abandon formal notions of differentiation. So we replace p by a more general function $\sigma : \mathbb{R}^d \times \mathbb{R}^d \to \mathbb{C}$. Informally, we may then define a *pseudo-differential operator* $\sigma(x, D)$ by the prescription:-

$$(\sigma(x, D)f)(x) = (2\pi)^{-\frac{d}{2}} \int_{\mathbb{R}^d} \sigma(x, u)\hat{f}(u)e^{i(u,x)} du,$$

and σ is then called the *symbol* of this operator. Of course we have been somewhat cavalier here and we should make some further assumptions on the symbol σ to ensure that $\sigma(x, D)$ really is a bona fide operator. There are various classes of symbols which may be defined to achieve this. One of the most useful is the *Hörmander class* $S^m_{\rho,\delta}$. This is defined to be the set of all $\sigma \in C^\infty(\mathbb{R}^d)$ such that for each multi-index α and β,

$$|D^\alpha_x D^\beta_u \sigma(x, u)| \le C_{\alpha,\beta}(1 + |u|^2)^{\frac{1}{2}(m - \rho|\alpha| + \delta|\beta|)},$$

for each $x, u \in \mathbb{R}^d$, where $C_{\alpha,\beta} > 0, m \in \mathbb{R}$ and $\rho, \delta \in [0, 1]$. In this case $\sigma(x, D) : S(\mathbb{R}^d, \mathbb{C}) \to S(\mathbb{R}^d, \mathbb{C})$ and extends to an operator $S'(\mathbb{R}^d, \mathbb{C}) \to S'(\mathbb{R}^d, \mathbb{C})$.

For those who hanker after operators in Banach spaces, note the following,

- If $\rho > 0$ and $m < -d + \rho(d-1)$, then $\sigma(x, D) : L^p(\mathbb{R}^d, \mathbb{C}) \to L^p(\mathbb{R}^d, \mathbb{C})$ for $1 \le p \le \infty$.
- If $m = 0$ and $0 \le \delta < \rho \le 1$, then $\sigma(x, D) : L^2(\mathbb{R}^d, \mathbb{C}) \to L^2(\mathbb{R}^d, \mathbb{C})$.

Proofs of these and more general results can be found in Taylor [79]. However, note that this book, like most on the subject, is written from the point of view of partial differential equations, where it is natural for the symbol to be smooth in both variables. For applications to Markov processes, this is too restrictive and we usually impose much weaker requirements on the dependence of σ in the x-variable (see Jacob [48]).

6 Generators of Lévy Processes

The next result is the key theorem of this lecture.

Theorem 6.1. *Let X be a Lévy process with Lévy symbol η and characteristics (b, a, ν). Let $(T_t, t \geq 0)$ be the associated Feller semigroup and A be its infinitesimal generator.*

1. *For each $t \geq 0, f \in S(\mathbb{R}^d), x \in \mathbb{R}^d$,*

$$(T_t f)(x) = (2\pi)^{-\frac{d}{2}} \int_{\mathbb{R}^d} e^{i(u,x)} e^{t\eta(u)} \hat{f}(u) du,$$

   so that T_t is a pseudo-differential operator with symbol $e^{t\eta}$.
2. *For each $f \in S(\mathbb{R}^d), x \in \mathbb{R}^d$,*

$$(Af)(x) = (2\pi)^{-\frac{d}{2}} \int_{\mathbb{R}^d} e^{i(u,x)} \eta(u) \hat{f}(u) du,$$

   so that A is a pseudo-differential operator with symbol η.
3. *For each $f \in S(\mathbb{R}^d), x \in \mathbb{R}^d$,*

$$(Af)(x) = b^i \partial_i f(x) + \frac{1}{2} a^{ij} \partial_i \partial_j f(x) +$$

$$+ \int_{\mathbb{R}^d - \{0\}} [f(x+y) - f(x) - y^i \partial_i f(x) \chi_{\hat{B}}(y)] \nu(dy). \quad (6.1)$$

Proof. (Sketch) I'll leave out all the analytic details and just present the "bare bones" of the calculations.

1. We apply Fourier inversion to find for all $t \geq 0, f \in S(\mathbb{R}^d), x \in \mathbb{R}^d$,

$$(T_t f)(x) = \mathbb{E}(f(X(t) + x)) = (2\pi)^{-\frac{d}{2}} \mathbb{E}\left[\int_{\mathbb{R}^d} e^{i(u, x + X(t))} \hat{f}(u) du\right].$$

Apply Fubini's theorem to obtain

$$(T_t f)(x) = (2\pi)^{-\frac{d}{2}} \int_{\mathbb{R}^d} e^{i(u,x)} \mathbb{E}(e^{i(u, X(t))}) \hat{f}(u) du$$

$$= (2\pi)^{-\frac{d}{2}} \int_{\mathbb{R}^d} e^{i(u,x)} e^{t\eta(u)} \hat{f}(u) du.$$

2. For each $f \in S(\mathbb{R}^d), x \in \mathbb{R}^d$, we have by the result of (1),

$$(Af)(x) = \lim_{t \to 0} \frac{1}{t}((T_t f)(x) - f(x))$$

$$= (2\pi)^{-\frac{d}{2}} \lim_{t \to 0} \int_{\mathbb{R}^d} e^{i(u,x)} \frac{e^{t\eta(u)} - 1}{t} \hat{f}(u) du.$$

Use dominated convergence to deduce the required result.

3. Applying the Lévy-Khinchine formula to the result of (2), we obtain for each $f \in S(\mathbb{R}^d), x \in \mathbb{R}^d$,

$$(Af)(x) = (2\pi)^{-\frac{d}{2}} \int_{\mathbb{R}^d} e^{i(x,u)} \left[i(b,u) - \frac{1}{2}(au,u) + \right.$$

$$\left. + \int_{\mathbb{R}^d-\{0\}} (e^{i(u,y)} - 1 - i(u,y)\chi_{\hat{B}}(y))\nu(dy) \right] \hat{f}(u)du.$$

The result now follows immediately from elementary properties of the Fourier transform.

$\square$

The results of Theorem 6.1 can be written in the convenient shorthand form,

$$\widehat{(T(t)f)}(u) = e^{t\eta(u)}\hat{f}(u) \quad \widehat{Af}(u) = \eta(u)\hat{f}(u),$$

for each $t \geq 0, f \in S(\mathbb{R}^d), u \in \mathbb{R}^d$.

Example 1: Standard Brownian Motion

Let X be a standard Brownian motion in $\mathbb{R}^d$. Then X has characteristics $(0, I, 0)$ and so we see from (6.1) that

$$A = \frac{1}{2}\sum_{i=1}^{d} \partial_i^2 = \frac{1}{2}\Delta,$$

where Δ is the usual Laplacian operator.

Example 2: Brownian Motion with Drift

Let X be a Brownian motion with drift in $\mathbb{R}^d$. Then X has characteristics $(b, a, 0)$ and A is a diffusion operator of the form

$$A = b^i \partial_i + \frac{1}{2}a^{ij}\partial_i\partial_j,$$

Example 3 : The Poisson Process

Let X be a Poisson process with intensity $\lambda > 0$. Then X has characteristics $(0, 0, \lambda\delta_1)$ and A is a difference operator

$$(Af)(x) = \lambda(f(x+1) - f(x)),$$

for all $f \in S(\mathbb{R}^d), x \in \mathbb{R}^d$. Note that $||Af|| \leq 2\lambda||f||$, so that A extends to a bounded operator on the whole of $C_0(\mathbb{R}^d)$.

Example 4: The Compound Poisson Process
Exercise 2.7. Verify that

$$(Af)(x) = \int_{\mathbb{R}^d} (f(x+y) - f(x))\nu(dy),$$

for all $f \in S(\mathbb{R}^d), x \in \mathbb{R}^d$, where ν is a finite measure. A again extends to a bounded operator on the whole of $C_0(\mathbb{R}^d)$.

Example 5 : Rotationally Invariant Stable Processes

Let X be a rotationally invariant stable process of index α where $0 < \alpha < 2$. Its symbol is given by $\eta(u) = -|u|^\alpha$ for all $u \in \mathbb{R}^d$(see section 1.2.5). It is instructive to pretend that η is the symbol of a legitimate differential operator, then using the usual correspondence $u_j \to -i\partial_j$, for $1 \leq j \leq d$, we would write

$$A = \eta(D) = -(\sqrt{-\partial_1^2 - \partial_2^2 - \ldots - \partial_d^2})^\alpha$$
$$= -(-\Delta)^{\frac{\alpha}{2}}.$$

In fact, it is very useful to interpret $\eta(D)$ as a "fractional power of the Laplacian". We will consider fractional powers of more general generators in the next section.

Example 6: Relativistic Schrödinger Operators

Fix $m, c > 0$ and consider the Lévy symbol $-E_{m,c}$ which represents (minus) the free energy of a particle of mass m moving at relativistic speeds (when $d = 3$),

$$E_{m,c}(u) = \sqrt{m^2c^4 + c^2|u|^2} - mc^2.$$

Arguing as above, we make the correspondence $u_j \to -i\partial_j$, for $1 \leq j \leq d$. Readers with a background in physics, will recognise that this is precisely the prescription for "quantisation" of the free energy, and the corresponding generator is then given by

$$A = -(\sqrt{m^2c^4 - c^2\Delta} - mc^2).$$

$-A$ is called a *relativistic Schrödinger operator* by physicists. Of course, it is more natural from the point of view of quantum mechanics to consider this as an operator in $L^2(\mathbb{R}^d)$ - see later. For more on relativistic Schrödinger operators from both a probabilistic and physical point of view, see Carmona et al. [25], and references therein.

The pseudo-differential operator representation of generators extends to a wide class of Markov processes - the symbols will, in general, be functions of x as well as u. They will still have a "Lévy-Khinchine type structure", but the characteristics are no longer constant. Results of this type are due to Courrège, and have been used in recent years by Jacob, Schilling and Hoh to study path properties of Feller processes (see e.g. Jacob [48]).

Note that if X is a Lévy process, we can also give the resolvent a probabilistic interpretation - for each $\lambda > 0$, $(R_\lambda(A)f)(x) = \mathbb{E}(f(x + X(\tau)))$, where τ is an exponentially distributed "random time", which is independent of the process X and has rate $\frac{1}{\lambda}$. It is not difficult to check that $R_\lambda(A)$ is a pseudo-differential operator with symbol $(\lambda - \eta(\cdot))^{-1}$.

6.1 Subordination of Semigroups

We now apply some of the ideas developed above to the subordination of semigroups.

In the following, X will always denote a Lévy process in $\mathbb{R}^d$ with symbol η_X, Feller semigroup $(T_t^X, t \geq 0)$ and generator A_X.

Let $S = (S(t), t \geq 0)$ be a subordinator, so that S is an one-dimensional, increasing Lévy process and for each $u, t > 0$,

$$\mathbb{E}(e^{-uS(t)}) = e^{-t\psi(u)},$$

where ψ is the Bernstein function given by

$$\psi(u) = bu + \int_0^\infty (1 - e^{-uy})\lambda(dy),$$

with $b \geq 0$ and $\int_0^\infty (y \wedge 1)\lambda(dy) < \infty$.

Recall from Theorem 3.5 and Proposition 3.6 that $Z = (Z(t), t \geq 0)$ is again a Lévy process where each $Z(t) = X(T(t))$ and the symbol of Z is $\eta_Z = -\psi \circ (-\eta_X)$. We write $(T_t^Z, t \geq 0)$ and A_Z for the semigroup and generator associated to Z, respectively.

Theorem 6.2. *1. For all $t \geq 0, f \in B_b(\mathbb{R}^d), x \in \mathbb{R}^d$,*

$$(T_t^Z f)(x) = \int_0^\infty (T_s^X f)(x) p_{S(t)}(ds).$$

2. For all $f \in S(\mathbb{R}^d)$,

$$A^Z f = bA^X f + \int_0^\infty (T_s^X f - f)\lambda(ds).$$

Proof.

1. By Exercise 1.13, for each $A \in \mathcal{B}(\mathbb{R}^d)$, $p_{Z(t)}(A) = \int_0^\infty p_{X(s)}(A)p_{S(t)}(ds)$. Hence for each $t \geq 0, f \in B_b(\mathbb{R}^d), x \in \mathbb{R}^d$, we obtain

$$\begin{aligned}
(T_t^Z f)(x) &= \mathbb{E}(f(Z(t) + x)) \\
&= \int_{\mathbb{R}^d} f(x + y) p_{Z(t)}(dy) \\
&= \int_0^\infty \left(\int_{\mathbb{R}^d} f(x + y) p_{X(s)}(dy) \right) p_{S(t)}(ds) \\
&= \int_0^\infty (T_s^X f)(x) p_{S(t)}(ds).
\end{aligned}$$

2.

$$\eta_Z(u) = -\psi \circ (-\eta_X) = b\eta_X(u) + \int_0^\infty (e^{s\eta_X(u)} - 1)\lambda(ds) \dots \text{(i)},$$

but by Theorem 6.1 (2), we have

$$(A_Z f)(x) = (2\pi)^{-\frac{d}{2}} \int_{\mathbb{R}^d} e^{i(u,x)} \eta_Z(u) \hat{f}(u) du \dots \text{(ii)}.$$

The required result now follows from substituting (i) into (ii), a straightforward application of Fubini's theorem and a further application of Theorem 6.1 (1) and (2). The details are left as an exercise.

$$\square$$

The formula $\eta_Z = -\psi \circ (-\eta_X)$ suggests a natural functional calculus wherein we define $A_Z = -\psi(-A_X)$ for any Bernstein function ψ. As an example, we may generalise the fractional power of the Laplacian, discussed in the last section, to define $(-A_X)^\alpha$ for any Lévy process X and any $0 < \alpha < 1$. To carry this out, we employ the α-stable subordinator. This has characteristics $(0, \lambda)$ where $\lambda(dx) = \frac{\alpha}{\Gamma(1-\alpha)} \frac{dx}{x^{1+\alpha}}$. Theorem 6.2 (2) then yields the beautiful formula

$$-(-A_X)^\alpha f = \frac{\alpha}{\Gamma(1-\alpha)} \int_0^\infty (T_s^X f - f) \frac{ds}{s^{1+\alpha}}, \tag{6.2}$$

for all $f \in S(\mathbb{R}^d)$.

Theorem 6.2 has a far-reaching generalisation which we quote without proof:-

Theorem 6.3 (Phillips). *Let $(T_t, t \geq 0)$ be a strongly continuous, contraction semigroup of linear operators on a Banach space B with infinitesimal generator A and let $(S(t), t \geq 0)$ be a subordinator with characteristics (b, λ).*

- *The prescription*

$$T_t^S \phi = \int_0^\infty (T_s \phi) p_{S(t)}(ds),$$

  for each $t \geq 0, \phi \in B$, defines a strongly continuous, contraction semigroup $(T_t^S, t \geq 0)$ in B.
- *If A^S is the infinitesimal generator of $(T_t^S, t \geq 0)$, then D_A is a core for A^S and for each $\phi \in D_A$,*

$$A^S \phi = bA\phi + \int_0^\infty (T_s^X \phi - \phi) \lambda(ds).$$

- *If $B = C_0(\mathbb{R}^d)$ and $(T_t, t \geq 0)$ is a Feller semigroup, then $(T_t^S, t \geq 0)$ is also a Feller semigroup.*

For a proof of this result see e.g. Sato [74] p.212-5.

This powerful theorem enables the extension of (6.2) to define fractional powers of arbitrary infinitesimal generators of semigroups.

7 L^p-Markov Semigroups and Lévy Processes

We fix $1 \leq p < \infty$ and let $(T_t, t \geq 0)$ be a strongly continuous, contraction semigroup of operators in $L^p(\mathbb{R}^d)$. We say that it is *sub-Markovian* if $f \in L^p(\mathbb{R}^d)$ and

$$0 \leq f \leq 1 \text{ (a.e.)} \Rightarrow 0 \leq T_t f \leq 1 \text{ (a.e.)},$$

for all $t \geq 0$.

Any semigroup on $L^p(\mathbb{R}^d)$ can be restricted to the dense subspace $C_c(\mathbb{R}^d)$. If this restriction can then be extended to a semigroup on $B_b(\mathbb{R}^d)$ which satisfies $T_t(1) = 1$ then the semigroup is said to be *conservative*.

A semigroup which is both sub-Markovian and conservative is said to be L^p-*Markov*.

Notes (a) Be mindful that the phrases "strongly continuous" and "contraction" in the above definition are now with respect to the L^p norm, given by $\|g\|_p = \left(\int_{\mathbb{R}^d} |g(x)|^p dx \right)^{\frac{1}{p}}$, for each $g \in L^p(\mathbb{R}^d)$.

(b) If $(T_t, t \geq 0)$ is sub-Markovian then it is L^p-*positivity preserving* in that $f \in L^p(\mathbb{R}^d)$ and $f \geq 0$ a.e. $\Rightarrow T_t f \geq 0$ a.e. for all $t \geq 0$.

Example Let $X = (X(t), t \geq 0)$ be a Markov process on $\mathbb{R}^d$ and define the usual stochastic evolution $(T_t f)(x) = \mathbb{E}(f(X(t))|X(0) = x)$ for each $f \in B_b(\mathbb{R}^d), x \in \mathbb{R}^d, t \geq 0$. Suppose that $(T_t, t \geq 0)$ also yields a strongly continuous, contraction semigroup on $L^p(\mathbb{R}^d)$, then it is clearly L^p-Markov.

Our good friends the Lévy processes provide a natural class for which the conditions of the last example hold, as the next theorem demonstrates.

Theorem 7.1. *If $X = (X(t), t \geq 0)$ is a Lévy process, then for each $1 \leq p < \infty$, the prescription $(T_t f)(x) = \mathbb{E}(f(X(t) + x))$ where $f \in L^p(\mathbb{R}^d), x \in \mathbb{R}^d, t \geq 0$ gives rise to an L^p-Markov semigroup $(T_t \geq 0)$.*

We omit the proof - but we should check that T_t is a bona fide operator in L^p. Let q_t be the law of $X(t)$, for each $t \geq 0$. For all $f \in L^p(\mathbb{R}^d), t \geq 0$, by Jensen's inequality (or Hölder's inequality if you prefer) and Fubini's theorem, we obtain

$$\|T_t f\|_p^p = \int_{\mathbb{R}^d} \left| \int_{\mathbb{R}^d} f(x+y) q_t(dy) \right|^p dx$$

$$\leq \int_{\mathbb{R}^d} \int_{\mathbb{R}^d} |f(x+y)|^p q_t(dy) dx$$

$$= \int_{\mathbb{R}^d} \left(\int_{\mathbb{R}^d} |f(x+y)|^p dx \right) q_t(dy)$$

$$= \int_{\mathbb{R}^d} \left(\int_{\mathbb{R}^d} |f(x)|^p dx \right) q_t(dy) = \|f\|_p^p,$$

and we have proved that each T_t is a contraction in $L^p(\mathbb{R}^d)$.

For the case, $p = 2$ we can explicitly compute the domain of the infinitesimal generator of a Lévy process. To establish this, let X be a Lévy process with Lévy symbol η and let A be the infinitesimal generator of the associated L^2-Markov semigroup. Define

$$\mathcal{H}_\eta(\mathbb{R}^d) = \left\{ f \in L^2(\mathbb{R}^d); \int_{\mathbb{R}^d} |\eta(u)|^2 |\hat{f}(u)|^2 du < \infty \right\}.$$

Theorem 7.2. $D_A = \mathcal{H}_\eta(\mathbb{R}^d)$.

See [13] or Berg and Forst [18], p.92. for the proof. Readers should note that the proof has also established the pseudo-differential operator representation

$$Af = (2\pi)^{-\frac{d}{2}} \int_{\mathbb{R}^d} e^{i(u,x)} \eta(u) \hat{f}(u) du,$$

for all $f \in \mathcal{H}_\eta(\mathbb{R}^d)$.

The space $\mathcal{H}_\eta(\mathbb{R}^d)$ is called an *anisotropic Sobolev space* by Jacob [48]. Note that if we take X to be a standard Brownian motion then $\eta(u) = -\frac{1}{2}|u|^2$, for all $u \in \mathbb{R}^d$ and

$$\mathcal{H}_\eta(\mathbb{R}^d) = \left\{ f \in L^2(\mathbb{R}^d); \int_{\mathbb{R}^d} |u|^4 |\hat{f}(u)|^2 du < \infty \right\}.$$

This is precisely the Sobolev space, usually denoted $\mathcal{H}_2(\mathbb{R}^d)$ which can equivalently be defined as the completion of $C_c^\infty(\mathbb{R}^d)$ with respect to the norm

$$\|f\|_2 = \left(\int_{\mathbb{R}^d} (1 + |u|^2)^2 |\hat{f}(u)|^2 du \right)^{\frac{1}{2}},$$

for each $f \in C_c^\infty(\mathbb{R}^d)$. By Theorem 7.2, $\mathcal{H}_2(\mathbb{R}^d)$ is the domain of the Laplacian Δ acting in $L^2(\mathbb{R}^d)$.

Exercise 2.8. Write down the domains of the fractional powers of the Laplacian $(-\Delta)^{\frac{\alpha}{2}}$, where $0 < \alpha < 2$.

7.1 Self-Adjoint Semigroups

We begin with some general considerations.

Let H be a Hilbert space and $(T_t, t \geq 0)$ be a strongly continuous, contraction semigroup in H. We say that $(T_t, t \geq 0)$ is *self-adjoint* if $T_t = T_t^*$, for each $t \geq 0$.

Theorem 7.3. *There is a one-to-one correspondence between the generators of self-adjoint semigroups in H and linear operators A in H such that $-A$ is positive, self-adjoint.*

See Davies [29] p. 99-100 for a proof.

Theorem 7.4. *If X is a Lévy process, then its associated semigroup $(T_t, t \geq 0)$ is self-adjoint in $L^2(\mathbb{R}^d)$ if and only if X is symmetric.*

Proof. We'll prove the easy part of this here. Suppose that X is symmetric, then $q_t(A) = q_t(-A)$ for each $A \in \mathcal{B}(\mathbb{R}^d), t \geq 0$, where q_t is the law of $X(t)$. Then for each $f \in L^2(\mathbb{R}^d), x \in \mathbb{R}^d, t \geq 0$,

$$(T_t f)(x) = \mathbb{E}(f(x + X(t)) = \int_{\mathbb{R}^d} f(x + y) q_t(dy)$$

$$= \int_{\mathbb{R}^d} f(x + y) q_t(-dy) = \int_{\mathbb{R}^d} f(x - y) q_t(dy)$$

$$= \mathbb{E}(f(x - X(t)).$$

So for each $f, g \in L^2(\mathbb{R}^d), t \geq 0$, using Fubini's theorem, we obtain

$$< T_t f, g > = \int_{\mathbb{R}^d} (T_t f)(x) g(x) dx$$

$$= \int_{\mathbb{R}^d} \mathbb{E}(f(x - X(t)) g(x) dx$$

$$= \int_{\mathbb{R}^d} \left(\int_{\mathbb{R}^d} f(x - y) g(x) dx \right) q_t(dy)$$

$$= \int_{\mathbb{R}^d} \left(\int_{\mathbb{R}^d} f(x) g(x + y) dx \right) q_t(dy)$$

$$= < f, T_t g > .$$

□

Corollary 7.5. *Let A be the infinitesimal generator of a Lévy process with Lévy symbol η, then $-A$ is positive, self-adjoint if and only if*

$$\eta(u) = -\frac{1}{2}(u, au) + \int_{\mathbb{R}^d - \{0\}} (\cos(u, y) - 1) \nu(dy),$$

for each $u \in \mathbb{R}^d$, where a is a positive definite symmetric matrix and ν is a symmetric Lévy measure.

Equivalently, we see that A is self-adjoint if and only if $\Im \eta = 0$.

In particular, we find that the discussion of this section has yielded a probabilistic proof of the self-adjointness of the following important operators in $L^2(\mathbb{R}^d)$.

Example 1 The Laplacian

In fact, we consider multiples of the Laplacian and let $a = 2\gamma I$ where $\gamma > 0$, then for all $u \in \mathbb{R}^d$,

$$\eta(u) = -\gamma|u|^2 \quad \text{and} \quad A = \gamma\Delta.$$

Example 2 Fractional Powers of the Laplacian

Let $0 < \alpha < 2$, and for all $u \in \mathbb{R}^d$,

$$\eta(u) = |u|^\alpha \quad \text{and} \quad A = -(-\Delta)^{\frac{\alpha}{2}}.$$

Example 3 Relativistic Schrödinger Operators

Let $m, c > 0$ and for all $u \in \mathbb{R}^d$,

$$E_{m,c}(u) = \sqrt{m^2 c^4 + c^2 |u|^2} - mc^2 \quad \text{and} \quad A = -(\sqrt{m^2 c^4 - c^2 \Delta} - mc^2).$$

Note that in all three of the above examples, the domain of each operator is the non-isotropic Sobolev space of Theorem 7.2.

Examples 1 and 3 are important in quantum mechanics as the observables (modulo a minus sign) which describe the kinetic energy of a particle moving at non-relativistic (for a suitable value of γ) and relativistic speeds, respectively. We emphasise that it is vital that we know that such operators really are self-adjoint (and not just symmetric, say) so that they legitimately satisfy the quantum-mechanical formalism.

Note that, in general, if A_X is the self-adjoint generator of a Lévy process and $(S(t), t \geq 0)$ is an independent subordinator then the generator A_Z of the subordinated process Z is also self-adjoint. This follows immediately from (i) in the proof of Theorem 6.2 (2).

Dirichlet Forms

Let A be the self-adjoint generator of a symmetric Lévy process and for each $f, g \in C_c^\infty(\mathbb{R}^d)$, define

$$\mathcal{E}(f, g) = - < f, Ag >,$$

then $\mathcal{E}$ extends to a *symmetric Dirichlet form* in $L^2(\mathbb{R}^d)$, i.e. a closed symmetric form in H with domain D, such that $f \in D \Rightarrow (f \vee 0) \wedge 1 \in D$ and

$$\mathcal{E}((f \vee 0) \wedge 1) \leq \mathcal{E}(f) \tag{7.1}$$

for all $f \in D$, where we have written $\mathcal{E}(f) = \mathcal{E}(f, f)$. A straightforward calculation (*Exercise* 2.9) yields

$$\mathcal{E}(f, g) = \frac{1}{2} a^{ij} \int_{\mathbb{R}^d} (\partial_i f)(x)(\partial_j g)(x) dx$$
$$+ \frac{1}{2} \int_{(\mathbb{R}^d \times \mathbb{R}^d) - D} (f(x) - f(x+y))(g(x) - g(x+y)) \nu(dy) dx$$

where D is the diagonal, $D = \{(x, x), x \in \mathbb{R}^d\}$. This is the prototype for the *Beurling-Deny formula* for symmetric Dirichlet forms. Remarkably, (7.1) encodes the Markov property and this has deep consequences (see e.g. Fukushima et. al. [38], Chapter 3 of [13] and references therein).

8 Lecture 3: Analysis of Jumps

"As a further precaution, to render any escape impossible, they passed a rope around his neck, ran the two ends between his legs and tied them to his wrists - the device know in prisons as 'the martingale' ".

Victor Hugo "Les Miserables"

Starting with this lecture, we enter the world of modern stochastic analysis. We begin by looking at some key concepts.

8.1 Martingales

From now on we will assume that our filtration $(\mathcal{F}_t, t \geq 0)$ satisfies the "usual hypotheses": -

1. (Completeness) $\mathcal{F}_0$ contains all sets of P-measure zero.
2. (Right continuity) $\mathcal{F}_t = \mathcal{F}_{t+}$ where $\mathcal{F}_{t+} = \bigcap_{\epsilon>0} \mathcal{F}_{t+\epsilon}$.

Given a filtration $(\mathcal{F}_t, t \geq 0)$, we can always enlarge it to satisfy the completeness property (1) by the following trick. Let $\mathcal{N}$ denote the collection of all sets of P-measure zero in $\mathcal{F}$ and define $\mathcal{G}_t = \mathcal{F}_t \vee \mathcal{N}$ for each $t \geq 0$, then $(\mathcal{G}_t, t \geq 0)$ is another filtration of $\mathcal{F}$ which we call the *augmented filtration*.
Now let X be an adapted process defined on a filtered probability space which also satisfies the integrability requirement $\mathbb{E}(|X(t)|) < \infty$ for all $t \geq 0$. We say that it is a *martingale* if for all $0 \leq s < t < \infty$,

$$\mathbb{E}(X(t)|\mathcal{F}_s) = X(s) \quad \text{a.s.}$$

Note that if X is a martingale, then the map $t \to \mathbb{E}(X(t))$ is constant.
Here's a nice example of a martingale built from a Lévy process.

Proposition 8.1. *If X is a Lévy process with Lévy symbol η, then for each $u \in \mathbb{R}^d, M_u = (M_u(t), t \geq 0)$ is a complex martingale with respect to $\mathcal{F}^X$ where each*

$$M_u(t) = e^{i(u,X(t))-t\eta(u)}.$$

Proof. $\mathbb{E}(|M_u(t)|) = e^{-t\eta(u)} < \infty$ for each $t \geq 0$.
For each $0 \leq s \leq t$, write $M_u(t) = M_u(s)e^{i(u,X(t)-X(s))-(t-s)\eta(u)}$; then by (L2) and Theorem 3.2,

$$\mathbb{E}(M_u(t)|\mathcal{F}_s^X) = M_u(s)\mathbb{E}(e^{i(u,X(t-s))})e^{-(t-s)\eta(u)}$$
$$= M_u(s) \quad \text{as required.}$$

$\square$

Exercise 3.1 Show that the following processes, whose values at each $t \geq 0$ are given below, are all martingales:-

(i) $C(t) = \sigma B(t)$ where $B(t)$ is a standard Brownian motion, and σ is an $r \times d$ matrix.

(ii) $|C(t)|^2 - \mathrm{tr}(A)t$ where $A = \sigma^T \sigma$.

(iii) $\exp((u, C(t)) - \frac{1}{2}(u, Au))$ where $u \in \mathbb{R}^d$.

(iv) $\tilde{N}(t)$ where $\tilde{N}$ is a compensated Poisson process with intensity λ.

(v) $\tilde{N}(t)^2 - \lambda t$.

(vi) $(\mathbb{E}(Y|\mathcal{F}_t), t \geq 0)$ where Y is an arbitrary random variable in a filtered probability space for which $\mathbb{E}(|Y|) < \infty$.

Martingales which are of the form (vi) above are called *closed*. Note that in (i) to (v), the martingales have mean zero. In general, martingales with this latter property are said to be *centered*. A martingale $M = (M(t), t \geq 0)$ is said to be L^2, if $\mathbb{E}(|M(t)|^2) < \infty$ for each $t \geq 0$ and is *continuous* if it has continuous sample paths.

A more wide-ranging concept than the martingale is the following:-

An adapted process X for which $\mathbb{E}(|X(t)|) < \infty$ for all $t \geq 0$ is a *submartingale* if for all $0 \leq s < t < \infty, 1 \leq i \leq d$,

$$\mathbb{E}(X_i(t)|\mathcal{F}_s) \geq X_i(s) \quad \text{a.s}$$

X is called a *supermartingale* if $-X$ is a submartingale.

By a straightforward application of the conditional form of Jensen's inequality (*Exercise* 3.2) we see that if X is a real-valued martingale and if $f : \mathbb{R} \to \mathbb{R}$ is convex with $\mathbb{E}(|f(X(t))|) < \infty$ for all $t \geq 0$, then $f(X)$ is a submartingale. In particular, if each $X(t) \geq 0$ (a.s.) then $(X(t)^p, t \geq 0)$ is a submartingale whenever $1 < p < \infty$ and $\mathbb{E}(|X(t)|^p) < \infty$ for all $t \geq 0$.

The following estimate is very useful for sharpening pointwise convergence to uniform convergence on compacta:-

Theorem 8.2 (Doob's Martingale Inequality). *If $(X(t), t \geq 0)$ is a positive submartingale, then for any $p > 1$,*

$$\mathbb{E}\left(\sup_{0 \leq s \leq t} X(s)^p\right) \leq q^p \mathbb{E}(X(t)^p),$$

where $\dfrac{1}{p} + \dfrac{1}{q} = 1$.

See e.g. Revuz and Yor [68] for a proof.

8.2 Càdlàg Paths

A function $f : \mathbb{R}^+ \to \mathbb{R}^d$ is *càdlàg* if it is *continue à droite et limité à gauche*, i.e. right continuous with left limits. Such a function has only jump discontinuities.

Define $f(t-) = \lim_{s \uparrow t} f(s)$ and $\Delta f(t) = f(t) - f(t-)$. If f is càdlàg, $\{0 \leq t \leq T, \Delta f(t) \neq 0\}$ is at most countable.

Here are some important facts about the paths of martingales and Lévy processes, see e.g. Revuz and Yor [68] for the proofs of the first, Protter [66] or [13] for proofs of the others.

- If M is a martingale, whose filtration satisfies the usual hypotheses, then M has a càdlàg modification.
- Every Lévy process has a càdlàg modification which is itself a Lévy process.
- If X is a Lévy process with càdlàg paths, then its augmented natural filtration is right continuous.

From now on, we will always make the following assumptions:-

- $(\Omega, \mathcal{F}, P)$ will be a fixed probability space equipped with a filtration $(\mathcal{F}_t, t \geq 0)$ which satisfies the usual hypotheses.
- Every Lévy process $X = (X(t), t \geq 0)$ will be assumed to be $\mathcal{F}_t$-adapted and have càdlàg sample paths.
- $X(t) - X(s)$ is independent of $\mathcal{F}_s$ for all $0 \leq s < t < \infty$.

8.3 Stopping Times

A *stopping time* is a random variable $T : \Omega \rightarrow [0, \infty]$ for which the event $(T \leq t) \in \mathcal{F}_t$, for each $t \geq 0$.
Any ordinary deterministic time is clearly a stopping time. A more interesting example which has many important applications is the *first hitting time of a process to a set*. Let X be an $\mathcal{F}_t$-adapted càdlàg process and $A \in \mathcal{B}(\mathbb{R}^d)$ then

$$T_A = \inf\{t \geq 0; X(t) \in A\}.$$

e.g. if A is open, $\{T_A \geq t\} = \bigcap_{r \in \mathbb{Q}, r \leq t}\{X(r) \in A^c\} \in \mathcal{F}_t$.
If X is an adapted process and T is a stopping time (with respect to the same filtration) then the *stopped random variable* $X(T)$ is defined by

$$X(T)(\omega) = X(T(\omega))(\omega),$$

and the *stopped σ-algebra* $\mathcal{F}_T$, by

$$\mathcal{F}_T = \{A \in \mathcal{F}; A \cap \{T \leq t\} \in \mathcal{F}_t, \forall t \geq 0\}.$$

If X is càdlàg, then $X(T)$ is $\mathcal{F}_T$-measurable.
A key application of these concepts is in providing the following "random time" version of the martingale notion

Theorem 8.3 (Doob's Optional Stopping Theorem). *If X is a càdlàg martingale and S and T are bounded stopping times for which $S \leq T$ (a.s.), then $X(S)$ and $X(T)$ are both integrable with*

$$\mathbb{E}(X(T)|\mathcal{F}_S) = X(S) \quad a.s.$$

For a proof, see e.g. Revuz and Yor [68]. An immediate corollary is that

$$\mathbb{E}(X(T)) = \mathbb{E}(X(0)),$$

for each bounded stopping time T.

Exercise 3.3 If S and T are stopping times and $\alpha \geq 1$ show that $S+T, \alpha T, S\wedge T$ and $S \vee T$ are also stopping times.

If T is an unbounded stopping time and one wants to employ Theorem 8.3, a useful trick is to replace T by the bounded stopping times $T \wedge n$ (where $n \in \mathbb{N}$) and then take the limit as $n \to \infty$ to obtain the required result. This procedure is sometimes called *localisation*.

A *local martingale*. This is an adapted process $M = (M(t), t \geq 0)$ for which there exists a sequence of stopping times $\tau_1 \leq \tau_2 \leq \ldots \leq \tau_n \to \infty$ (a.s.), such that each of the processes $(M(t \wedge \tau_n), t \geq 0)$ is a martingale. Any martingale is clearly a local martingale.

Here's a nice application of stopping times.

Theorem 8.4. *Let $B = (B(t), t \geq 0)$ be a one-dimensional standard Brownian motion and for each $t \geq 0$ define*

$$T(t) = \inf\{s > 0; B(s) = \frac{t}{\sqrt{2}}\};$$

then $T = (T(t), t \geq 0)$ is the Lévy subordinator.

Proof. (cf. Rogers and Williams [69] p.18). Clearly each $T(t)$ is a stopping time. By Exercise 3.1(ii), the process given for each $\theta \in \mathbb{R}$, by $M_\theta(t) = \exp(\theta B(t) - \frac{1}{2}\theta^2 t)$ is a continuous martingale with respect to the augmented natural filtration for Brownian motion. Now by Theorem 8.3, for each $t \geq 0, n \in \mathbb{N}$, we have

$$1 = \mathbb{E}(\exp\{\theta B(T(t) \wedge n) - \frac{1}{2}\theta^2(T(t) \wedge n)\}).$$

In this case (see [13]), the limiting argument works and we have,

$$1 = \mathbb{E}(\exp\{\theta B(T(t)) - \frac{1}{2}\theta^2 T(t)\})$$
$$= e^{\frac{\theta t}{\sqrt{2}}} \mathbb{E}\exp\{-\frac{1}{2}\theta^2 T(t)\}.$$

On substituting $\theta = \sqrt{2u}$, we obtain

$$\mathbb{E}(\exp\{-uT(t)\}) = \exp(-t\sqrt{u}).$$

$\square$

Exercise 3.4 Generalise the proof given above to find the characteristic function for the inverse Gaussian subordinator.

If X is an $\mathcal{F}_t$-adapted process and T is a stopping time then we may define a new process $X_T = (X_T(t), t \geq 0)$ by the procedure

$$X_T(t) = X(T + t) - X(T),$$

for each $t \geq 0$. The following result is called the strong Markov property for Lévy processes.

Theorem 8.5 (Strong Markov Property). *If X is a Lévy process and T is a stopping time, then on the set $(T < \infty)$*

1. *X_T is again a Lévy process which is independent of $\mathcal{F}_T$.*
2. *For each $t \geq 0, X_T(t)$ has the same law as $X(t)$.*
3. *X_T has càdlàg paths and is $\mathcal{F}_{T+t}$-adapted.*

See Protter [66] or [13] for the proof.

8.4 The Jumps of A Lévy Process - Poisson Random Measures

The *jump process* $\Delta X = (\Delta X(t), t \geq 0)$ associated to a Lévy process is defined by

$$\Delta X(t) = X(t) - X(t-),$$

for each $t \geq 0$.

Theorem 8.6. *If N is a Lévy process which is increasing (a.s.) and is such that $(\Delta N(t), t \geq 0)$ takes values in $\{0, 1\}$, then N is a Poisson process.*

Proof. Define a sequence of stopping times recursively by $T_0 = 0$ and $T_n = \inf\{t > T_{n-1}; N(t + T_{n-1}) - N(T_{n-1})) \neq 0\}$ for each $n \in \mathbb{N}$. It follows from (L2) that the sequence $(T_1, T_2 - T_1, \ldots, T_n - T_{n-1}, \ldots)$ is i.i.d.
By (L2) again, we have for each $s, t \geq 0$,

$$P(T_1 > s + t) = P(N(s) = 0, N(t + s) - N(s) = 0)$$
$$= P(T_1 > s)P(T_1 > t).$$

From the fact that N is increasing (a.s.), it follows easily that the map $t \to P(T_1 > t)$ is decreasing and by a straightforward application of stochastic continuity (L3) we find that the map $t \to P(T_1 > t)$ is continuous at $t = 0$. Hence there exists $\lambda > 0$ such that $P(T_1 > t) = e^{-\lambda t}$ for each $t \geq 0$. So T_1 has an exponential distribution with parameter λ and

$$P(N(t) = 0) = P(T_1 > t) = e^{-\lambda t},$$

for each $t \geq 0$.
Now assume as an inductive hypothesis that $P(N(t) = n) = e^{-\lambda t}\frac{(\lambda t)^n}{n!}$, then

$$P(N(t) = n + 1) = P(T_{n+2} > t, T_{n+1} \leq t) = P(T_{n+2} > t) - P(T_{n+1} > t).$$

But $T_{n+1} = T_1 + (T_2 - T_1) + \cdots + (T_{n+1} - T_n)$

is the sum of $(n+1)$ i.i.d. exponential random variables, and so has a gamma distribution with density $f_{T_{n+1}}(s) = e^{-\lambda s} \frac{\lambda^{n+1} s^n}{n!}$ for $s > 0$.
The required result follows on integration. □

The following result shows that ΔX is not a straightforward process to analyse.

Lemma 8.7. *If X is a Lévy process, then for fixed $t > 0, \Delta X(t) = 0$ (a.s.).*

Proof. Let $(t(n), n \in \mathbb{N})$ be a sequence in $\mathbb{R}^+$ with $t(n) \uparrow t$ as $n \to \infty$, then since X has càdlàg paths, $\lim_{n \to \infty} X(t(n)) = X(t-)$. However, by (L3) the sequence $(X(t(n)), n \in \mathbb{N})$ converges in probability to $X(t)$, and so has a subsequence which converges almost surely to $X(t)$. The result follows by uniqueness of limits. □

Much of the analytic difficulty in manipulating Lévy processes arises from the fact that it is possible for them to have

$$\sum_{0 \leq s \leq t} |\Delta X(s)| = \infty \quad \text{a.s.}$$

and the way in which these difficulties are overcome exploits the fact that we always have

$$\sum_{0 \leq s \leq t} |\Delta X(s)|^2 < \infty \quad \text{a.s.}$$

We will gain more insight into these ideas as the discussion progresses.

Exercise 3.5 Show that $\sum_{0 \leq s \leq t} |\Delta X(s)| < \infty$ (a.s.) if X is a compound Poisson process.

Rather than exploring ΔX itself further, we will find it more profitable to count jumps of specified size. More precisely, let $0 \leq t < \infty$ and $A \in \mathcal{B}(\mathbb{R}^d - \{0\})$. Define

$$N(t, A) = \#\{0 \leq s \leq t; \Delta X(s) \in A\}$$
$$= \sum_{0 \leq s \leq t} \chi_A(\Delta X(s)).$$

Note that for each $\omega \in \Omega, t \geq 0$, the set function $A \to N(t, A)(\omega)$ is a counting measure on $\mathcal{B}(\mathbb{R}^d - \{0\})$ and hence

$$\mathbb{E}(N(t, A)) = \int N(t, A)(\omega) dP(\omega)$$

is a Borel measure on $\mathcal{B}(\mathbb{R}^d - \{0\})$. We write $\mu(\cdot) = \mathbb{E}(N(1, \cdot))$ and call it the *intensity measure* associated to X.
We say that $A \in \mathcal{B}(\mathbb{R}^d - \{0\})$ is *bounded below* if $0 \notin \bar{A}$.

Lemma 8.8. *If A is bounded below, then* $N(t, A) < \infty$ *(a.s.) for all* $t \geq 0$.

Proof. Define a sequence of stopping times $(T_n^A, n \in \mathbb{N})$ by $T_1^A = \inf\{t > 0; \Delta X(t) \in A\}$, and for $n > 1, T_n^A = \inf\{t > T_{n-1}^A; \Delta X(t) \in A\}$. Since X has càdlàg paths, we have $T_1^A > 0$ (a.s.) and $\lim_{n \to \infty} T_n^A = \infty$ (a.s.). Indeed if either of these were not the case, then the set of all jumps in A would have an accumulation point, and this is not possible if X is càdlàg. Hence, for each $t \geq 0$,

$$N(t, A) = \sum_{n \in \mathbb{N}} \chi_{\{T_n^A \leq t\}} < \infty \text{ a.s.}$$

$\square$

Be aware that if A fails to be bounded below, then Lemma 8.8 may no longer hold, because of the accumulation of large numbers of small jumps.

The following result should at least be plausible, given Theorem 8.6 and Lemma 8.8. See [13] for a proof.

Theorem 8.9. *1. If A is bounded below, then* $(N(t, A), t \geq 0)$ *is a Poisson process with intensity* $\mu(A)$.

    2. If $A_1, \ldots, A_m \in \mathcal{B}(\mathbb{R}^d - \{0\})$ *are disjoint, then the random variables* $N(t, A_1), \ldots, N(t, A_m)$ *are independent.*

It follows immediately that $\mu(A) < \infty$ whenever A is bounded below, hence the measure μ is σ-finite.

The main properties of N, which we will use extensively in the sequel, are summarised below:-.

1. For each $t > 0, \omega \in \Omega, N(t, .)(\omega)$ is a counting measure on $\mathcal{B}(\mathbb{R}^d - \{0\})$.
2. For each A bounded below, $(N(t, A), t \geq 0)$ is a Poisson process with intensity $\mu(A) = \mathbb{E}(N(1, A))$.
3. The *compensator* $(\tilde{N}(t, A), t \geq 0)$ is a martingale-valued measure where $\tilde{N}(t, A) = N(t, A) - t\mu(A)$, for A bounded below, i.e.
   For fixed A bounded below, $(\tilde{N}(t, A), t \geq 0)$ is a martingale.
   For fixed $t \geq 0, \omega \in \Omega, \tilde{N}(t, \cdot)(\omega)$ is a σ-finite measure (almost surely).

8.5 Poisson Integration

Let f be a Borel measurable function from $\mathbb{R}^d$ to $\mathbb{R}^d$ and let A be bounded below then for each $t > 0, \omega \in \Omega$, we may define the *Poisson integral* of f as a random finite sum by

$$\int_A f(x) N(t, dx)(\omega) = \sum_{x \in A} f(x) N(t, \{x\})(\omega).$$

Note that each $\int_A f(x) N(t, dx)$ is an $\mathbb{R}^d$-valued random variable and gives rise to a càdlàg stochastic process, as we vary t.

Now since $N(t, \{x\}) \neq 0 \Leftrightarrow \Delta X(u) = x$ for at least one $0 \leq u \leq t$, we have

$$\int_A f(x)N(t, dx) = \sum_{0 \leq u \leq t} f(\Delta X(u))\chi_A(\Delta X(u)). \tag{8.1}$$

In the sequel, we will sometimes use μ_A to denote the restriction to A of the measure μ. In the following theorem, Var stands for variance.

Theorem 8.10. *Let A be bounded below, then*

1. $\left(\int_A f(x)N(t, dx), t \geq 0\right)$ *is a compound Poisson process, with characteristic function*

$$\mathbb{E}\left(e^{i\left(u, \int_A f(x)N(t, dx)\right)}\right) = e^{t \int_A (e^{i(u,x)} - 1)\mu_f(dx)},$$

   for each $u \in \mathbb{R}^d$, where $\mu_f = \mu \circ f^{-1}$.
2. *If $f \in L^1(A, \mu_A)$, then*

$$\mathbb{E}\left(\int_A f(x)N(t, dx)\right) = t \int_A f(x)\mu(dx).$$

3. *If $f \in L^2(A, \mu_A)$, then*

$$Var\left(\left\|\int_A f(x)N(t, dx)\right\|\right) = t \int_A |f(x)|^2 \mu(dx).$$

Proof. - part of it!

1. For simplicity, we will prove this result in the case where $f \in L^1(A, \mu_A)$. The general proof for arbitrary measurable f can be found in Sato [74] p.124. First let f be a simple function and write $f = \sum_{j=1}^n c_j \chi_{A_j}$ where each $c_j \in \mathbb{R}^d$. We can assume, without loss of generality, that the A_j's are disjoint Borel subsets of A. By Theorem 8.9, we find that

$$\mathbb{E}\left(e^{i\left(u, \int_A f(x)N(t, dx)\right)}\right) = \mathbb{E}\left(e^{i\left(u, \sum_{j=1}^n c_j N(t, A_j)\right)}\right)$$

$$= \prod_{j=1}^n \mathbb{E}\left(e^{i(u, c_j N(t, A_j))}\right)$$

$$= \prod_{j=1}^n e^{t(e^{i(u,c_j)} - 1)\mu(A_j)}$$

$$= e^{t \int_A (e^{i(u, f(x))} - 1)\mu(dx)}.$$

Now for an arbitrary $f \in L^1(A, \mu_A)$, we can find a sequence of simple functions converging to f in L^1 and hence a subsequence which converges to f almost surely. Passing to the limit along this subsequence in the above yields the required result, via dominated convergence.

(2) and (3) follow from (1) by differentiation.   □

It follows from Theorem 8.10 (2) that a Poisson integral will fail to have a finite mean if $f \notin L^1(A, \mu)$.

Exercise 3.6 Show that if $\int_A |f(x)| \mu(dx) < \infty$ then

$$\sum_{0 \leq u \leq t} |f(\Delta X(u))| \chi_A(\Delta X(u)) < \infty \qquad \text{(a.s.)}.$$

For each $f \in L^1(A, \mu_A), t \geq 0$, we define the *compensated Poisson integral* by

$$\int_A f(x) \tilde{N}(t, dx) = \int_A f(x) N(t, dx) - t \int_A f(x) \mu(dx).$$

A straightforward argument, as in Exercise 3.1(iv), shows that $\left(\int_A f(x) \tilde{N}(t, dx), t \geq 0 \right)$ is a martingale and we will use this fact extensively in the sequel. Note that by Theorem 8.10 (2) and (3), we can easily deduce the following two important facts:

$$\mathbb{E} \left(e^{i \left(u, \int_A f(x) \tilde{N}(t, dx) \right)} \right) = e^{t \int_A (e^{i(u,x)} - 1 - i(u,x)) \mu_f(dx)}, \qquad (8.2)$$

for each $u \in \mathbb{R}^d$, and for $f \in L^2(A, \mu_A)$,

$$\mathbb{E} \left(\left| \int_A f(x) \tilde{N}(t, dx) \right|^2 \right) = t \int_A |f(x)|^2 \mu(dx). \qquad (8.3)$$

8.6 Processes of Finite Variation

We begin by introducing a useful class of functions. Let $\mathcal{P} = \{a = t_1 < t_2 < \cdots < t_n < t_{n+1} = b\}$ be a partition of the interval $[a, b]$ in $\mathbb{R}$, and define its mesh to be $\delta = \max_{1 \leq i \leq n} |t_{i+1} - t_i|$. We define the *variation* $\text{Var}_\mathcal{P}(g)$ of a càdlàg mapping $g : [a, b] \to \mathbb{R}^d$ over the partition $\mathcal{P}$ by the prescription

$$\text{Var}_\mathcal{P}(g) = \sum_{i=1}^{n} |g(t_{i+1}) - g(t_i)|.$$

If $V(g) = \sup_\mathcal{P} \text{Var}_\mathcal{P}(g) < \infty$, we say that g has *finite variation* on $[a, b]$. If g is defined on the whole of $\mathbb{R}$ (or $\mathbb{R}^+$), it is said to have *finite variation* if it has finite variation on each compact interval.

It is a trivial observation that every non-decreasing g is of finite variation. Conversely if g is of finite variation, then it can always be written as the difference of two non-decreasing functions (to see this, just write $g = \frac{V(g)+g}{2} - \frac{V(g)-g}{2}$, where $V(g)(t)$ is the variation of g on $[a, t]$).

Functions of finite variation are important in integration, for suppose that we are given a function g which we are proposing as an integrator, then as a minimum we will want to be able to define the Stieltjes integral $\int_I f \, dg$, for all continuous functions f (where I is some finite interval). It is shown on p.40-41 of Protter [66], that a necessary and sufficient condition for obtaining such an integral as a limit of Riemann sums is that g has finite variation.

Exercise 3.7 Show that all the functions of finite variation on $[a, b]$ (or on $\mathbb{R}$) form a vector space.

A stochastic process $(X(t), t \geq 0)$ is of *finite variation* if the paths $(X(t)(\omega), t \geq 0)$ are of finite variation for almost all $\omega \in \Omega$. The following is an important example for us.

Example *Poisson Integrals*

Let N be a Poisson random measure with intensity measure μ and let $f : \mathbb{R}^d \to \mathbb{R}^d$ be Borel measurable. For A bounded below, let $Y = (Y(t), t \geq 0)$ be given by $Y(t) = \int_A f(x) N(t, dx)$, then Y is of finite variation on $[0, t]$ for each $t \geq 0$. To see this, we observe that for all partitions $\mathcal{P}$ of $[0, t]$, we have

$$\mathrm{Var}_{\mathcal{P}}(Y) \leq \sum_{0 \leq s \leq t} |f(\Delta X(s))| \chi_A(\Delta X(s)) < \infty \quad \text{a.s.} \tag{8.4}$$

where $X(t) = \int_A x N(t, dx)$, for each $t \geq 0$.

Exercise 3.8 Let Y be a Poisson integral as above and let η be its Lévy symbol. For each $u \in \mathbb{R}^d$, consider the martingales $M_u = (M_u(t), t \geq 0)$ where each

$$M_u(t) = e^{i(u, Y(t)) - t\eta(u)}.$$

Show that M_u is of finite variation. (Hint: Use the mean value theorem.)

Exercise 3.9 Show that every subordinator is of finite variation.

In fact, a necessary and sufficient condition for a Lévy process to be of finite variation is that there is no Brownian part (i.e. $a = 0$ in the Lévy-Khinchine formula), and $\int_{|x|<1} |x| \nu(dx) < \infty$, see e.g. Bertoin [19] p.15.

8.7 The Lévy-Itô Decomposition

This is the key result of this lecture.
First, note that for A bounded below, for each $t \geq 0$

$$\int_A x N(t, dx) = \sum_{0 \leq u \leq t} \Delta X(u) \chi_A(\Delta X(u))$$

is the sum of all the jumps taking values in the set A up to the time t. Since the paths of X are càdlàg, this is clearly a finite random sum. In particular,

$\int_{|x|\geq 1} xN(t,dx)$ is the sum of all jumps of size bigger than one. It is a compound Poisson process, has finite variation but may have no finite moments. Conversely it can be shown that $X(t) - \int_{|x|\geq 1} xN(t,dx)$ is a Lévy process having finite moments to all orders.

Now lets turn our attention to the small jumps. We study compensated integrals, which we know are martingales. Introduce the notation $M(t,A) = \int_A x\tilde{N}(t,dx)$, for $t \geq 0$ and A bounded below. For each $m \in \mathbb{N}$, let $B_m = \left\{ x \in \mathbb{R}^d, \frac{1}{m+1} < |x| \leq \frac{1}{m} \right\}$ and for each $n \in \mathbb{N}$, let $A_n = \bigcup_{m=1}^{n} B_m$. It can be shown that

$$\int_{|x|<1} x\tilde{N}(t,dx) = L^2 - \lim_{n\to\infty} M(t,A_n),$$

and hence it is a martingale. Moreover, on taking limits in (8.2), we get

$$\mathbb{E}\left(\exp i \left(u, \int_{|x|<1} x\tilde{N}(t,dx) \right) \right) = e^{t\int_{|x|<1}(e^{i(u,x)}-1-i(u,x))\mu(dx)}.$$

Consider

$$B_a(t) = X(t) - bt - \int_{|x|<1} x\tilde{N}(t,dx) - \int_{|x|\geq 1} xN(t,dx),$$

where $b = \mathbb{E}\left(X(1) - \int_{|x|\geq 1} xN(1,dx) \right)$. The process B_a is a centred martingale with continuous sample paths. With a little more work, we can show that $\mathrm{Cov}(B_a^i(t)B_a^j(t)) = a^{ij}t$. From this and Lévy's celebrated martingale characterisation of Brownian motion (to be proved in the next lecture) we have that B_a is a Brownian motion with covariance a. Hence we have

Theorem 8.11 (The Lévy-Itô Decomposition).

If X is a Lévy process, then there exists $b \in \mathbb{R}^d$, a Brownian motion B_a with covariance matrix a in $\mathbb{R}^d$ and an independent Poisson random measure N on $\mathbb{R}^+ \times (\mathbb{R}^d - \{0\})$ such that for each $t \geq 0$,

$$X(t) = bt + B_a(t) + \int_{|x|<1} x\tilde{N}(t,dx) + \int_{|x|\geq 1} xN(t,dx) \qquad (8.5)$$

Note that the three processes in (8.5) are all independent.

Exercise 3.10 Write down the Lévy-Itô decompositions for the cases where X is (a) α-stable, (b) a subordinator, (c) a subordinated process.

Exercise 3.11 Deduce that if X is a Lévy process then for each $t \geq 0$,

$$\sum_{0 \leq s \leq t} (\Delta X(s))^2 < \infty \qquad \text{(a.s.)}.$$

An interesting by-product of the Lévy-Itô decomposition is the Lévy-Khintchine formula, which follows easily by independence in the Lévy-Itô decomposition:-

Corollary 8.12. *If X is a Lévy process, then for each $u \in \mathbb{R}^d, t \geq 0$,*

$$\mathbb{E}(e^{i(u,X(t))}) = \tag{8.6}$$

$$\exp\left(t\left[i(b,u) - \frac{1}{2}(u,Au) + \int_{\mathbb{R}^d - \{0\}} (e^{i(u,y)} - 1 - i(u,y)\chi_B(y))\mu(dy)\right]\right)$$
$$\tag{8.7}$$

so the intensity measure μ is the Lévy measure for X.

The process $\int_{|x|<1} x\tilde{N}(t,dx)$ is the *compensated sum of small jumps*. The compensation takes care of the analytic complications in the Lévy-Khintchine formula in a probabilistically pleasing way, since it is an L^2-martingale.

The process $\int_{|x|\geq 1} xN(t,dx)$ describes the "large jumps" - it is a compound Poisson process, but may have no finite moments.

H.Geman, D.Madan and M.Yor [39] have proposed a nice financial interpretation for the jump terms in the Lèvy-Itô decomposition:- where the intensity measure is infinite, the stock price manifests "infinite activity" and this is the mathematical signature of the jitter arising from the interaction of pure supply shocks and pure demand shocks. On the other hand, where the intensity measure is finite, we have "finite activity", and this corresponds to sudden shocks that can cause unexpected movements in the market, such as a terrorist atrocity or a major earthquake.

Semimartingales A stochastic process X is a *semimartingale* if it is an adapted process such that for each $t \geq 0$,

$$X(t) = X(0) + M(t) + C(t),$$

where $M = (M(t), t \geq 0)$ is a local martingale and $C = (C(t), t \geq 0)$ is an adapted process of finite variation. In particular

Every Lévy process is a semimartingale.

To see this, use the Lévy-Itô decomposition to write

$$M(t) = B_a(t) + \int_{|x|<1} x\tilde{N}(t,dx) \text{ - a martingale},$$

$$C(t) = bt + \int_{|x|\geq 1} xN(t,dx).$$

8.8 The Interlacing Construction

The interlacing technique gives greater insight into the Lévy-Itô decomposition.

Let $Y = (Y(t), t \geq 0)$ be a Lévy process with Lévy measure ν whose jumps are bounded by 1 so that we have the Lévy-Itô decomposition

$$Y(t) = bt + B_a(t) + \int_{|x|<1} x \tilde{N}(t, dx),$$

for each $t \geq 0$. For the following construction to be non-trivial we will find it convenient to assume that Y may have jumps of arbitrarily small size, i.e. there exists no $0 < a < 1$ such that $\nu((-a, a)) = 0$. Now define a sequence $(\epsilon_n, n \in \mathbb{N})$ which decreases monotonically to zero by

$$\epsilon_n = \sup \left\{ y \geq 0, \int_{0<|x|<y} x^2 \nu(dx) \leq 8^{-n} \right\}.$$

We define an associated sequence of Lévy processes $Y_n = (Y_n(t), t \geq 0)$ wherein the size of each jump is bounded below by ϵ_n and above by 1 as follows:

$$Y_n(t) = bt + B_a(t) + \int_{\epsilon_n \leq |x|<1} x \tilde{N}(t, dx)$$

$$= C_n(t) + \int_{\epsilon_n \leq |x|<1} x N(t, dx),$$

where for each $n \in \mathbb{N}$, C_n is the Brownian motion with drift given by

$$C_n(t) = B_a(t) + t \left(b - \int_{\epsilon_n \leq |x|<1} x \nu(dx) \right),$$

for each $t \geq 0$.

Now $\int_{\epsilon_n \leq |x|<1} x N(t, dx)$ is a compound Poisson process with jumps $\Delta Y(t)$ taking place at times $(T_n^m, m \in \mathbb{N})$. We can thus build the process Y_n by interlacing:

$$
\begin{aligned}
Y_n(t) &= C_n(t) \quad \text{for } 0 \leq t < T_n^1, \\
&= C_n(T_n^1) + \Delta Y(T_n^1) \quad \text{when } t = T_n^1, \\
&= Y_n(T_n^1) + C_n(t) - C_n(T_n^1) \quad \text{for } T_n^1 < t < T_n^2, \\
&= Y_n(T_n^2-) + \Delta Y(T_n^2) \quad \text{when } t = T_2,
\end{aligned}
$$

and so on recursively.

Our main result is the following

Theorem 8.13. *For each $t \geq 0$,*

$$\lim_{n \to \infty} Y_n(t) = Y(t) \quad a.s.$$

and the convergence is uniform on compact intervals of $\mathbb{R}^+$.

The proof can be found in [13], section 2.5.2.
Now let X be an arbitrary Lévy process then by the Lévy-Itô decomposition, for each $t \geq 0$

$$X(t) = Y(t) + \int_{|x| \geq 1} x N(t, dx).$$

But $\int_{|x| \geq 1} x N(t, dx)$ is a compound Poisson process and so the paths of X can be obtained by a further interlacing with jumps of size bigger than 1.

9 Lecture 4: Stochastic Integration

In this lecture, we give a rather rapid account of classical stochastic integration in a form suitable for application to Lévy processes.
Let $X = M + C$ be a semimartingale. The problem of stochastic integration is to make sense of objects of the form

$$\int_0^t F(s) dX(s) = \int_0^t F(s) dM(s) + \int_0^t F(s) dC(s).$$

The second integral can be well-defined using the usual Lebesgue-Stieltjes approach. The first one cannot - indeed if M is a continuous martingale of finite variation, then M is a.s. constant (see Revuz and Yor [68]).
Refer to the martingale part of the Lévy-Itô decomposition (8.11). Define a "martingale-valued measure" by

$$M(t, E) = B(t)\delta_0(E) + \tilde{N}(t, E - \{0\}),$$

for $E \in \mathcal{B}(\mathbb{R}^d)$, where $B = (B(t), t \geq 0)$ is a one-dimensional Brownian motion. The following key properties then hold:-

- $M((s, t], E) = M(t, E) - M(s, E)$ is independent of $\mathcal{F}_s$, for $0 \leq s < t < \infty$.
- $\mathbb{E}(M((s, t], E)) = 0$.
- $\mathbb{E}(M((s, t], E)^2) = \rho((s, t], E)$
  where $\rho((s, t], E) = (t - s)(\delta_0(E) + \nu(E - \{0\}))$.

We're going to unify the usual stochastic integral with the Poisson integral, by defining:

$$\int_0^t \int_E F(s, x) M(ds, dx) = \int_0^t G(s) dB(s) + \int_0^t \int_{E - \{0\}} F(s, x) \tilde{N}(ds, dx).$$

where $G(s) = F(s, 0)$. Of course, we need some conditions on the class of integrands:-

Fix $E \in \mathcal{B}(\mathbb{R}^d)$ and $0 < T < \infty$ and let $\mathcal{P}$ denote the smallest σ-algebra with respect to which all mappings $F : [0, T] \times E \times \Omega \to \mathbb{R}$ satisfying (1) and (2) below are measurable.

1. For each $0 \leq t \leq T$, the mapping $(x, \omega) \to F(t, x, \omega)$ is $\mathcal{B}(E) \otimes \mathcal{F}_t$ measurable,
2. For each $x \in E, \omega \in \Omega$, the mapping $t \to F(t, x, \omega)$ is left continuous.

We call $\mathcal{P}$ the *predictable σ-algebra*. A $\mathcal{P}$-measurable mapping $G : [0, T] \times E \times \Omega \to \mathbb{R}$ is then said to be *predictable*. The definition clearly extends naturally to the case where $[0, T]$ is replaced by $\mathbb{R}^+$.

Note that by (1), if G is predictable then the process $t \to G(t, x, \cdot)$ is adapted, for each $x \in E$. If G satisfies (1) and is left continuous then it is clearly predictable.

Define $\mathcal{H}_2(T, E)$ to be the linear space of all equivalence classes of mappings $F : [0, T] \times E \times \Omega \to \mathbb{R}$ which coincide almost everywhere with respect to $\rho \times P$ and which satisfy the following conditions:

- F is predictable,
-
$$\int_0^T \int_E \mathbb{E}(|F(t, x)|^2) \rho(dt, dx) < \infty.$$

It can be shown that $\mathcal{H}_2(T, E)$ is a real Hilbert space with respect to the inner product $< F, G >_{T, \rho} = \int_0^T \int_E \mathbb{E}((F(t, x), G(t, x))) \rho(dt, dx)$.

Define $S(T, E)$ to be the linear space of all simple processes in $\mathcal{H}_2(T, E)$, where F is *simple* if for some $m, n \in \mathbb{N}$, there exists $0 \leq t_1 \leq t_2 \leq \cdots \leq t_{m+1} = T$ and there exists a family of disjoint Borel subsets $A_1, A_2, \ldots, A_n$ of E with each $\mu(A_i) < \infty$ such that

$$F = \sum_{j=1}^{m} \sum_{k=1}^{n} F_k(t_j) \chi_{(t_j, t_{j+1}]} \chi_{A_k},$$

where each $F_k(t_j)$ is a bounded $\mathcal{F}_{t_j}$-measurable random variable. Note that F is left continuous and $\mathcal{B}(E) \otimes \mathcal{F}_t$ measurable, hence it is predictable. An important fact is that

$$S(T, E) \text{ is dense in } \mathcal{H}_2(T, E),$$

and this is proved in [13] - see also Steele [78] for a very careful treatment of the Brownian case.

One of Itô's greatest achievements was the definition of the stochastic integral $I_T(F)$, for F simple, by separating the "past" from the "future" within the Riemann sum:-

$$I_T(F) = \sum_{j=1}^{m} \sum_{k=1}^{n} F_k(t_j) M((t_j, t_{j+1}], A_k). \qquad (9.1)$$

Exercise 4.1 Deduce that, if $F, G \in S(T, E)$ and $\alpha, \beta \in \mathbb{R}$, then $\alpha F + \beta G \in S(T, E)$ and

$$I_T(\alpha F + \beta G) = \alpha I_T(F) + \beta I_T(G).$$

Lemma 9.1. *For each* $T \geq 0, F \in S(T, E)$,

$$\mathbb{E}(I_T(F)) = 0, \quad \mathbb{E}(I_T(F)^2) = \int_0^T \int_E \mathbb{E}(|F(t, x)|^2) \rho(dt, dx).$$

Proof. $\mathbb{E}(I_T(F)) = 0$ is a straightforward application of linearity and independence. The second result is quite messy - we lose nothing important by just looking at the Brownian case, with $d = 1$. So let $F(t) = \sum_{j=1}^{m} F(t_j) \chi_{(t_j, t_{j+1}]}$, then $I_T(F) = \sum_{j=1}^{m} F(t_j)(B(t_{j+1}) - B(t_j))$, and

$$I_T(F)^2 = \sum_{j=1}^{m} \sum_{p=1}^{m} F(t_j) F(t_p)(B(t_{j+1}) - B(t_j))(B(t_{p+1}) - B(t_p)).$$

Now fix j and split the second sum into three pieces - corresponding to $p < j, p = j$ and $p > j$. When $p < j$, $F(t_j)F(t_p)(B(t_{p+1}) - B(t_p)) \in \mathcal{F}_{t_j}$ which is independent of $B(t_{j+1}) - B(t_j)$,

$$\mathbb{E}[F(t_j)F(t_p)F(t_j)(B(t_{j+1}) - B(t_j))(B(t_{p+1}) - B(t_p))]$$
$$= \mathbb{E}[F(t_j)F(t_p)F(t_j)(B(t_{p+1}) - B(t_p))]\mathbb{E}(B(t_{j+1}) - B(t_j)) = 0.$$

Exactly the same argument works when $p > j$. What remains is the case $p = j$, and by independence again,

$$\mathbb{E}(I_T(F)^2) = \sum_{j=1}^{m} \mathbb{E}(F(t_j)^2)\mathbb{E}(B(t_{j+1}) - B(t_j))^2$$

$$= \sum_{j=1}^{m} \mathbb{E}(F(t_j)^2)(t_{j+1} - t_j). \qquad \square$$

We deduce from Lemma 9.1 and Exercise 4.1, that I_T is a linear isometry from $S(T, E)$ into $L^2(\Omega, \mathcal{F}, P)$, and hence it extends to an isometric embedding of the whole of $\mathcal{H}_2(T, E)$ into $L^2(\Omega, \mathcal{F}, P)$. We continue to denote this extension as I_T and we call $I_T(F)$ the *(Itô) stochastic integral* of $F \in \mathcal{H}_2(T, E)$. When convenient, we will use the Leibniz notation $I_T(F) = \int_0^T \int_E F(t, x)M(dt, dx)$. The following theorem summarises some useful properties of the stochastic integral.

Theorem 9.2. *If* $F, G \in \mathcal{H}_2(T, E)$ *and* $\alpha, \beta \in \mathbb{R}$, *then :*

1. $I_T(\alpha F + \beta G) = \alpha I_T(F) + \beta I_T(G)$.
2. $\mathbb{E}(I_T(F)) = 0, \quad \mathbb{E}(I_T(F)^2) = \int_0^T \int_E \mathbb{E}(|F(t,x)|^2)\rho(dt, dx)$.
3. $(I_t(F), t \geq 0)$ is $\mathcal{F}_t$-adapted.
4. $(I_t(F), t \geq 0)$ is a square-integrable martingale.

Proof. (1) and (2) are *Exercise 4.2*
For (3), let $(F_n, n \in \mathbb{N})$ be a sequence in $S(T, E)$ converging to F; then each process $(I_t(F_n), t \geq 0)$ is clearly adapted. Since each $I_t(F_n) \to I_t(F)$ in L^2 as $n \to \infty$, we can find a subsequence $(F_{n_k}, n_k \in \mathbb{N})$ such that $I_t(F_{n_k}) \to I_t(F)$ a.s. as $n_k \to \infty$, and the required result follows.

(4) Let $F \in S(T, E)$ and (without loss of generality) choose $0 < s = t_l < t_{l+1} < t$. Then it is easy to see that $I_t(F) = I_s(F) + I_{s,t}(F)$ and hence $\mathbb{E}_s(I_t(F)) = I_s(F) + \mathbb{E}_s(I_{s,t}(F))$ by (3). However,

$$\mathbb{E}_s(I_{s,t}(F)) = \mathbb{E}_s \left(\sum_{j=l+1}^m \sum_{k=1}^n F_k(t_j) M((t_j, t_{j+1}], A_k) \right)$$

$$= \sum_{j=l+1}^n \sum_{k=1}^n \mathbb{E}_s(F_k(t_j)) \mathbb{E}(M((t_j, t_{j+1}], A_k)) = 0.$$

The result now follows by the continuity of $\mathbb{E}_s$ in L^2. Indeed, let $(F_n, n \in \mathbb{N})$ be a sequence in $S(T, E)$ converging to F; then we have

$$||\mathbb{E}_s(I_t(F)) - \mathbb{E}_s(I_t(F_n))||_2 \leq ||I_t(F) - I_t(F_n)||_2$$
$$= ||F - F_n||_{T,\rho} \to 0 \text{ as } n \to \infty.$$

□

We can extend the stochastic integral $I_T(F)$ to integrands in $\mathcal{P}_2(T, E)$. This is the linear space of all equivalence classes of mappings $F : [0, T] \times E \times \Omega \to \mathbb{R}$ which coincide almost everywhere with respect to $\rho \times P$, and which satisfy the following conditions:

- F is predictable.
- $P \left(\int_0^T \int_E |F(t,x)|^2 \rho(dt, dx) < \infty \right) = 1$.

If $F \in \mathcal{P}_2(T, E), (I_t(F), t \geq 0)$ is always a local martingale, but not necessarily a martingale. See [13], section 4.2.2 for details.

Poisson Stochastic Integrals

Let A be an arbitrary Borel set in $\mathbb{R}^d - \{0\}$ which is bounded below, and introduce the compound Poisson process $P = (P(t), t \geq 0)$, where each $P(t) = \int_A x N(t, dx)$. Let K be a predictable mapping, then generalising equation (8.1), we define

$$\int_0^T \int_A K(t,x)N(dt,dx) = \sum_{0 \le u \le T} K(u, \Delta P(u))\chi_A(\Delta P(u)), \qquad (9.2)$$

as a random finite sum.

In particular, if H satisfies the square-integrability condition given above, we may then define, for each $1 \le i \le d$,

$$\int_0^T \int_A H^i(t,x)\tilde{N}(dt,dx) = \int_0^T \int_A H^i(t,x)N(dt,dx) - \int_0^T \int_A H^i(t,x)\nu(dx)dt.$$

The definition (9.2) can, in principle, be used to define stochastic integrals for a more general class of integrands than we have been considering. For simplicity, let $N = (N(t), t \ge 0)$ be a Poisson process of intensity 1 and let $f : \mathbb{R} \to \mathbb{R}$, then we may define

$$\int_0^t f(N(s))dN(s) = \sum_{0 \le s \le t} f(N(s-) + \Delta N(s))\Delta N(s).$$

Exercise 4.3. Show that for each $t \ge 0$,

$$\int_0^t N(s)d\tilde{N}(s) - \int_0^t N(s-)d\tilde{N}(s) = N(t).$$

Hence deduce that the process whose value at time t is $\int_0^t N(s)d\tilde{N}(s)$ cannot be a local martingale.

Within any theory of stochastic integration, it is highly desirable that the stochastic integral of a process against a martingale as integrator should at least be a local martingale. The last example illustrates the perils of abandoning the requirement of predictability on our integrands, which ensures that this is the case.

Lévy-type stochastic integrals

We take $E = \hat{B} - \{0\}$ throughout this subsection. We say that an $\mathbb{R}^d$-valued stochastic process $Y = (Y(t), t \ge 0)$ is a *Lévy-type stochastic integral* if it can be written in the following form for each $1 \le i \le d, t \ge 0$,

$$Y^i(t) = Y^i(0) + \int_0^t G^i(s)ds + \int_0^t F_j^i(s)dB^j(s) + \int_0^t \int_{|x| < 1} H^i(s,x)\tilde{N}(ds,dx)$$

$$+ \int_0^t \int_{|x| \ge 1} K^i(s,x)N(ds,dx), \qquad (9.3)$$

where for each $1 \le i \le d, 1 \le j \le m, t \ge 0, |G^i|^{\frac{1}{2}}, F_j^i \in \mathcal{P}_2(T), H^i \in \mathcal{P}_2(T, E)$ and K is predictable. B is an m-dimensional standard Brownian motion and

N is an independent Poisson random measure on $\mathbb{R}^+ \times (\mathbb{R}^d - \{0\})$ with compensator $\tilde{N}$ and intensity measure ν, which we will assume is a Lévy measure. We will assume that the random variable $Y(0)$ is $\mathcal{F}_0$-measurable, and then it is clear that Y is an adapted process.

We can often simplify complicated expressions by employing the notation of *stochastic differentials* to represent Lévy-type stochastic integrals. We then write (9.3) as

$$dY(t) = G(t)dt + F(t)dB(t) + H(t,x)\tilde{N}(dt,dx) + K(t,x)N(dt,dx).$$

When we want to particularly emphasise the domains of integration with respect to x, we will use an equivalent notation

$$dY(t) = G(t)dt + F(t)dB(t) + \int_{|x|<1} H(t,x)\tilde{N}(dt,dx) + \int_{|x|\geq 1} K(t,x)N(dt,dx).$$

Clearly Y is a semimartingale.

Let $M = (M(t), t \geq 0)$ be an adapted process which is such that $MJ \in \mathcal{P}_2(t,A)$ whenever $J \in \mathcal{P}_2(t,A)$ (where $A \in \mathcal{B}(\mathbb{R}^d)$ is arbitrary) . For example, it is sufficient to take M to be adapted and left-continuous.

For these processes we can define an adapted process $Z = (Z(t), t \geq 0)$ by the prescription that it has the stochastic differential

$$\begin{aligned} dZ(t) = {}& M(t)G(t)dt + M(t)F(t)dB(t) + M(t)H(t,x)\tilde{N}(dt,dx) \\ &+ M(t)K(t,x)N(dt,dx), \end{aligned}$$

and we will adopt the natural notation,

$$dZ(t) = M(t)dY(t).$$

Example (Lévy Stochastic Integrals)

Let X be a Lévy process with characteristics (b, a, ν) and Lévy-Itô decomposition given by equation (8.5):

$$X(t) = bt + B_a(t) + \int_{|x|<1} x\tilde{N}(t,dx) + \int_{|x|\geq 1} xN(t,dx),$$

for each $t \geq 0$. Let $L \in \mathcal{P}_2(t)$ for all $t \geq 0$ and in (9.3), choose each $F_j^i = a_j^i L, H^i = K^i = x^i L$. Then we can construct processes with the stochastic differential

$$dY(t) = L(t)dX(t) \tag{9.4}$$

We call Y a *Lévy stochastic integral.*

In the case where X has finite variation, the Lévy stochastic integral Y can also be constructed as a Lebesgue-Stieltjes integral, and this coincides, up to set of measure zero, with the prescription (9.4).

For many applications of interest, X is α-stable - for an alternative approach to stochastic integration in this case, see [71].

Example: The Ornstein Uhlenbeck Process (OU Process)

$$Y(t) = e^{-\lambda t} Y_0 + \int_0^t e^{-\lambda(t-s)} dX(s) \tag{9.5}$$

where Y_0 is a $\mathcal{F}_0$-measurable random variable. The condition $\int_{|x|>1} \log(1 + |x|)\nu(dx) < \infty$ is necessary and sufficient for there to be a choice of distribution for Y_0 such that it is stationary. There are important applications to finance which have recently been developed by Ole Barndorff-Nielsen and Neil Sheppard [16]. Intriguingly, every self-decomposable random variable can be naturally embedded in an OU process whose Lévy measure satisfies the logarithmic moment condition given above [80].

Exercise 4.4 If X is a standard Brownian motion show that each $Y(t)$ is Gaussian with mean $e^{-\lambda t} y_0$ and variance $\frac{1}{2\lambda}(1 - e^{-2\lambda t} I)$.

When X is a Brownian motion, the OU process is a good model of the physical phenomenon of Brownian motion (see Nelson [61]).

9.1 Itô's Formula

We begin with the easy case - Itô's formula for Poisson stochastic integrals of the form

$$W^i(t) = W^i(0) + \int_0^t \int_A K^i(t,x) N(dt, dx) \tag{9.6}$$

for $1 \leq i \leq d$, where $t \geq 0, A$ is bounded below and each K^i is predictable. Itô's formula for such processes takes a particularly simple form.

Lemma 9.3. *If W is a Poisson stochastic integral of the form (9.6) then for each $f \in C(\mathbb{R}^d)$, and for each $t \geq 0$, with probability one, we have*

$$f(W(t)) - f(W(0)) = \int_0^t \int_A [f(W(s-) + K(s,x)) - f(W(s-))] N(ds, dx).$$

Proof. Let $Y(t) = \int_A x N(dt, dx)$ and recall that the jump times for Y are defined recursively as $T_0^A = 0$ and for each $n \in \mathbb{N}, T_n^A = \inf\{t > T_{n-1}^A; \Delta Y(t) \in A\}$. We then find that,

$$f((W(t)) - f(W(0))$$

$$= \sum_{0 \le s \le t} f(W(s)) - f(W(s-))$$

$$= \sum_{n=1}^{\infty} f(W(t \wedge T_n^A)) - f(W(t \wedge T_{n-1}^A))$$

$$= \sum_{n=1}^{\infty} [f(W(t \wedge T_n^A -)) + K(t \wedge T_n^A, \Delta Y(t \wedge T_n^A)) - f(t \wedge W(T_n^A -))]$$

$$= \int_0^t \int_A [f(W(s-) + K(s,x)) - f(W(s-))] N(ds, dx).$$

$\square$

The celebrated Itô formula for Brownian motion is probably well-known to you so I'll briefly outline the proof. Let $(\mathcal{P}_n, n \in \mathbb{N})$ be a sequence of partitions of the form $\mathcal{P}_n = \{0 = t_0^{(n)} < t_1^{(n)} < \ldots < t_{m(n)}^{(n)} < t_{m(n)+1}^{(n)} = T\}$ and suppose that $\lim_{n \to \infty} \delta(\mathcal{P}_n) = 0$, where the mesh, $\delta(\mathcal{P}_n) = \max_{0 \le j \le m(n)} |t_{j+1}^{(n)} - t_j^{(n)}|$. As a preliminary - you need the following:-

Lemma 9.4. *If $W_{kl} \in \mathcal{H}_2(T)$ for each $1 \le k, l \le m$, then*

$$L^2 - \lim_{n \to \infty} \sum_{j=0}^{n} W_{kl}(t_j^{(n)})(B^k(t_{j+1}^{(n)}) - B^k(t_j^{(n)}))(B^l(t_{j+1}^{(n)}) - B^l(t_j^{(n)}))$$

$$= \sum_{k=1}^{m} \int_0^T W_{kk}(s) ds.$$

The proof is similar to that of Lemma 9.1 - but you will need the Gaussian moment $\mathbb{E}(B(t)^4) = 3t^2$ (see e.g. [13], section 4.4.1 for details). Now let M be a Brownian integral with drift of the form

$$M^i(t) = \int_0^t F_j^i(s) dB^j(s) + \int_0^t G^i(s) ds, \tag{9.7}$$

where each $F_j^i, (G^i)^{\frac{1}{2}} \in \mathcal{P}_2(t)$, for all $t \ge 0, 1 \le i \le d, 1 \le j \le m$. For each $1 \le i \le j$, we introduce the *quadratic variation process* denoted as $([M^i, M^j](t), t \ge 0)$ by

$$[M^i, M^j](t) = \sum_{k=1}^{m} \int_0^T F_k^i(s) F_k^j(s) ds.$$

We will explore quadratic variation in greater depth in the sequel. The following slick method of proving Itô's formula is based on the proof in Kunita [53], pp.64-5.

Theorem 9.5 (Itô's Theorem 1).

If $M = (M(t), t \geq 0)$ is a Brownian integral with drift of the form (9.7), then for all $f \in C^2(\mathbb{R}^d), t \geq 0$, with probability 1, we have

$$f(M(t)) - f(M(0)) = \int_0^t \partial_i f(M(s)) dM^i(s) + \frac{1}{2} \int_0^t \partial_i \partial_j f(M(s)) d[M^i, M^j](s).$$

Proof. Let $(\mathcal{P}_n, n \in \mathbb{N})$ be a sequence of partitions of $[0, t]$ as above. By Taylor's theorem, we have, for each such partition (where we suppress the index n).

$$f(M(t)) - f(M(0)) = \sum_{k=0}^m f(M(t_{k+1})) - f(M(t_k))$$

$$= J_1(t) + \frac{1}{2} J_2(t),$$

where

$$J_1(t) = \sum_{k=0}^m \partial_i f(M(t_k))(M^i(t_{k+1}) - M^i(t_k)),$$

$$J_2(t) = \sum_{k=0}^m \partial_i \partial_j f(N_{ij}^k)(M^i(t_{k+1}) - M^i(t_k))(M^j(t_{k+1}) - M^j(t_k)),$$

and where the N_{ij}^k's are each $\mathcal{F}(t_{k+1})$-adapted $\mathbb{R}^d$-valued random variables satisfying $|N_{ij}^k - M(t_k)| \leq |M(t_{k+1}) - M(t_k)|$.
We write each $J_2(t) = K_1(t) + K_2(t)$, where

$$K_1(t) = \sum_{k=0}^m \partial_i \partial_j f(M(t_k))(M^i(t_{k+1}) - M^i(t_k))(M^j(t_{k+1}) - M^j(t_k)),$$

$$K_2(t) = \sum_{k=0}^m [\partial_i \partial_j f(N_{ij}^k) - \partial_i \partial_j f(M(t_k))](M^i(t_{k+1}) - M^i(t_k))(M^j(t_{k+1}) - M^j(t_k)).$$

Now take limits as $n \to \infty$. It turns out that $K_2(t) \to 0$, in probability and the result follows. □

Itô's formula for general Lévy-type stochastic integrals is obtained essentially by combining the Poisson and Brownian results and making sure you take good care of the compensators for small jumps. You should be able to guess the right result. See e.g. [13] or Ikeda and Watanabe [47] for a proof.

To give a precise statement, consider a Lévy-type stochastic integral of the form

$$dY(t) = G(t)dt + F(t)dB(t) + H(t, x)\tilde{N}(dt, dx) + K(t, x)N(dt, dx). \quad (9.8)$$

Theorem 9.6 (Itô's Theorem 2).
If Y is a Lévy-type stochastic integral of the form (9.8), then for each $f \in C^2(\mathbb{R}^d), t \geq 0$, with probability 1, we have

$$f(Y(t)) - f(Y(0)) = \int_0^t \partial_i f(Y(s-)) dY_c^i(s) + \frac{1}{2} \int_0^t \partial_i \partial_j f(Y(s-)) d[Y_c^i, Y_c^j](s)$$

$$+ \int_0^t \int_{|x| \geq 1} [f(Y(s-) + K(s,x)) - f(Y(s-))] N(ds, dx)$$

$$+ \int_0^t \int_{|x| < 1} [f(Y(s-) + H(s,x)) - f(Y(s-))] \tilde{N}(ds, dx)$$

$$+ \int_0^t \int_{|x| < 1} [f(Y(s-) + H(s,x)) - f(Y(s-))$$

$$- H^i(s,x) \partial_i f(Y(s-))] \nu(dx) ds.$$

Tedious but straightforward algebra (*Exercise* 4.6) yields the following form, which is important since it extends to general semimartingales:-

Theorem 9.7 (Itô's Theorem 3). *If Y is a Lévy-type stochastic integral of the form (9.8), then for each $f \in C^2(\mathbb{R}^d), t \geq 0$, with probability 1, we have*

$$f(Y(t)) - f(Y(0)) = \int_0^t \partial_i f(Y(s-)) dY^i(s) + \frac{1}{2} \int_0^t \partial_i \partial_j f(Y(s-)) d[Y_c^i, Y_c^j](s)$$

$$+ \sum_{0 \leq s \leq t} [f(Y(s)) - f(Y(s-)) - \Delta Y^i(s) \partial_i f(Y(s-))].$$

Here Y_c denotes the *continuous* part of Y defined by $Y_c^i(t) = \int_0^t G^i(s) ds + \int_0^t F_j^i(s) dB^j(s)$.
A current fascinating area of investigation involves extending Itô's formula to fractional Brownian motion, which is not a semimartingale, see e.g. [4].

9.2 Quadratic Variation and Itô's Product Formula

We extend the definition of quadratic variation to the more general case of Lévy-type stochastic integrals $Y = (Y(t), t \geq 0)$ of the form (9.8). So for each $t \geq 0$ we define a $d \times d$ matrix-valued adapted process $[Y, Y] = ([Y, Y](t), t \geq 0)$ by the following prescription for its (i,j)th entry $(1 \leq i, j \leq d)$,

$$[Y^i, Y^j](t) = [Y_c^i, Y_c^j](t) + \sum_{0 \leq s \leq t} \Delta Y^i(s) \Delta Y^j(s). \tag{9.9}$$

Each $[Y^i, Y^j](t)$ is almost surely finite, and we have

$$[Y^i, Y^j](t) = \sum_{k=1}^{m} \int_0^T F_k^i(s) F_k^j(s) ds + \int_0^t \int_{|x|<1} H^i(s,x) H^j(s,x) N(ds,dx)$$

$$+ \int_0^t \int_{|x|\geq 1} K^i(s,x) K^j(s,x) N(ds,dx), \tag{9.10}$$

so that we clearly have each $[Y^i, Y^j](t) = [Y^j, Y^i](t)$. Note that the integral over small jumps in this case is always finite (Why ?)

Exercise 4.7 Show that for each $\alpha, \beta \in \mathbb{R}$ and $1 \leq i, j, k \leq d, t \geq 0$,

$$[\alpha Y^i + \beta Y^j, Y^k](t) = \alpha[Y^i, Y^k](t) + \beta[Y^j, Y^k](t).$$

The importance of $[Y, Y]$ is that it measures the deviation in the stochastic differential of products from the usual Leibniz formula. The following result makes this precise

Theorem 9.8 (Itô's Product Formula). *If Y^1 and Y^2 are real-valued Lévy-type stochastic integrals of the form (9.8), then for all $t \geq 0$, with probability one, we have that*

$$Y^1(t)Y^2(t) = Y^1(0)Y^2(0) + \int_0^t Y^1(s-)dY^2(s)$$

$$+ \int_0^t Y^2(s-)dY^1(s) + [Y^1, Y^2](t).$$

Proof. We consider Y^1 and Y^2 as components of a vector $Y = (Y^1, Y^2)$ and we take f in Theorem 9.7 to be the smooth mapping from $\mathbb{R}^2$ to $\mathbb{R}$ given by $f(x^1, x^2) = x^1 x^2$.
By Theorem 9.7, we then obtain, for each $t \geq 0$, with probability one,

$$Y^1(t)Y^2(t) = Y^1(0)Y^2(0) + \int_0^t Y^1(s-)dY^2(s)$$

$$+ \int_0^t Y^2(s-)dY^1(s) + [Y_c^1, Y_c^2](t)$$

$$+ \sum_{0 \leq s \leq t} [Y^1(s)Y^2(s) - Y^1(s-)Y^2(s-)$$

$$- (Y^1(s) - Y^1(s-))Y^2(s-) - (Y^2(s) - Y^2(s-))Y^1(s-)],$$

from which the required result easily follows. $\square$

Exercise 4.8 Extend this result to the case where Y^1 and Y^2 are d-dimensional.

We can learn much about the way our Itô formulae work by writing the product formula in differential form:-

$$d(Y^1(t)Y^2(t)) = Y^1(t-)dY^2(t) + Y^2(t-)dY^1(t) + d[Y^1, Y^2](t).$$

By equation (9.10), we see that the term $d[Y^1, Y^2](t)$, which is sometimes called an *Itô correction*, arises as a result of the following formal product relations between differentials:-

$$dB^i(t)dB^j(t) = \delta^{ij}dt \quad ; \quad N(dt, dx)N(dt, dy) = N(dt, dx)\delta(x - y),$$

for $1 \leq i, j \leq m$, with all other products of differentials vanishing and if you have little previous experience of this game, these relations are a very valuable guide to intuition.

For completeness, we will give another characterisation of quadratic variation which is sometimes quite useful. We recall the sequence of partitions $(\mathcal{P}_n, n \in \mathbb{N})$, with mesh tending to zero which were introduced earlier.

Theorem 9.9. *If X and Y are real-valued Lévy-type stochastic integrals of the form (9.8), then for each $t \geq 0$, with probability one, we have*

$$[X, Y](t) = \lim_{n \to \infty} \sum_{j=0}^{m_n} (X(t_{j+1}^{(n)}) - X(t_j^{(n)}))(Y(t_{j+1}^{(n)}) - Y(t_j^{(n)})),$$

where the limit is taken in probability.

Proof. By polarisation, it is sufficient to consider the case $X = Y$. Using the identity
$$(x - y)^2 = x^2 - y^2 - 2y(x - y)$$
for $x, y \in \mathbb{R}$, we deduce that

$$\sum_{j=0}^{m_n} (X(t_{j+1}^{(n)}) - X(t_j^{(n)}))^2 = \sum_{j=0}^{m_n} X(t_{j+1}^{(n)})^2 - \sum_{j=0}^{m_n} X(t_j^{(n)})^2$$
$$- 2\sum_{j=0}^{m_n} X(t_j^{(n)})(X(t_{j+1}^{(n)}) - X(t_j^{(n)})),$$

and the required result follows from Itô's product formula (Theorem 9.8).   □

Many of the results of this lecture extend from Lévy-type stochastic integrals to arbitrary semimartingales and full details can be found in Jacod-Shiryaev [51] and Protter [66]. In particular, if F is a simple process and X is a semimartingale we can again use Itô's prescription to define

$$\int_0^t F(s)dX(s) = \sum F(t_j)(X(t_{j+1}) - X(t_j)),$$

and then pass to the limit to obtain more general stochastic integrals. Itô's formula can be established in the form given in Theorem 9.7 and the quadratic variation of semimartingales defined as the correction term in the corresponding Itô product formula.

Although stochastic calculus for general semimartingales is not the subject of this book, we do require one result - the famous Lévy characterisation of Brownian motion.

Theorem 9.10 (Lévy's characterisation). *Let $M = (M(t), t \geq 0)$ be a continuous centered martingale, which is adapted to a given filtration $(\mathcal{F}_t, t \geq 0)$. If $[M_i, M_j](t) = a_{ij}t$ for each $t \geq 0, 1 \leq i, j \leq d$ where $a = (a_{ij})$ is a positive definite symmetric matrix, then M is an $\mathcal{F}_t$-adapted Brownian motion with covariance a.*

Proof. Fix $u \in \mathbb{R}^d$ and define the process $(Y_u(t), t \geq 0)$ by $Y_u(t) = e^{i(u, M(t))}$, then by Itô's formula, we obtain

$$dY_u(t) = iu^j Y_u(t) dM_j(t) - \frac{1}{2} u^i u^j Y_u(t) d[M_i, M_j](t)$$

$$= iu^j Y_u(t) dM_j(t) - \frac{1}{2}(u, au) Y_u(t) dt.$$

Upon integrating from s to t, we obtain

$$Y_u(t) = Y_u(s) + iu^j \int_s^t Y_u(\tau) dM_j(\tau) - \frac{1}{2}(u, au) \int_s^t Y_u(\tau) d\tau.$$

Now take conditional expectations of both sides with respect to $\mathcal{F}_s$, and use the conditional Fubini Theorem to obtain

$$\mathbb{E}(Y_u(t)|\mathcal{F}_s) = Y_u(s) - \frac{1}{2}(u, au) \int_s^t \mathbb{E}(Y_u(\tau)|\mathcal{F}_s) d\tau.$$

Hence $\mathbb{E}(e^{i(u, M(t))}|\mathcal{F}_s) = e^{-\frac{1}{2}(u, au)(t-s)}$.

Exercise 4.8 Confirm that this is enough to make M a Brownian motion. $\square$

Note: A number of interesting propositions which are equivalent to the Lévy characterisation can be found in Kunita [53], p.67.

9.3 Stochastic Differential Equations

Using Picard iteration one can show the existence of a unique solution to

$$dY(t) = b(Y(t-))dt + \sigma(Y(t-))dB(t) + \tag{9.11}$$
$$+ \int_{|x| < c} F(Y(t-), x)\tilde{N}(dt, dx) + \int_{|x| \geq c} G(Y(t-), x)N(dt, dx),$$

which is a convenient shorthand for the system of SDE's:-

$$dY^i(t) = b^i(Y(t-))dt + \sigma^i_j(Y(t-))dB^j(t) + \tag{9.12}$$
$$+ \int_{|x| \leq c} F^i(Y(t-), x)\tilde{N}(dt, dx) + \int_{|x| > c} G^i(Y(t-), x)N(dt, dx),$$

where each $1 \leq i \leq d$. The conditions under which this holds are:-

(1) **Lipschitz Condition**

There exists $K_1 > 0$ such that for all $y_1, y_2 \in \mathbb{R}^d$,

$$|b(y_1) - b(y_2)|^2 + ||a(y_1, y_1) - 2a(y_1, y_2) + a(y_2, y_2)|| \qquad (9.13)$$
$$+ \int_{|x|<c} |F(y_1, x) - F(y_2, x)|^2 \nu(dx) \leq K_1 |y_1 - y_2|^2.$$

(2) **Growth Condition**

There exists $K_2 > 0$ such that for all $y \in \mathbb{R}^d$,

$$|b(y)|^2 + ||a(y, y)|| + \int_{|x|<c} |F(y, x)|^2 \nu(dx) \leq K_2(1 + |y|^2). \qquad (9.14)$$

(3) **Big Jumps Condition**

G is jointly measurable and $y \to G(y, x)$ is continuous for all $|x| \geq 1$.

Here, $|| \cdot ||$ is the matrix seminorm $||a|| = \sum_{i=1}^{d} |a_i^i|$, and $a(x, y) = \sigma(x)\sigma(y)^T$.

We also impose the *standard initial condition* $Y(0) = Y_0$ (a.s.) for which Y_0 is independent of $(\mathcal{F}_t, t > 0)$. Full details and proofs can be found in section 6.2 of [13].

A special case of considerable interest is

$$dY(t) = L(Y(t-))dX(t).$$

You can check that the conditions given above boil down to the single requirement that L be globally Lipshitz, in order to get existence and uniqueness.

Example:- **Stochastic Exponentials**

$$dY(t) = Y(t-)dX(t),$$

i.e. $dY^i(t) = Y^i(t-)dX^i(t)$ for each $1 \leq i \leq d$. This has a unique solution given (in the case $d = 1$, with $Y_0 = 1$(a.s.)) by the *stochastic (Doléans-Dade) exponential*

$$Y(t) = \mathcal{E}_X(t) = \exp\left\{ X(t) - \frac{1}{2}[X_c, X_c](t) \right\} \prod_{0 \leq s \leq t} (1 + \Delta X(s))e^{-\Delta X(s)},$$

for each $t \geq 0$.

Solutions of SDEs are Markov processes and, in the case where there are no jumps, diffusion processes. In general, we obtain a Feller semigroup $(T_t f)(y) = \mathbb{E}(f(Y(t))|Y(0) = y)$ with generator $\mathcal{L}$. We have $C_0^2(\mathbb{R}^d) \subseteq \text{Dom}(\mathcal{L})$ and

$$(\mathcal{L}f)(y) = b^i(y)(\partial_i f)(y) + \frac{1}{2}a^{ij}(y,y)(\partial_i \partial_j f)(y) \qquad (9.15)$$

$$+ \int_{|x|<c} (f(y + F(y,x)) - f(y) - F^i(y,x)(\partial_i f)(y))\nu(dx)$$

$$+ \int_{|x|\geq c} (f(y + G(y,x)) - f(y))\nu(dx),$$

for each $f \in C_0^2(\mathbb{R}^d), y \in \mathbb{R}^d$.

Sometimes it useful to study solutions of SDE's as two-parameter processes corresponding to a "starting time" s and a "finishing time" t. We also consider solutions as functions of the initial condition as well as of chance i.e. each $\Phi_{s,t} : \mathbb{R}^d \times \Omega \to \mathbb{R}^d$.

$$d\Phi_{s,t}(y) = b(\Phi_{s,t-}(y))dt + \sigma(\Phi_{s,t-}(y))dB(t) \qquad (9.16)$$

$$+ \int_{|x|<c} F(\Phi_{s,t-}(y),x)\tilde{N}(dt,dx) + \int_{|x|\geq c} G(\Phi_{s,t-}(y),x)N(dt,dx)$$

with initial condition $\Phi_{s,s}(y) = y$ (a.s.). These form a *stochastic flow* i.e.

(i) $\Phi_{r,t} = \Phi_{s,t} \circ \Phi_{r,s}$ (a.s.), for all $0 \leq r < s < t < \infty$,

(ii) $\Phi_{s,s}(y) = y$ (a.s.), for all $s \geq 0, y \in \mathbb{R}^d$.

If, in addition, each $\Phi_{s,t}$ is almost surely a homeomorphism (C^k-diffeomorphism) of $\mathbb{R}^d$, we say that Φ is a *stochastic flow of homeomorphisms* (C^k-*diffeomorphisms*, respectively).

If, in addition to (i) and (ii), we have that

(iii) For each $n \in \mathbb{N}, 0 \leq t_1 < t_2 < \cdots < t_n < \infty, y \in \mathbb{R}^d$, the random variables $\{\Phi_{t_j,t_{j+1}}(y); 1 \leq j \leq n-1\}$ are independent.

(iv) The mappings $t \to \Phi_{s,t}(y)$ are càdlàg, for each $y \in \mathbb{R}^d, 0 \leq s < t$,

we say that Φ is a *Lévy flow*.

If (iv) can be strengthened from "càdlàg" to "continuous", we say that Φ is a *Brownian flow*.

It is not difficult to show that solutions to (9.16) are a Lévy flow. The systematic study of Lévy flows was initiated by Fujiwara and Kunita in the important paper [37]. A review of progress in finding conditions which guarantee the diffeomorphism property is in [13], see also [54].

10 Lecture 5: Lévy Processes in Groups

10.1 Lévy Processes in Locally Compact Groups - The Basics

Let G be a topological group with identity e, so G is a group which is also a topological space in which the composition $G \times G \to G$, given by $(\sigma, \tau) \to \sigma\tau$

and the inverse $G \to G$ given by $\sigma \to \sigma^{-1}$ are continuous. For each $\sigma \in G$, *left translation* $l_\sigma : G \to G$ is defined by $l_\sigma(\tau) = \sigma\tau$. Each l_σ is an homeomorphism of G. r_σ is defined similarly.

We assume throughout that G is Hausdorff and locally compact. Every such group is equipped with a non-zero left-invariant regular Borel measure, called *Haar measure m*, which is unique up to multiplication by a positive constant, so $m(\sigma A) = m(A)$ for all $\sigma \in G, A \in \mathcal{B}(G)$. We often write $m(d\tau) = d\tau$. We may thus equip G with the Banach spaces $L^p(G, m) = L^p(G)$, for $1 \leq p \leq \infty$. We'll also need $C_0(G)$, the Banach space (under the sup norm) of continuous functions $f : G \to \mathbb{R}$ which vanish at infinity, i.e. given any $\epsilon > 0$, there exists a compact $K \subset G$ such that $|f(\sigma)| < \epsilon$, whenever $x \in G - K$. For each $\sigma \in G, f \in C_0(G)$, define $L_\sigma f = f \circ l_\sigma$, then L_σ is an isometric isomorphism of $C_0(G)$ (and also of each $L^p(G)$). $R_\sigma f = f \circ r_\sigma$ has similar properties.

Note that if G is compact, then m is always finite and also right-invariant. $\dfrac{m(\cdot)}{m(G)}$ is then a probability measure on G, which is called *normalised Haar measure*.

Now consider a stochastic process $(Y(t), t \geq 0)$ taking values in G. The group structure allows us to construct left increments $Y(s)^{-1}Y(t)$ and right increments $Y(t)Y(s)^{-1}$ of the process Y for $0 \leq s < t < \infty$ and unless the group is abelian there is no reason why these should coincide.

We say that a process Y has *stationary and independent left increments* if

1. for each $n \in \mathbf{N}$ and $0 \leq t_1 \ldots \leq t_n < \infty$,
   the random variables $Y(t_1)^{-1}Y(t_2), \ldots, Y(t_{n-1})^{-1}Y(t_n)$ are independent,
2. for each $0 \leq s < t < \infty, Y(s)^{-1}Y(t)$ has the same law as $Y(t - s)$.

Now we can define a *left Lévy process in G* to be a process Y satisfying the following

1. Y has stationary and independent left increments
2. $Y(0) = e$ (a.s.)
3. Y is (left) stochastically continuous i.e.

$$\lim_{s \to t} P(Y(s)^{-1}Y(t) \in A) = 0$$

for all $A \in \mathcal{B}(G)$ with $e \notin \bar{A}$

We can similarly define a *right* Lévy process by replacing "left" with "right" in (1) and (3) above.

Exercise 5.1 Deduce that there is a one to one correspondence between left and right Lévy processes in G wherein the right Lévy process corresponding to the left Lévy process Y is $Y^{-1} = (Y^{-1}(t), t \geq 0)$.

In the light of the above we will drop the left/right distinction and concentrate on left Lévy processes which we will call *Lévy processes* from now on. Note that when $G = \mathbb{R}^d$, then our processes are precisely the usual ones.

What mathematical tools can we use to investigate Lévy processes in Lie groups ? We cannot define a characteristic function in general so Fourier methods are not obviously available (we'll have plenty more to say about this later). What about the semigroup approach ?

Let $(p_t, t \geq 0)$ be the law of the Lévy process Y, then it follows from the definition that $(p_t, t \geq 0)$ is a weakly continuous convolution semigroup of probability measures on G where the convolution operation is defined for measures μ and ν on G by

$$(\mu * \nu)(A) = \int_G \mu(d\tau)\nu(\tau^{-1}A)$$

for each $A \in \mathcal{B}(G)$. So that in particular we have, for all $s, t \geq 0$,

$$p_{s+t} = p_s * p_t \quad \text{and} \quad \text{wklim}_{t \downarrow 0} p_t = \delta_e \tag{10.1}$$

where δ_e is Dirac measure concentrated at e. Define a family of linear operators $(T_t, t \geq 0)$ on $C_0(G)$ by the prescription

$$(T_t f)(\tau) = \mathbb{E}(f(\tau Y(t))) = \int_G f(\tau\sigma)p_t(d\sigma) \tag{10.2}$$

for each $t \geq 0, f \in C_0(G), \tau \in G$.

Exercise 5.2. Show that $(T_t, t \geq 0)$ is a Feller semigroup.

Note that $L_\tau T_t = T_t L_\tau$, for all $\tau \in G, t \geq 0$.

We would like to be able to characterise the generator of this semigroup. It will help to look at some sub-categories of "locomp groups" and return to the general case later.

10.2 Lévy Processes in LCA Groups

In this section, we assume that G is *abelian* and write all group operations additively. So G is a LCA group (locally compact, abelian). An excellent reference for all the group theory developed below is Rudin [72].

Let $\widehat{G}$ be the set of all continuous homomorphisms from G into the one-torus $\mathbb{T} = \{z \in \mathbb{C}; |z| = 1\}$. Then $\widehat{G}$ is also a locally compact abelian group called the *dual group* of G. Elements of $\widehat{G}$ are called *characters* of G. We emphasise the duality between G and $\widehat{G}$ by writing $\gamma(\tau) = \langle \gamma, \tau \rangle$, for each $\tau \in G, \gamma \in \widehat{G}$. So we have

$$\langle \gamma, \tau_1 + \tau_2 \rangle = \langle \gamma, \tau_1 \rangle \langle \gamma, \tau_2 \rangle, \quad \langle \gamma, -\tau \rangle = \overline{\langle \gamma, \tau \rangle},$$

for each $\gamma \in \widehat{G}, \tau_1, \tau_2, \tau \in G$.
Useful facts:

$$G \text{ discrete} \Rightarrow \widehat{G} \text{ compact.}$$

$$G \text{ compact} \Rightarrow \widehat{G} \text{ discrete.}$$

$$\text{e.g. } \widehat{\mathbb{R}} = \mathbb{R}, \quad \widehat{\mathbb{T}} = \mathbb{Z}, \quad \widehat{\mathbb{Z}} = \mathbb{T}.$$

In general, we have *Pontryagin duality* - $\widehat{\widehat{G}} \cong G$.

The Lévy-Khintchine Formula

Let μ be a probability measure on G. Define its *characteristic function* $\phi_\mu : \widehat{G} \to \mathbb{C}$ by

$$\phi_\mu(\gamma) = \int_G \langle \gamma, \tau \rangle \mu(d\tau).$$

Lemma 10.1. *Let* μ_1, μ_2 *be probability measures on* G. *For each* $\gamma \in G$,

$$\phi_{\mu_1 * \mu_2}(\gamma) = \phi_{\mu_1}(\gamma)\phi_{\mu_2}(\gamma).$$

Proof. By Fubini's theorem,

$$\phi_{\mu_1 * \mu_2}(\gamma) = \int_G \langle \gamma, \tau \rangle (\mu_1 * \mu_2)(d\tau)$$

$$= \int_G \int_G \langle \gamma, \tau + \sigma \rangle \mu_1(d\tau)\mu_2(d\sigma)$$

$$= \int_G \int_G \langle \gamma, \tau \rangle \langle \gamma, \sigma \rangle \mu_1(d\tau)\mu_2(d\sigma)$$

$$= \phi_{\mu_1}(\gamma)\phi_{\mu_2}(\gamma). \qquad \square$$

Exercise 5.3 Show that ϕ_μ is positive definite, i.e. $\sum_{i,j} c_i \bar{c}_j \phi_\mu(\gamma_i - \gamma_j) \geq 0$, for all $c_i \in \mathbb{C}, \gamma_i \in \widehat{G}, 1 \leq i, j \leq n, n \in \mathbb{N}$.

We can now follow much of the path developed in Lecture 1 - there are generalisations of Bochner's theorem, Schoenberg's correspondence, infinite divisibility etc. For details see Chapter IV of Parthasarathy [64]. We will pass straight to Lévy processes $Y = (Y(t), t \geq 0)$.

The main result about these in LCA groups is the generalised Lévy-Khintchine formula, a proof of which can again be found in Parthasarathy [64]. For each $t \geq 0, \gamma \in \widehat{G}$, consider the characteristic function:-

$$\mathbb{E}(\langle \gamma, Y(t) \rangle) = \int_G \langle \gamma, \tau \rangle p_t(d\tau).$$

Theorem 10.2 (Lévy Khinchine formula - LCA case). *For each* $t \geq 0, \gamma \in \widehat{G}$,

$$\mathbb{E}(\langle \gamma, Y(t) \rangle) = e^{t\eta(\gamma)},$$

where $\eta : \widehat{G} \to \mathbb{C}$ *is of the form*

$$\eta(\gamma) = il(\gamma) + q(\gamma) + \int_{G-\{0\}} [\langle \gamma, \tau \rangle - 1 - ig(\tau, \gamma)]\nu(d\tau), \quad \text{where}$$

- $l : \widehat{G} \to \mathbb{R}$ is a continuous homomorphism,
- $q : \widehat{G} \to \mathbb{R}$ is a continuous non-negative quadratic form, i.e. $q(\gamma_1 + \gamma_2) + q(\gamma_1 - \gamma_2) = 2q(\gamma_1) + 2q(\gamma_2)$,
- ν is a σ-finite measure on $G - \{0\}$ for which $\nu(V^c) < \infty$, for every neighborhood of the identity $V \in \mathcal{B}(G)$ and $\int_{G-\{0\}}(1 - \Re\langle \gamma, \tau \rangle)\nu(d\tau) < \infty$ for all $\gamma \in \widehat{G}$.
- $g : G \times \widehat{G} \to \mathbb{R}$ is continuous, bounded on compact sets and is subject to other technical conditions which are listed in [64], lemma 5.3.

Note The general theory of infinite divisibility in LCA groups is complicated by the existence of *idempotents*, i.e. probability measures μ for which $\mu * \mu = \mu$.

Exercise 5.3 Show that if G is compact, then its normalised Haar measure is idempotent.

If Y is a Lévy process with laws $(p_t, t \geq 0)$, then p_t cannot be idempotent and this simplifies matters for us.

Semigroups and The Fourier Transform

If $f \in L^1(G)$, we define its *Fourier transform* $\widehat{f}$ by

$$\widehat{f}(\gamma) = \int_G \langle \gamma, -\tau \rangle f(\tau)d\tau, \quad \text{for each } \gamma \in \widehat{G}.$$

Then $\widehat{f} \in C_0(\widehat{G})$. In fact $L^1(G)$ is a commutative Banach algebra under convolution, $\widehat{G}$ is its maximal ideal space and $f \to \widehat{f}$ is the Gelfand transform. We also have the *Plancherel theorem* whereby the mapping $f \to \widehat{f}$ extends from an isometric embedding of $L^1(G) \cap L^2(G)$ into $L^2(\widehat{G})$ to a unitary isomorphism between $L^2(G)$ and $L^2(\widehat{G})$. If $f \in L^1(G) \cap L^2(G)$, we can, by taking adjoints, use Fourier inversion to write $f(\tau) = \int_{\widehat{G}} \langle \gamma, \tau \rangle \widehat{f}(\gamma)d\gamma$. Now let $(T_t, t \geq 0)$ be the Feller semigroup of Y acting in $L^2(G)$, then we can imitate the argument of Lecture 2 to write, via Fubini's theorem and the Lévy-Khintchine formula :-

$$(T_t f)(\tau) = \mathbb{E}(f(\tau + Y(t)))$$
$$= \int_{\widehat{G}} \mathbb{E}(\langle \gamma, \tau + Y(t) \rangle)\widehat{f}(\gamma)d\gamma$$
$$= \int_{\widehat{G}} \langle \gamma, \tau \rangle \mathbb{E}(\langle \gamma, Y(t) \rangle)\widehat{f}(\gamma)d\gamma$$
$$= \int_{\widehat{G}} \langle \gamma, \tau \rangle e^{t\eta(\gamma)}\widehat{f}(\gamma)d\gamma$$

If $\mathcal{A}$ is the infinitesimal generator of $(T_t, t \geq 0)$, then formal differentiation yields the pseudo-differential operator representation:-

$$(\mathcal{A}f)(\tau) = \int_{\widehat{G}} \langle \gamma, \tau \rangle \eta(\gamma) \widehat{f}(\gamma) d\gamma.$$

In Berg and Forst [18], this argument is made precise and it is shown that this representation holds on the non-isotropic Sobolev space $\mathrm{Dom}(\mathcal{A}) = \mathcal{H}_\eta(G) = \{f \in L^2(G); \eta\widehat{f} \in L^2(\widehat{G})\}$. [18] is also a good source for extending other aspects of Lévy processes in Euclidean space to LCA's, such as subordination.

10.3 Lévy Processes in Lie Groups

Background on Lie Groups

If you are new to Lie theory, a nice introduction can be found in the articles by Segal and Carter in [26] and the second half of Simon [77].

A *Lie group* is a group G which is also a C^∞ manifold in which the composition $G \times G \to G$, given by $(\sigma, \tau) \to \sigma\tau$ and the inverse $G \to G$ given by $\sigma \to \sigma^{-1}$ are C^∞.

Examples The Classical Groups - e.g. $GL(n, \mathbb{C})$, $SL(n, \mathbb{R})$, $U(n)$, $O(n)$, $SU(n)$, $SO(n)$, $Sp(n)$, $Spin(n)$ etc.

The Heisenberg Group $\mathbb{H}^n$ is $\mathbb{R}^{2n+1}$ equipped with the composition law

$$(a_1, q_1, p_1)(a_2, q_2, p_2) = (a_1 + a_2 + \frac{1}{2}(p_1 \cdot q_2 - q_1 \cdot p_2), q_1 + q_2, p_1 + p_2),$$

where each $a_i \in \mathbb{R}, q_i, p_i \in \mathbb{R}^n (i = 1, 2)$.

$O(m, n)$ is the group of linear transformations in $\mathbb{R}^{m+n}$ which leave invariant the pseudo-metric

$$x_1^2 + \cdots + x_m^2 - x_{m+1}^2 - \cdots x_{m+n}^2,$$

e.g. $O(3, 1)$ is the Lorentz group.

If M is any d-dimensional C^∞-manifold then for each $p \in M, T_p(M)$ denotes the *tangent space* at p. It is a d-dimensional linear space which consists precisely of all the point derivations at p, i.e. $X_p \in T_p(M)$ if and only if X_p is a linear map from $C^\infty(M)$ to $\mathbb{R}$ and

$$X_p(fg) = X_p(f)g(p) + f(p)X_p(g),$$

for all $f, g \in C^\infty(M)$.

In local co-ordinates, each $X_p = a_p^i \partial_i$, where $a_p^i \in \mathbb{R}(1 \leq i \leq d)$. The *tangent bundle* $T(M) = \bigcup_{p \in M} T_p(M)$ inherits a differential structure from M and becomes a $2d$-dimensional C^∞-manifold. X is a *smooth vector field* if it is a smooth section of $T(M)$ i.e. $X : M \rightarrow T(M)$ is C^∞ and $X(p) \in T_p(M)$, for each $p \in M$. In local co-ordinates, $X(p) = a^i(p)\partial_i$ where each $a^i : \mathbb{R}^d \rightarrow \mathbb{R}$ is C^∞.

If M and N are both C^∞ manifolds and $\phi : M \rightarrow N$ is C^∞, then we can "linearise" ϕ to obtain its *differential* which is a linear map from $T_p(M)$ to $T_{\phi(p)}(N)$. This is defined as follows:- let $X_p \in T_p(M), f \in C^\infty(N)$ - define $Y_p \in T_{\phi(p)}(N)$, by

$$(Y_p f)(\phi(p)) = (X_p(f \circ \phi))(p),$$

$$\text{then } d\phi(X_p) = Y_p.$$

Note that if ϕ is bijective then so is $d\phi$.

Let G be a Lie group. Fix $X \in T_e(G)$. Define a vector field by $X^L(\tau) = dl_\tau(X)$. $X^L(\cdot)$ is called a *left invariant vector field* since each $X^L(\sigma) = dl_{\sigma\tau^{-1}}X^L(\tau)$. Right invariant vector fields X^L are defined similarly, using right instead of left translation. The linear space of all left invariant vector fields induces a d-dimensional *Lie algebra* structure onto $\mathbf{g} = T_e(G)$. You should think of $\mathbf{g}$ as a "linearisation" of G. The fact that $\mathbf{g}$ is a (real) Lie algebra means that $\mathbf{g}$ is a finite-dimensional real vector space equipped with a bilinear map $[\cdot,\cdot] : \mathbf{g} \times \mathbf{g} \rightarrow \mathbf{g}$ for which

1. $[X,Y] = -[Y,X]$
2. (The Jacobi Identity) $[X,[Y,Z]] + [Y,[Z,X]] + [Z,[X,Y]] = 0$

for all $X,Y,Z \in \mathbf{g}$.

Examples: $G = SO(n), \mathbf{g} = so(n)$ - the space of all skew-symmetric $n \times n$ matrices having zero trace.

$G = \mathbb{H}^n$. A basis for the Lie algebra of left-invariant vector fields is $\{T, L_1, \ldots, L_n, M_1, \ldots, M_n\}$, where for $1 \leq j \leq n$,

$$T = \frac{\partial}{\partial t}, L_j = \frac{\partial}{\partial q_j} + \frac{1}{2}p_j\frac{\partial}{\partial t}, M_j = \frac{\partial}{\partial p_j} - \frac{1}{2}q_j\frac{\partial}{\partial t}$$

and we have the commutation relations, for $1 \leq j, k \leq n$,

$$[L_j, L_k] = [M_j, M_k] = [M_j, T] = [L_j, T] = 0, [M_j, L_k] = \delta_{jk}T.$$

Fix $X \in \mathbf{g}$ and consider the differential equation $\dfrac{d\psi(t)}{dt} = X(\psi(t))$, with initial condition $\psi(0) = e$. In local co-ordinates, if $X(p) = a^i(p)\partial_i$, then $\dfrac{d\psi(t)^i}{dt} = a^i(\psi(t))(1 \leq i \leq d)$. We write $\psi(t) = \exp(tX)$, then $(\exp(tX), t \in \mathbb{R})$ is a one-parameter subgroup of G i.e.

$$\exp((s+t)X) = \exp(sX)\exp(tX), \quad [\exp(tX)]^{-1} = \exp(-tX).$$

The mapping $\mathbf{g} \rightarrow G$ given by $X \rightarrow \exp(X)$ is called the *exponential map*. exp has some nice properties, e.g. we can always find a neighborhood V of 0 in $\mathbf{g}$

which is mapped diffeomorphically by exp to a neighbourhood N of e in G. Fix a basis $X_1, \ldots, X_n$ of $\mathbf{g}$. Smooth functions $x_1, \ldots, x_n : N \to \mathbb{R}$ are called *canonical co-ordinates* for G at e (with respect to $X_1, \ldots, X_n$) if

$$x_i \left(\exp \left(\sum_{i=1}^{n} a^i X_i \right) \right) = a_i,$$

whenever $\sum_{i=1}^{n} a^i X_i \in V$.

The following formulae can be useful: - for each $X \in \mathbf{g}, f \in C^\infty(G), \tau \in G$,

$$(X^L f)(\tau) = \frac{d}{da} f(\tau \exp(aX)) \Big|_{a=0}, \quad (X^R f)(\tau) = \frac{d}{da} f(\exp(aX)\tau) \Big|_{a=0}$$

If $\mathbf{h}_1, \mathbf{h}_2 \subseteq \mathbf{g}$, define $[\mathbf{h}_1, \mathbf{h}_2] = \{[X, Y], X \in \mathbf{h}_1, Y \in \mathbf{h}_2\}$. We obtain a decreasing sequence of subsets (in fact these are *ideals*) of $\mathbf{g}, (\mathbf{g}_n, n \in \mathbb{N})$ by $\mathbf{g}_1 = [\mathbf{g}, \mathbf{g}]$ and for $n > 2, \mathbf{g}_n = [\mathbf{g}, \mathbf{g}_{n-1}]$. We say that $\mathbf{g}$ and G are *nilpotent* if $\mathbf{g}_n = 0$ for some $n \in \mathbb{N}$ (and hence for all $m > n$), e.g. $\mathbb{H}_n$ is nilpotent.

Hunt's Representation Formula

Now let $Y = (Y(t), t \geq 0)$ be a Lévy process with laws $(p_t, t \geq 0)$ in a Lie group G with Lie algebra $\mathbf{g}$. Let $(T(t), t \geq 0)$ be the associated Feller semigroup acting in $C_0(G)$. The starting point for studying Lévy processes in Lie groups is a wonderful formula due to Hunt [45] who effectively generalised (6.1) from Lecture 2 to give a Lévy-Khintchine type decomposition for the infinitesimal generator $\mathcal{A}$ of Y. In the sequel, we'll write $\mathcal{D} = \text{Dom}(\mathcal{A})$.

We first fix a basis $(X_j, 1 \leq j \leq n)$ of $\mathbf{g}$ and define the dense linear manifold $C_2^L(G)$ by

$$C_2^L(G) = \{f \in C_0(G); X_i^L(f) \in C_0(G), X_i^L X_j^L(f) \in C_0(G) \text{ for all } 1 \leq i, j \leq n\}.$$

$C_2^L(G)$ is a Banach space with respect to the norm

$$\|f\|_{2,L} = \|f\| + \sum_{i=1}^{n} \|X_i^L f\| + \sum_{j,k=1}^{n} \|X_j^L X_k^L f\|.$$

The space $C_2^R(G)$ and the norms $\| \cdot \|_{2,R}$ are defined similarly. Note that the smooth functions of compact support $C_c^\infty(G) \subseteq C_2^L(G) \cap C_2^R(G)$.

There exist functions $x_i \in C_c^\infty(G), 1 \leq i \leq n$ so that $(x_1, \ldots, x_n)$ are a system of canonical co-ordinates for G at e.

Furthermore, there exists a map $h \in \mathcal{D}$ which is such that

1. $h > 0$ on $G - \{e\}$,
2. There exists a compact neighbourhood of the identity V such that for all $\tau \in V$,

$$h(\tau) = \sum_{i=1}^{n} x_i(\tau)^2.$$

Any such function is called a *Hunt function* in G. A positive measure ν defined on $\mathcal{B}(G - \{e\})$ is called a *Lévy measure* whenever

$$\int_{G-\{e\}} h(\sigma)\nu(d\sigma) < \infty$$

for some Hunt function h.

Hunt's theorem, which is given below is the main result of [45]. The arguments were later simplified for lesser mortals by S.Ramaswami - a student of K.R.Parthasarathy. These were then incorporated into the proof given by Herbert Heyer in his seminal 1977 treatise [43]. A quite slick proof will appear shortly in Liao [57].

Theorem 10.3 (Hunt's theorem). *Let Y be a Lévy process in G with infinitesimal generator $\mathcal{A}$ then*

1. $C_2^L(G) \subseteq \mathrm{Dom}(\mathcal{A})$.
2. *For each $\tau \in G, f \in C_2^L(G)$,*

$$\mathcal{A}f(\tau) = b^i X_i^L f(\tau) + a^{ij} X_i^L X_j^L f(\tau) \tag{10.3}$$
$$+ \int_{G-\{e\}} (f(\tau\sigma) - f(\tau) - y^i(\sigma)X_i^L f(\tau))\nu(d\sigma),$$

where $b = (b^1, \ldots b^n) \in \mathbb{R}^n, a = (a^{ij})$ is a non-negative-definite, symmetric $n \times n$ real-valued matrix and ν is a Lévy measure on $G - \{e\}$.

Conversely, any linear operator with a representation as in (10.3) is the restriction to $C_2(G)$ of the infinitesimal generator of a unique convolution semigroup of probability measures.

Proof.(Sketch) This is quite long and involved. Of necessity we'll take a crude approach so as to get across the main ideas as to how the first part is obtained. The essence of Hunt's proof resides in a careful analysis of the *generating functional*

$$\mathcal{B}f = \lim_{t\to\infty} \frac{1}{t}[(T_t f)(e) - f(e)] = \lim_{t\to\infty} \frac{1}{t}\int_G (f(\sigma) - f(e))p_t(d\sigma).$$

Note that if $f \in \mathcal{D}$, then $(\mathcal{A}f)(\tau) = \mathcal{B}(L_\tau f)$.
Ramaswami proved the following two useful results. Given any Borel neighborhood N of e,

(i) $\sup_{t>0} \dfrac{p_t(N^c)}{t} < \infty$ (ii) $\sup_{t>0} \dfrac{1}{t}\int_N \left(\sum_{i=1}^n x_i(\sigma)^2\right) p_t(d\sigma) < \infty.$

Now $\lim_{t\to 0} \dfrac{(T_t f)(\sigma) - f(\sigma)}{t}$ certainly exists uniformly in $\sigma \in G$, for all $f \in C_2^L(G) \cap \mathcal{D}$. After some argument (and using Ramamswami's lemmas), we have that there exists $C > 0$ such that

$$\left| \frac{(T_t f)(\sigma) - f(\sigma)}{t} \right| \leq C \|f\|_{2,R},$$

for all $f \in C_2^L(G)$. Since $C_2^L(G) \cap \mathcal{D}$ is dense in $C_2^L(G)$, we conclude that the $\lim_{t \to 0} \frac{(T_t f)(\sigma) - f(\sigma)}{t}$ exists uniformly in $\sigma \in G$ for all $f \in C_2^L(G)$. Hence $C_2^L(G) \subseteq \mathcal{D}$.

Let $f \in C_2^L$, then $\mathcal{B}f$ exists. For each $\tau \in G$ define,

$$g(\tau) = f(\tau) - f(e) - \alpha_i x^i(\tau) - \beta_{ij} x^i(\tau) x^j(\tau),$$

where $\alpha_i - X_i^L f(e)$ and $\beta_{ij} = X_i^L X_j^L f(e)$. Hence

$$\mathcal{B}(f) = \lim_{t \to 0} \frac{1}{t} \int_G g(\sigma) p_t(d\sigma) + \alpha_i \mathcal{B}(x^i) + \beta_{ij} \mathcal{B}(x^i x^j).$$

After some work, we get

$$\lim_{t \to 0} \frac{1}{t} \int_G g(\sigma) p_t(d\sigma) = \int_{G-\{e\}} g(\sigma) \nu(d\sigma)$$

$$= \int_{G-\{e\}} (f(\sigma) - f(e) - x^i(\sigma)(X_i^L f)(e)) \nu(d\sigma)$$

$$- \beta_{ij} \int_{G-\{e\}} x^i(\sigma) x^j(\sigma) \nu(d\sigma).$$

Now rearrange terms to find (10.3) at $\tau = e$. You can now get the general result on replacing f by $L_\sigma f$ □

Martingale Representation

More insight into the nature of the paths of Lévy processes in Lie groups can be obtained from the following result, due to Applebaum and Kunita [8].

In the following, we take $\mathcal{F}_t = \sigma\{Y(s), 0 \leq s \leq t\}$, for each $t \geq 0$. We will need the *Doob-Meyer decomposition* for real valued martingales $(M(t), t \geq 0)$ and $(N(t), t \geq 0)$, which asserts that there is a unique predictable process - denoted as $(\langle M(t), N(t) \rangle, t \geq 0)$ which is such that $M(t)N(t) - M(0)N(0) - \langle M(t), N(t) \rangle$ is a local martingale. If M and N are continuous, then $\langle M(t), N(t) \rangle = [M, N](t)$, for each $t \geq 0$. $\langle \cdot, \cdot \rangle$ is sometimes called the *Meyer angle bracket*.

Theorem 10.4. *If $Y = (Y(t), t \geq 0)$ is a càdlàg Lévy process in G with infinitesimal generator $\mathcal{A}$ of the form (10.3), then there exists*

- *an $\mathcal{F}_t$-adapted Poisson random measure N on $\mathbb{R}^+ \times (G - \{e\})$,*
- *an n-dimensional $\mathcal{F}_t$-adapted Brownian motion $B = (B(t), t \geq 0)$ with mean zero and covariance $Cov(B_i(t), B_j(t)) = 2t a_{ij}$, for each $t \geq 0$, which is independent of N,*

such that for each $f \in C_2^L(G), t \geq 0,$

$$f(Y(t)) = f(e) + \int_0^t (X_i^L f)(Y(s-))dB^i(s) + \int_0^t (\mathcal{A}f)(Y(s-))ds +$$

$$+ \int_0^t \int_{G-\{e\}} [f(Y(s-)\sigma) - f(Y(s-))]\tilde{N}(ds, d\sigma),$$

where $\tilde{N}(ds, d\sigma) = N(ds, d\sigma) - ds\nu(d\sigma).$
Furthermore, Y *is uniquely determined by* B *and* N *and*

$$\mathcal{F}_t = \sigma\{B(s), N((s,t] \times E); 0 \leq s \leq t, E \in \mathcal{B}(G - \{e\})\},$$

for each $t \geq 0.$

Proof (Sketch). For each $0 \leq s \leq t < \infty, \tau \in G$, we introduce the notation $Y_{s,t}(\tau) = \tau Y(s)^{-1} Y(t)$. Now fix $s \geq 0$. For each $f \in C_2(G)^L, \tau \in G$, define $M_s^{f,\tau} = (M_{s,t}^{f,\tau}, t \geq s)$ by

$$M_{s,t}^{f,\tau} = f(Y_{s,t}(\tau)) - f(e) - \int_s^t \mathcal{A}f(Y_{s,u}(\tau))du.$$

Then $M_s^{f,\tau}$ is a centred L^2-martingale. We can compute the associated Meyer angle bracket to obtain

$$\langle M_{s,t}^{f,\tau_1}, M_{s,t}^{g,\tau_2} \rangle = \int_s^t B(f,g)(Y_{s,u}(\tau_1), Y_{s,u}(\tau_2))du,$$

for each $f, g \in C_2^L(G), \tau_1, \tau_2 \in G$, where B is the "carré de champ". This is defined by polarisation, from

$$B(f,f)(\rho, \rho) = (\mathcal{A}f^2)(\rho) - 2f(\rho)(\mathcal{A}f)(\rho),$$

and a calculation then yields,

$$B(f,g)(\rho_1, \rho_2) = 2 \sum_{i,j=1}^n a_{ij}(X_i^L f)(\rho_1)(X_j^L g)(\rho_2)$$

$$+ \int_{G-\{e\}} (f(\rho_1\tau) - f(\rho_1))(g(\rho_2\tau) - g(\rho_2))\nu(d\tau),$$

for each $\rho_1, \rho_2 \in G$. Now let $\mathcal{P} = \{0 = t_0 < t_1 < t_2 < \cdots\}$ be a partition of $\mathbb{R}^+$ with mesh $\delta(\mathcal{P}) = \max_{n \in \mathbb{N}}(t_n - t_{n-1}) < \infty$. We define a centred L^2-martingale $(Z_t^{\mathcal{P},f,\tau}, t \geq 0)$ by

$$Z_t^{\mathcal{P},f,\tau} = \sum_{n \in \mathbb{N}} M_{t \wedge t_{n-1}, t \wedge t_n}^{f,\tau},$$

for each $t \geq 0$. Then we obtain another centred L^2-martingale $(Z_t^{f,\tau}, t \geq 0)$ by

$$Z_t^{f,\tau} = L^2 - \lim_{\delta(\mathcal{P}) \to 0} Z_t^{\mathcal{P},f,\tau},$$

for each $t \geq 0$. Moreover, for each $f, g \in C_2(G), \tau_1, \tau_2 \in G, t \geq 0$, we have

$$\langle Z_t^{f,\tau_1}, Z_t^{g,\tau_2} \rangle = tB(f,g)(\tau_1, \tau_2) \text{ and } M_{s,t}^{f,\tau} = \int_s^t dZ_u^{f,Y_{s,u^-}(\tau)},$$

in the sense of the non-linear stochastic integral of Fujiwara and Kunita [37], lemma 4.2.

For each $t \geq 0$, let $Z_t^{f,\tau} = Z_t^{(c),f,\tau} + Z_t^{(d),f,\tau}$ be the unique decomposition into continuous and discontinuous centred martingales. For each $0 \leq s \leq t < \infty, E \in \mathcal{B}(G - \{e\})$, define

$$N((s,t],E) = \#\{0 \leq s < u \leq t; \Delta Y(u) \in E\}.$$

Then N extends to a Poisson random measure on $\mathbb{R}^+ \times (G - \{e\})$ with intensity measure ν, and for each $t \geq 0$,

$$Z_t^{(d),f,\tau} = \int_0^t \int_{G-\{e\}} (f(\tau\sigma) - f(\tau)) \tilde{N}(ds, d\sigma).$$

For each $1 \leq i \leq n, t \geq 0$, define $B_i(t) = Y_t^{(c),x_i,e}$. Then for each $1 \leq i, j \leq n, \langle B_i(t), B_j(t) \rangle = 2ta_{ij}$. Hence $(B(t), t \geq 0)$ is a d-dimensional Brownian motion, by Lévy's characterisation, and for each $t \geq 0$,

$$Z_t^{(c),f,\tau} = \sum_{1 \leq i \leq n} X_i^L f(\tau) B^i(t),$$

from which the required result follows.   □

One of the motivations for proving this result was to obtain a class of Lévy flows on manifolds - briefly if there is an action of a Lie group G on a manifold M (i.e. a homomorphism from G into the group Diff(M)), then the right increment $X(t)X(s)^{-1}$ is mapped to a stochastic flow of diffeomorphisms acting on M.

In the case where G is simply connected and nilpotent, Pap [62] has given a recursive formula for the construction of Y. In particular, if G is the Heisenberg group $\mathbb{H}^n$, then there exists a Lévy process $(A(t), Q_1(t), \ldots Q_n(t), P_1(t), \ldots, P_n(t))$ on $\mathbb{R}^{2n+1}$ which is such that for each $t \geq 0$, we have

$$Y(t) = (C_{A,Q,P}(t), Q_1(t), \ldots Q_n(t), P_1(t), \ldots, P_n(t)) \text{ a.s.},$$

where $C_{A,Q,P}(t) = A(t) + \frac{1}{2} \sum_{j=1}^n \int_0^t (P_j(s-)dQ_j(s) - Q_j(s-)dP_j(s))$.

For further work on Lévy processes on $\mathbb{H}^n$, see [6].

Examples of Lévy Processes in G

Example 1: Brownian Motion on a Lie Group

A left-invariant Brownian motion on G is the unique solution of the Stratono-vitch SDE

$$dY(t) = X_i^L(Y(t)) \circ dB^i(t) \tag{10.4}$$

with $Y(0) = e$ (a.s.) and we have taken $a = \frac{1}{2}I$.

If we write each $X_i^L(\cdot) = c_i^j(\cdot)\partial_i$, in local co-ordinates, we have for $Y(t) = (Y^1(t), \ldots, Y^n(t))$,

$$dY^j(t) = c_i^j(Y(t))dB^i(t) + \frac{1}{2}\sum_{i=1}^n \partial_k(c_i^j(Y(t))(c_i^k(Y(t))dt.$$

We have $2\mathcal{A} = \Delta_G = \sum_{j=1}^n (X_j^L)^2$ which is a left-invariant Laplacian in G. In the case, where $a = cI$, with $c \neq \frac{1}{2}$ we call the solution to (10.4) a c-Brownian motion.

As we have defined it above, Brownian motion depends upon the choice of basis $(X_1^L, \ldots, X_n^L)$. If we equip G with a left-invariant Riemannian metric m and require that $(X_1^L, \ldots, X_n^L)$ is orthonormal with respect to the corresponding inner product, then Y is a geometrically intrinsic object and Δ_G is the Laplace-Beltrami operator associated to (G, m).

Brownian motion has been the most intensively studied Lévy process in Lie groups. Recently it has played a key role in the development of analysis and geometry in path groups and loop groups (see e.g. Chapter XI of [60]).

Example 2: The Compound Poisson Process

Let $(\gamma_n, n \in \mathbf{N})$ be a sequence of i.i.d random variables taking values in G with common law μ and let $(N(t), t \geq 0)$ be an independent Poisson process with intensity $\lambda > 0$. We define the compound Poisson process in G by

$$Y(t) = \gamma_1\gamma_2 \cdots \gamma_{N(t)} \tag{10.5}$$

for $t > 0$.

Exercise 5.5 Show that in this case the generator is bounded and is given as

$$\mathcal{A}f(\tau) = \int_G (f(\tau\sigma) - f(\tau))\nu(d\sigma)$$

for each $f \in C_0(G)$ where the Lévy measure $\nu(\cdot) = \lambda\mu(\cdot)$ is finite.

Example 3: Stable Processes in Nilpotent Lie Groups

A *dilation* of a Lie group G is a family of automorphisms $\delta = (\delta(r), r > 0)$ for which

1. $\delta(r)\delta(s) = \delta(rs)$ for all $r, s > 0$

2. The map from $(0, \infty) \to G$ given by $r \to \delta(r)(\tau)$ is continuous for all $\tau \in G$

3. $\delta(r)(\tau) \to e$ as $r \to 0$, for all $\tau \in G$.

Let Y be a Lévy process in G. We say that it is *stable* with respect to the dilation δ if

$$\delta(r)Y(s) \text{ has the same law as } Y(rs) \text{ for each } r, s > 0.$$

Dilations (and hence stable Lévy processes) can only exist on simply connected nilpotent groups. For more on this topic, see the survey article [10] and references therein to works of H.Kunita.

Example 4: Subordinated Processes

Let $Y = (Y(t), t \geq 0)$ be a Lévy process on G and $T = (T(t), t \geq 0)$ be a subordinator which is independent of Y. Just as in the Euclidean case, we can construct a new Lévy process $Z = (Z(t), t \geq 0)$ by the prescription $Z(t) = Y(T(t))$, for each $t \geq 0$. For example, suppose that Y is an c-Brownian motion, with $c = 1$, and T is an independent α-stable subordinator. In this case, we can employ Phillip's theorem (theorem 6.3) to see that the generator of Z is $-(-\Delta_G)^{\alpha}$, on $\text{Dom}(\Delta_G)$. Such processes are called *pseudo-stable* by Cohen [27].

Lévy processes in Lie groups is a subject which is currently undergoing intense development - see the survey article [10] and the forthcoming book by Liao [57]. The latter contains a lot of interesting material on the asymptotics of Lévy processes on non-compact semi-simple Lie groups, as $t \to \infty$.

In a recent fascinating paper, Liao [58] has found some classes of Lévy processes on compact Lie groups which have L^2-densities. The density then has a "non-commutative Fourier series" expansion via the Peter-Weyl theorem. In the special case of c-Brownian motion on $SU(2)$, Liao obtains the following beautiful formula for its density ρ_t at time t:

$$\rho_t(\theta) = \sum_{n=1}^{\infty} n \exp\left(-\frac{c(n^2 - 1)t}{32\pi^2}\right) \frac{\sin(2\pi n\theta)}{\sin(2\pi\theta)},$$

where $\theta \in (0, 1]$ parameterises the maximal torus. $\{\text{diag}\left(e^{2\pi i\theta}, e^{-2\pi i\theta}\right), \theta \in [0, 1)\}$.

Another important theme, originally due to Gangolli in the 1960's, is to study *spherically symmetric Lévy processes* on semi-simple Lie groups G (i.e. those whose laws are bi-invariant under the action of a fixed compact subgroup K.) Using Harish-Chandra's theory of spherical functions, one can do "Fourier analysis" and obtain a Lévy-Khintchine type formula. One of the reasons why this is interesting is that G/K is a Riemannian (globally) symmetric space and all such spaces can be obtained in this way. The Lévy process in G projects to a Lévy process in G/K and this is the prototype for constructions of Lévy processes in more general Riemannian manifolds (see [10], [7]).

10.4 Lévy Processes on Locally Compact Groups - Reprise

Now that we know about Lévy processes in Lie groups, we can return to the problem of trying to understand these in general locally compact groups. First, a little more background:-

Let $(I, <)$ be a partially ordered set. Suppose that for every $i \in I$, there exists a locally compact group G_i, such that for every $i, j \in I$ with $i < j$, there is a continuous open homomorphism $\pi_{ij} : G_j \to G_i$, such that $\pi_{ik} = \pi_{ij} \circ \pi_{jk}$, for all $i < j < k$. The *projective limit* $\varprojlim_{i \in I} G_i$ is the closed subgroup $\{(x_i, i \in I) \in \prod_{i \in I} G_i; x_i = \pi_{ij}(x_j) \text{ for all } i, j \in I, i < j\}$. In the 1950's, Yamabe proved that every connected locally compact group can be represented as a projective limit of Lie groups.

Let $(H_i, i \in I)$ be a family of compact, normal subgroups of a locally compact group G. We say that they form a *Lie system* if

1. $i < j \Rightarrow H_j \subseteq H_i$.
2. $\bigcap_{i \in I} H_i = \{e\}$.
3. G/H_i is a Lie group for all $i \in I$.

A locally compact group G is said to be *Lie projective* if there exists a Lie system $(H_i, i \in I)$ such that $G = \varprojlim_{i \in I} G/H_i$. Gluškov proved that in every locally compact group G there exists an open Lie projective subgroup G_1 which contains the connected component of the identity.

A *topological Lie algebra* is a (not necessarily finite dimensional) Lie algebra for which the Lie bracket is jointly continuous in the vector topology. Projective limits of Lie algebras were introduced by Lashoff. Suppose that for every $i \in I$, there exists a topological Lie algebra $\mathbf{g}_i$, such that for every $i, j \in I$ with $i < j$, there is a continuous open Lie algebra homomorphism $p_{ij} : \mathbf{g}_j \to \mathbf{g}_i$, such that $p_{ik} = p_{ij} \circ p_{jk}$, for all $i < j < k$. The *projective limit* $\varprojlim_{i \in I} \mathbf{g}_i$ is the closed subalgebra $\{(X_i, i \in I) \in \prod_{i \in I} \mathbf{g}_i; X_i = p_{ij}(X_j) \text{ for all } i, j \in I, i < j\}$.

The relationship between projective Lie groups and projective Lie algebras is straightforward when $G = \varprojlim_{i \in I} G_i$, with each G_i a Lie group. In this case $\mathbf{g} = \varprojlim_{i \in I} \mathcal{L}(G_i)$ is a topological Lie algebra wherein $p_{ij} = d\pi_{ij}$, for each $i, j \in I, i < j$. We then call $\mathbf{g}$ the *Lie algebra* of the locally compact group G and sometimes denote it by $\mathcal{L}(G)$. There is a natural notion of exponential map from $\mathcal{L}(G)$ to G which works as follows. If $X = (X_i, i \in I) \in \mathcal{L}(G)$, then

$$\exp(X) = (\exp(X_i), i \in I).$$

For each $X \in \mathcal{L}(G)$, the map $t \to \exp(tX)$ is a continuous homomorphism from $\mathbb{R}$ to G. We define the *left invariant vector field* X^L associated to X in the obvious way, i.e.

$$(X^L f)(\sigma) = \lim_{h \to 0} \frac{f(\sigma \exp(hX)) - f(\sigma)}{h},$$

where $f \in C(G)$ is such that the limit on the right hand side exists for all $\sigma \in G$. If G is an arbitrary locally compact group, we can apply Gluškov's theorem to define the Lie algebra $\mathcal{L}(G)$ of G to be that of G_1, so that

$$\mathcal{L}(G) = \lim_{\longleftarrow i \in I} \mathcal{L}(G_1/H_i).$$

For each $i \in I$, π_i is the canonical surjection from G_1 onto G_1/H_j and $d\pi_i$ is then the canonical surjection from $\mathcal{L}(G)$ onto $\mathcal{L}(G_1/H_j)$.

We will briefly summarize some recent probabilistic progress. In the late 1980's E.Born (a student of Siebert) showed that the Lie algebra of a locally compact group always has a *projective basis* - to be more precise, let S be a set for which $I \subseteq S$. A family $(X_i, i \in S)$ in $\mathcal{L}(G) - \{0\}$ is called a *projective basis* if for each $j \in I$, there is a finite subset $S_j \subset S$, such that $(d\pi_j(X_i), i \in S_j)$ is a basis for $\mathcal{L}(G/H_j)$ and $d\pi_j(X_i) = 0$ whenever $i \notin S_j$. In [22] he uses this to obtain an explicit generalisation of Hunt's formula (10.3) in the general locally compact case, building on earlier more general results described in Heyer [43]. More recently, Born's formula was applied to establish a generalisation of Theorem 10.4 to this setting [12].

Brownian motion on compact (non-Lie) groups is currently a topic of intense investigation by A.Bendikov and L.Saloffe-Coste at Cornell. They have made a case-study of the infinite torus $\mathbb{T}^\infty = \{(z_n, n \in \mathbb{N}); z_n \in \mathbb{T}\}$. A fascinating investigation of sample paths can be found in [17]. It will be interesting to see if such nice results can also be obtained for more general Lévy processes.

11 Lecture 6: Two Lévy Paths to Quantum Stochastics

11.1 Path 1: - Unitary Representations of Lévy Processes in Lie Groups

An important role is played in quantum stochastic calculus by unitary operator-valued processes $(U(t), t \geq 0)$, which satisfy SDEs of the form

$$dU(t) = U(t)dM(t), \quad U(0) = I$$

acting in $\mathcal{H}_0 \otimes \Gamma(L^2(\mathbb{R}^+), \mathcal{H}_1)$, where $\mathcal{H}_1, \mathcal{H}_0$ are complex separable Hilbert spaces and Γ is the symmetric Fock functor. Here $(M(t), t \geq 0)$ is a suitable operator-valued semimartingale built from annihilation, creation and conservation processes, as described in Martin Lindsay's lectures. Although not generally considered by probabilists, such equations also arise classically in a natural way. To see one approach to this, we will need to dabble in group representations

The Non-Commutative Fourier-Stieltjes Transform

Let G be a locally compact group and $\mathcal{H}$ a complex, separable Hilbert space. A *unitary representation* of G in $\mathcal{H}$ is a strongly continuous homomorphism

π from G into $\mathcal{U}(\mathcal{H})$, the group of all unitary operators in $\mathcal{H}$, equipped with the strong operator topology. So we have the following:-

- For each $g \in G, \pi(g)$ is a unitary operator in $\mathcal{H}$.
- For each $g, h \in G, \pi(gh) = \pi(g)\pi(h)$.
- For each $g \in G, \pi(e) = I, \quad \pi(g^{-1}) = \pi(g)^*$.
- For each $\psi \in \mathcal{H}$, the mapping from G to $\mathcal{H}$ given by $g \to \pi(g)\psi$ is continuous.

A closed subspace $\mathcal{H}_1$ of $\mathcal{H}$ is *invariant* for π if $\pi(\mathcal{H}_1) \subseteq \mathcal{H}_1$. A representation is *irreducible* if the only invariant subspaces are $\{0\}$ and $\mathcal{H}$.

Later on we will need the *direct sum* of representations, so if π_i are unitary representations of G in $\mathcal{H}_i$, for $i = 1, 2$, then $\pi_1 \oplus \pi_2$ is the unitary representation of G in $\mathcal{H}_1 \oplus \mathcal{H}_2$ for which $(\pi_1 \oplus \pi_2)(\psi_1, \psi_2) = (\pi_1\psi_1, \pi_2\psi_2)$.

Now suppose that we want to try to do Fourier analysis in G. If G is abelian, we have seen that we can build Fourier transforms by using the dual group $\hat{G}$. But $\hat{G}$ is precisely the set of all irreducible representations of G.

Now take G to be an arbitrary locally compact group, let μ be a probability measure on G and fix a representation π (not necessarily irreducible) of G in some $\mathcal{H}$. Following Heyer [42], we define the *non-commutative Fourier-Stieltjes transform* of μ at π to be the Bochner integral:-

$$\hat{\mu}(\pi)\psi = \int_G (\pi(\sigma)\psi)\mu(d\sigma),$$

where $\psi \in \mathcal{H}$. $\hat{\mu}(\pi) \in B(\mathcal{H})$ and is in fact a contraction, since

$$||\hat{\mu}(\pi)\psi|| = \left|\left| \int_G (\pi(\sigma)\psi)\mu(d\sigma) \right|\right|$$

$$\leq \int_G ||\pi(\sigma)\psi||\mu(d\sigma)$$

$$= \int_G ||\psi||\mu(d\sigma) = ||\psi||.$$

The next result shows that $\hat{\mu}(\pi)$ is a good generalisation of the characteristic function of a probability measure on abelian G.

Theorem 11.1. *If μ_1 and μ_2 are probability measures in G, then*

$$\widehat{(\mu_1 * \mu_2)}(\pi) = \widehat{\mu_1}(\pi).\widehat{\mu_2}(\pi).$$

Proof For each $\psi \in \mathcal{H}$,

$$(\widehat{\mu_1 * \mu_2})(\pi)\psi = \int_G (\pi(\sigma)\psi)(\mu_1 * \mu_2)(d\sigma)$$

$$= \int_G \int_G (\pi(\sigma)\psi)\mu_1(d\tau)\mu_2(\tau^{-1}d\sigma)$$

$$= \int_G \int_G (\pi(\tau\sigma)\psi)\mu_1(d\tau)\mu_2(d\sigma)$$

$$= \int_G \int_G \pi(\tau)\pi(\sigma)\psi\mu_1(d\tau)\mu_2(d\sigma)$$

$$= \widehat{\mu_1}(\pi).\widehat{\mu_2}(\pi)\psi.$$

Exercise 6.1 Define the dual measure $\tilde{\mu}$ to μ by $\tilde{\mu}(A) = \mu(A^{-1})$, for each $A \in \mathcal{B}(G)$. Show that $\widehat{\tilde{\mu}}(\pi) = (\widehat{\mu})(\pi)^*$.

Now let $X = (X(t), t \geq 0)$ be a Lévy process in G. We follow [11] and define a unitary operator valued process $U = (U(t), t \geq 0)$ by

$$U^\pi(t) = \pi(X(t)).$$

We further define $T_t^\pi = \mathbb{E}(U^\pi(t))$, for each $t \geq 0$. If p_t is the law of $X(t)$ observe that

$$T_t^\pi = \widehat{p_t}(\pi).$$

Theorem 11.2. $(T_t^\pi, t \geq 0)$ *is a one-parameter semigroup of linear operators acting in* $\mathcal{H}$.

Proof By Theorem 11.1,

$$T_{t+s}^\pi = \widehat{p_{t+s}}(\pi) = \widehat{p_t * p_s}(\pi) = \widehat{p_t}(\pi)\widehat{p_s}(\pi) = T_t^\pi T_s^\pi.$$

The proof of strong continuity is *Exercise 6.2*   □

Exercise 6.3 Define a bounded continuous function f on G by $f(\tau) = \langle \psi_1, \pi(\tau)\psi_2 \rangle$, where $\psi_1, \psi_2 \in \mathcal{H}$. Deduce that $(T_t f)(e) = \langle \psi_1, T_t^\pi \psi_2 \rangle$.

Question: What can we say about the generator of this semigroup?

First we need some more background on the "infinitesimal structure" of group representations. Let $V^\omega = \{\psi \in \mathcal{H}, g \to \pi(g)\psi$ is analytic $\}$. V^ω is the set of *analytic vectors* for π in G. It is a dense linear manifold in $\mathcal{H}$. For each $Y \in \mathbf{g}$, define a linear operator $d\pi(Y)$ on V^ω by

$$d\pi(Y)\psi = \frac{d}{da}\pi(\exp(aY))\psi\Big|_{a=0}.$$

$d\pi(Y)$ is essentially skew-adjoint on V^ω. To see that, at least, skew-symmetry holds, let $\psi_1, \psi_2 \in V^\omega$, then

$$\langle d\pi(Y)\psi_1, \psi_2 \rangle = \frac{d}{da} \langle \pi(\exp(aY))\psi_1, \psi_2 \rangle \Big|_{a=0}$$

$$= \frac{d}{da} \langle \psi_1, \pi(\exp(aY)^{-1})\psi_2 \rangle \Big|_{a=0}$$

$$= \frac{d}{da} \langle \psi_1, \pi(\exp(-aY))\psi_2 \rangle \Big|_{a=0}$$

$$= -\langle \psi_1, d\pi(Y)\psi_2 \rangle$$

In the sequel we will require the following linear operator L^π in $\mathcal{H}$:

$$L^\pi \psi = b^i d\pi(X_i)\psi + a^{ij} d\pi(X_i)d\pi(X_j)\psi +$$

$$+ \int_{G-\{e\}} (\pi(\sigma) - I - x^i(\sigma)d\pi(X_i))\psi\nu(d\sigma) \qquad (11.1)$$

where $\psi \in V^\omega \subset \mathrm{Dom}(L^\pi)$ (see [11]).

The following result shows that the unitary process satisfies an operator-valued SDE of the type we encounter in quantum stochastic calculus:-

Theorem 11.3. *For each $t \geq 0$ the following operator-valued SDE holds on the domain $\hat{V}^\omega$:*

$$U^\pi(t) = I + \int_0^t U^\pi(s-)d\pi(X_i)dB^i(s) + \int_0^t U^\pi(s-)L^\pi ds \qquad (11.2)$$

$$+ \int_0^t \int_{G-\{e\}} U^\pi(s-)(\pi(\sigma) - I)\hat{N}(ds, d\sigma)$$

Proof Let $\psi_1, \psi_2 \in V^\omega$ and consider the map $f : G \to \mathbb{C}$ defined by

$$f(\sigma) = <\psi_1, \pi(\sigma)\psi_2>$$

Although f is not necessarily in $C_2^L(G)$, it is bounded and uniformly continuous (for the left uniform structure in G) and we can still apply Theorem 10.4 to such functions to obtain

$$< \psi_1, (U^\pi(t) - I)\psi_2 > = \int_0^t X_i^L < \psi_1, U^\pi(s-)\psi_2 > dB^i(s)$$

$$+ \int_0^t \mathcal{A} < \psi_1, U^\pi(s-)\psi_2 > ds$$

$$+ \int_0^t \int_{G-\{e\}} [f(X(s-)\sigma) - f(X(s-))]\hat{N}(ds, d\sigma)$$

The weak form of the required result for U now follows from the fact that for any $\tau \in G, Y \in \mathbf{g}$,

$$Y^L < \psi_1, \pi(\tau)\psi_2 > = \frac{d}{da} < \psi_1, \pi(\tau)\pi(\exp(aY))\psi_2 > \Big|_{a=0}$$

$$= < \psi_1, \pi(\tau)d\pi(Y)\psi_2 >,$$

and by a similar argument

$$\int_0^t \int_{G-\{e\}} [f(X(s-)\sigma) - f(X(s-))]\tilde{N}(ds, d\sigma)$$

$$= \int_0^t \int_{G-\{e\}} \langle \psi_1, [\pi(X(s-)\sigma) - \pi(X(s-))]\psi_2 \rangle \tilde{N}(ds, d\sigma)$$

$$= \int_0^t \int_{G-\{e\}} \langle \psi_1, [U^\pi(s-)\pi(\sigma) - U^\pi(s-)]\psi_2 \rangle \tilde{N}(ds, d\sigma).$$

You also need *Exercise* 6.4

$$\mathcal{A}\langle \psi_1, \pi(\tau)\psi_2 \rangle = \langle \psi_1, \pi(\tau)L^\pi \psi_2 \rangle$$

The strong result follows by a density argument, once you've checked that the stochastic integrals yield well-defined linear operators.   □

Taking expectations in (11.2), we obtain for all $\psi \in V^\omega$,

$$T_t^\pi \psi - \psi = \int_0^t T_s^\pi L^\pi \psi ds.$$

Hence the action of the infinitesimal generator of $(T_t^\pi, t \geq 0)$ is given by L^π. Formally, we can think of

$$T_t^\pi = e^{tL^\pi}$$

as a *non-commutative Lévy-Khintchine formula*, where we think of L^π as a "function" of irreducible representations. This insight is made precise by Siebert [76] who effectively showed that the convolution semigroup $(p_t, t \geq 0)$ is uniquely determined by the actions of L^π on a suitable domain, for each irreducible representation π of G.

From now on, we drop the π superscript to simplify notation. Following the philosophy of quantum stochastics, we obtain a stochastic process $j = (j(t), t \geq 0)$ taking values in the automorphism group of $B(\mathcal{H})$ if we define

$$j(t)(a) = U(t)aU(t)^*,$$

for all $t \geq 0, a \in B(\mathcal{H})$. To examine this infinitesimally, we need to take care with unbounded operators. Fix a (left) Haar measure on G and consider $L^1(G)$ which is a commutative Banach *-algebra, with respect to convolution and the involution $\tilde{f}(\tau) \to \overline{f(\tau^{-1})}$ (c.f. Example 5.16 in J. Kustermans' lectures). We define the "non-commutative Fourier transform" of $f \in L^1(G)$ by

$$\pi(f) = \int_G \pi(\tau)f(\tau)d\tau$$

Clearly $\pi(f) \in B(\mathcal{H})$, and in fact π is a homomorphism from $L^1(G)$ into $B(\mathcal{H})$, and a *-homomorphism when G is unimodular- (*Exercise* 6.5 - prove this).

In fact we will restrict yet further to the $*$-subalgebra $C_c^\infty(G)$ of $L^1(G)$ and work on $\mathcal{B} = \pi(C_c^\infty(G))$, (so that we can differentiate!). We then find that for each $t \geq 0, f \in C_c^\infty(G)$, the following operator-valued SDE holds on the domain V^ω, where $[\cdot, \cdot]$ now denotes the commutator:

$$j(t)\hat{\pi}(f) = \hat{\pi}(f) + \int_0^t j(s)(\mathcal{M}(\hat{\pi}(f)))ds + \int_0^t j(s)([d\hat{\pi}(X_i), \hat{\pi}(f)])dB^i(s)$$

$$+ \int_0^t \int_{G-\{e\}} j(s)(\hat{\pi}(\sigma)\hat{\pi}(f)\hat{\pi}(\sigma^{-1}) - \hat{\pi}(f))\hat{N}(ds, d\sigma). \qquad (11.3)$$

Here the (unbounded) linear operator $\mathcal{M}$ on $B(\mathcal{H})$ is given by

$$\mathcal{M}(\pi(f)) = b^i[d\pi(X_i), \pi(f)] + a^{ij}[d\pi(X_i), [d\pi(X_j), \pi(f)]] +$$

$$+ \int_{G-\{e\}} (\pi(\sigma)\pi(f)\pi(\sigma^{-1}) - \pi(f) - x^i(\sigma)[d\pi(X_i), \pi(f)])\nu(d\sigma).$$

We have $V^\omega \in \text{Dom}(\mathcal{M}(\pi(f))$ for each $f \in C_K^\infty(G)$.
If we define $S(t)(\pi(f)) = \mathbb{E}(j(t)(\pi(f))$, then $(S(t), t \geq 0)$ extends to a quantum dynamical semigroup on the von Neumann algebra generated by $\mathcal{B}$. The action of the infinitesimal generator on $\mathcal{B}$ is given by the linear operator $\mathcal{M}$. For full details see [11].

The monograph by Diaconis [28] is an excellent source for other applications of group representations within probability theory.

11.2 Lévy Processes in Fock Space

In this last part, we touch on a beautiful area of mathematics which involves the interaction between factorisable representations of current groups, cohomology of groups, infinite divisibility and Fock space. This was developed in the 1960s by H.Araki, R.F.Streater, K.R.Parthasarathy and K.Schmidt and continued through the 1970s and 80s via the work of I.M.Gelf'and, M.J.Graev and A.M.Vershik and also S.Albeverio, R.Høegh-Krohn and their collaborators.
We will only touch on this subject. Our aim is to answer the question -

Can we naturally represent Lévy processes as operator-valued processes in Fock space?

Good references for the general theory are Erven and Falkowski [30] and Guichardet [41], and for the probabilistic developments - Parthasarathy [65] - particularly Chapter II, section 21.
We need some simple ideas from cohomology:-

Cohomology of Groups

Let G be a Lie group and M a Hausdorff space which is a left G-module. For each $n \in \mathbb{N}$, define

$$C_n(G, M) = \{f : \overbrace{G \times \cdots \times G}^{n \text{ times}} \to M, f \text{ is continuous }\}.$$

We also need $C_0(G, M) = C_1(\{e\}, M)$ which we can just identify with M itself. Elements of $C_n(G, M)$ are called n *cochains* taking values in M. The sequence of *coboundary operators* $\{\delta_n, n \in \mathbb{N} \cup \{0\}\}$, where each $\delta_n : C_n(G, M) \to C_{n+1}(G, M)$ is given by

$$(\delta_n f)(\sigma_1, \ldots, \sigma_{n+1}) = \sigma_1 f(\sigma_2, \ldots, \sigma_{n+1}) + \sum_{i=1}^{n} (-1)^i f(\sigma_1, \ldots, \sigma_i \sigma_{i+1}, \ldots, \sigma_{n+1})$$

$$+ (-1)^{n+1} f(\sigma_1, \ldots, \sigma_n),$$

for each $f \in C_n(G, M), \sigma_1, \ldots, \sigma_{n+1} \in G$. A tedious calculation yields, for all $n \in \mathbb{N} \cup \{0\}$,

$$\delta_{n+1} \circ \delta_n = 0 \quad \Rightarrow \quad \text{Im}(\delta_n) \subseteq \text{Ker}(\delta_{n+1})$$

Elements of $\text{Ker}(\delta_n)$ are called n-cocycles.
Elements of $\text{Im}(\delta_{n-1})$ are called n-coboundaries.
$H^n(G, M) = \text{Ker}(\delta_n)/\text{Im}(\delta_{n-1})$ is called the nth *cohomology group* of G with coefficients in M.

Exercise 6.6 Check that all 1-coboundaries are of the form $\sigma \to \sigma\psi - \psi$, where $\psi \in M$ is fixed. Check that f is a 1-cocycle if and only if $\sigma_1 f(\sigma_2) = f(\sigma_1\sigma_2) - f(\sigma_1)$ for all $\sigma_1, \sigma_2 \in G$.
If f is a 1-cocycle, show that

1. $f(e) = 0$,
2. $\sigma^{-1} f(\sigma) = -f(\sigma^{-1})$, for all $\sigma \in G$.

In all the situations we will be interested in, M will be a complex separable Hilbert space $\mathcal{H}$. We will fix a unitary representation π of G in $\mathcal{H}$. The left action $G \times \mathcal{H} \to \mathcal{H}$ is $(g, \psi) \to \pi(g)\psi$.

Exercise 6.7 Suppose that U is a projective unitary representation of a Lie group G in a Hilbert space $\mathcal{H}$, so that there exists a continuous map $g : G \times G \to \mathbb{T}$, such that for each $\sigma_1, \sigma_2 \in G, g(\sigma_1, e) = g(e, \sigma_2) = 1$ and

$$U(\sigma_1\sigma_2) = g(\sigma_1, \sigma_2)U(\sigma_1)U(\sigma_2).$$

Write each $g(\sigma_1, \sigma_2) = e^{is(\sigma_1, \sigma_2)}$, and show that s is a 2-cocycle where $M = \mathbb{R}$ and π is the trivial action on $M = \mathbb{R}$ (i.e. each $\pi(\sigma) = 1$).

The Lévy-Khintchine Formula - Fock Style

We need some "basic Focklore" - details of which are covered in Martin Lindsay's notes.

- $\Gamma(h)$ is symmetric Fock space over a complex separable Hilbert space h.
- $\{e(f), f \in h\}$ are the total set of exponential vectors in $\Gamma(h)$. The Fock vacuum is $e(0)$.
- The Weyl operators - $\{W(f), f \in h\}$ are unitary in $\Gamma(h)$ - their action on exponential vectors is given by

$$W(f)e(g) = e^{-\frac{1}{2}\|f\|^2 - <f,g>}e(f+g),$$

for each $g \in h$.
- The second quantisation of a unitary operator V in h is $\Gamma(V)$ which is unitary in $\Gamma(h)$. For each $f \in h$,

$$\Gamma(V)e(f) = e(Vf) \qquad \Gamma(V)W(f)\Gamma(V)^{-1} = W(Vf).$$

- Let $u, v \in h$ and $T = T^*$ have domain D in h. We denote the annihilation operator $a(u)$, the creation operator $a^\dagger(v)$ and the conservation operator $\lambda(T)$. "By their matrix elements, shall ye know them"-

$$< e(f), a(u)e(g) >=< u, g > e^{<f,g>}, \qquad < e(f), a^\dagger(v)e(g) >=< f, v >$$
$$e^{<f,g>}$$
$$< e(f), \lambda(T)e(g) >=< f, Tg > e^{<f,g>}.$$

Note that in the first two relations, f and g are arbitrary vectors in h - in the third $g \in \text{Dom}(T)$.

The *Euclidean group* of h is the semidirect product $\mathcal{U}(h) \odot h$ which is $\mathcal{U}(h) \times h$ equipped with the composition law,

$$(V_1, f_1) \odot (V_2, f_2) = (V_1 V_2, f_1 + V_1 f_2).$$

Define the extended Weyl operator $W(V, f) = W(f)\Gamma(V)$.

Exercise 6.8 Check that the extended Weyl operators yield a projective unitary representation of $\mathcal{U}(h) \odot h$ in $\Gamma(h)$, i.e.

$$W(V_1, f_1)W(V_2, f_2) = e^{-i\Im <f_1, V_1 f_2>}W((V_1, f_1) \odot (V_2, f_2)).$$

Now let $\pi : G \to \mathcal{U}(G)$ be a unitary representation of a Lie group G in h and let ρ be a 1-cocycle for G acting on h. We will also assume that we can find a continuous map $\beta : G \to \mathbb{R}$ for which

$$\beta(\sigma_1\sigma_2) - \beta(\sigma_1) - \beta(\sigma_2) = \Im < \rho(\sigma_1^{-1}), \rho(\sigma_2) >,$$

for all $\sigma_1, \sigma_2 \in G$. We will show that such maps β can exist below. Now for each $\sigma \in G$, define the following unitary operator in $\Gamma(h)$:

$$U(\sigma) = e^{i\beta(\sigma)}W(\pi(\sigma), \rho(\sigma)). \tag{11.4}$$

Exercise 6.9 Show that $U : G \to \mathcal{U}(\Gamma(h))$ is a continuous unitary representation of G in $\Gamma(h)$, so that - in particular each $U(\sigma_1\sigma_2) = U(\sigma_1)U(\sigma_2)$.

U is called a *type S representation* by Guichardet [41].

All of this seems quite abstract and a long way from probability theory. However, the following example (due to K.R.Parthasarathy) will yield some familiar-looking expressions.

Take $G = \mathbb{R}$. We write $h = h_1 \oplus h_2$. Fix two vectors $\psi_i \in h_i, i = 1, 2$. We define a unitary representation $\pi' = I \oplus \pi$ of $\mathbb{R}$ in h. Observe that by Stone's theorem, there exists a self-adjoint operator $T = \int_{\mathbb{R}} \lambda P(d\lambda)$ in h (which may be unbounded), such that for each $y \in \mathbb{R}$,

$$\pi(y) = e^{iyT} = \int_{\mathbb{R}} e^{iy\lambda} P(d\lambda) \tag{11.5}$$

Now ρ is a 1-cocycle for the action of π' in h, where

$$\rho(y) = y\psi_0 + (\pi(y) - I)\psi_1.$$

You can check that we can then take

$$\beta(y) = my + \Im < \psi_1, (\pi(y) - I)\psi_1 >,$$

where $m \in \mathbb{R}$ is arbitrary.
From (11.4), we get a continuous one-parameter unitary group $(U(y), y \in \mathbb{R})$ in $\Gamma(h)$. The fun really starts when we compute the vacuum expectation of U, i.e.

$$< e(0), U(y)e(0) > = \exp\left(i\beta(y) - \frac{1}{2}||\rho(y)||^2\right).$$

Easy computations yield

$$||\rho(y)||^2 = y^2||\psi_0||^2 + ||(\pi(y) - I)\psi_1||^2$$
$$= y^2||\psi_0||^2 - 2\Re < \psi_1, (\pi(y) - I)\psi_1 > .$$

Now combining this with the expression for β given above, we obtain

$$< e(0), U(y)e(0) >$$
$$= \exp\left\{imy - \frac{1}{2}y^2||\psi_0||^2 + < \psi_1, (\pi(y) - I)\psi_1 >\right\}$$
$$= \exp\left(imy - \frac{1}{2}y^2||\psi_0||^2 + \int_{\mathbb{R}}(e^{i\lambda y} - 1)\nu(d\lambda)\right), \tag{11.6}$$

where $\nu(d\lambda) = < \psi_1, P(d\lambda)\psi_1 >$. Applying the spectral theorem again, we can write $U(y) = e^{iyX}$ and $< e(0), U(y)e(0) > = \int_{\mathbb{R}} e^{iyx}p(dx)$ where p is a probability measure on $\mathbb{R}$. Thus we see that (11.6) is a Lévy -Khintchine formula, and X is a realisation of an infinitely divisible random variable, with a finite Lévy measure.

One way of extending these ideas to get the most general form of the Lévy -Khintchine formula, is to set up a partition $\mathbb{R} = \bigcup_{n \in \mathbb{N}} E_n$, where each $E_n \in \mathcal{B}(\mathbb{R})$ with $E_0 = \{0\}, E_1 = [-1, 1]^c$. Next pick a sequence $(\psi_n, n \in \mathbb{N})$ in h such that each $P(E_n)\psi_n = \psi_n$ and $\sum_{n=2}^{\infty} ||T\psi_n||^2 < \infty$. (e.g. You could work in $h = L^2(\mathbb{R} - \{0\}, \nu)$, where ν is any Lévy measure. Then take each $E_n = (-\frac{1}{n-1}, -\frac{1}{n}) \cup (\frac{1}{n}, \frac{1}{n-1}), \psi_n = \chi_{E_n}, T(x) = \text{multiplication by } x$.)
Now replace the prescriptions given above for ρ and β by

$$\rho(y) = y\psi_0 + \sum_{n=1}^{\infty} (\pi(y) - I)\psi_n,$$

$$\beta(y) = my + \Im < \psi_1, (\pi(y) - I)\psi_1 > + \sum_{n=2}^{\infty} \Im < \psi_n, (\pi(y) - I - iyT)\psi_n > .$$

Exercise 6.10 Check that this really gives you the most general form of the Lévy-Khintchine formula.

Finally, we must bring time back into the picture. Return to the case of a general Lie group G. We replace G with the current group $C(\mathbb{R}^+, G)$ of all Borel measurable functions from $\mathbb{R}^+$ to G which have compact support. This is a group under pointwise operations. Now if π is a unitary representation of G in a complex separable Hilbert space, $\mathcal{H}$, then $\tilde{\pi}$ is a unitary representation of $C(\mathbb{R}^+, G)$ in $L^2(\mathbb{R}^+, \mathcal{H})$, where for each $f \in C(\mathbb{R}^+, G), \psi \in L^2(\mathbb{R}^+, \mathcal{H}), t \geq 0$, we have

$$((\tilde{\pi}(f))\psi)(t) = \pi(f(t))\psi(t).$$

To get a type S representation $\tilde{U}$ of $C(\mathbb{R}^+, G)$, we simply replace $\Gamma(h)$ with $\Gamma(L^2(\mathbb{R}^+, h))$ and instead of the previous "characteristics" (π, ρ, β), we employ $(\tilde{\pi}, \tilde{\rho}, \tilde{\beta})$, where for each $f \in C(\mathbb{R}^+, G), t \geq 0$,

$$(\tilde{\rho}(f))(t) = \rho(f(t)), \quad \tilde{\beta}(f) = \int_0^t \beta(f(s))ds.$$

Exercise 6.11. Check that $\tilde{U}$ really is a type S representation.

Note. If we define the *expectation functional* $\mathcal{E}(f) = < e(0), \tilde{U}(f)e(0) >$, for each $f \in C(\mathbb{R}^+, G)$, then $\mathcal{E}(fg) = \mathcal{E}(f)\mathcal{E}(g)$, whenever f and g have disjoint support, i.e. the representation $\tilde{U}$ is *factorisable* in the sense of H.Araki [14].

From now on, take $G = \mathbb{R}$. First observe (*Exercise* 6.12), that if you take $f_{t,y} \in C(\mathbb{R}^+, \mathbb{R})$ to be of the form $f_{t,y}(s) = \chi_{[0,t)}(s)y$, for fixed $t \geq 0, y \in \mathbb{R}$, then on replacing U with $\tilde{U}$ in (11.6), we obtain the Lévy-Khintchine formula for a Lévy process $(X(t), t \geq 0)$. Now we will find the analogue of the Lévy-Itô decomposition in $\Gamma(L^2(\mathbb{R}^+, h))$. For simplicity, we work in the case where the

Lévy measure ν is finite. If we hold t constant and vary y, then $\tilde{U}_t(y) = \tilde{U}(f_{t,y})$ is a continuous one-parameter unitary group and by Stone's theorem, we may write $\tilde{U}_t(y) = e^{iyX(t)}$, where $(X(t), t \geq 0)$ is the Fock space realisation of a Lévy process. Choose $f_1, f_2 \in L^2(\mathbb{R}^+, h)$ where f_2 takes values in $\text{Dom}(H)$. We also require that $\psi_1 \in \text{Dom}(H)$.

Now (*Exercise* 6.13), compute

$$< e(f_1), X(t)e(f_2) > = \frac{d}{dy} < e(f_1), \tilde{U}_t(y)e(f_2) > \Big|_{y=0},$$

to obtain

$$X(t) = mt + i(a(\chi_{[0,t)} \otimes \psi_0) - a^\dagger(\chi_{[0,t)} \otimes \psi_0))$$
$$+ a^\dagger(\chi_{[0,t)} \otimes H\psi_1) + \lambda(\chi_{[0,t)} \otimes H) + a(\chi_{[0,t)} \otimes H\psi_1) + t < \psi_1, H\psi_1 >.$$

Comparing with (8.11) - we can identify the drift term as mt, the Brownian motion part (in momentum form) as $i(a(\chi_{[0,t)} \otimes \psi_0) - a^\dagger(\chi_{[0,t)} \otimes \psi_0))$ and the Poisson part as $a^\dagger(\chi_{[0,t)} \otimes T\psi_1) + \lambda(\chi_{[0,t)} \otimes T) + a(\chi_{[0,t)} \otimes T\psi_1) + t < \psi_1, H\psi_1 >$. Making the analogy with (8.11) more precise, we should really write the latter as an integral over jumps - we can do this, at least formally, by utilising the spectral decomposition of $H = \int_{\mathbb{R}} \lambda P(d\lambda)$ so that e.g.

$$a^\dagger(\chi_{[0,t)} \otimes H\psi_1) = \int_{\mathbb{R}} \lambda a^\dagger(\chi_{[0,t)} \otimes P(d\lambda)\psi_1).$$

More generally, to build a quantum stochastic calculus based on the noise generated by a Poisson random measure - we need to consider *quantum stochastic spectral integrals* based on $A^\dagger(dt, P(d\lambda)\psi_1), A(dt, P(d\lambda)\psi_1)$ and $\Lambda(dt, P(d\lambda))$. For details of this, see [5] and references therein, where you will find full "quantised" generalisations of (11.2) and (11.3).

It would be nice to be able to use Fock space methods to learn more about Lévy processes in Lie groups and this would involve extending the ideas given above to the current group $C(\mathbb{R}^+, G)$. This only seems to work well when there is a natural analogue of the Fourier transform available, e.g. in [9], Harish-Chandra's spherical transform was used to obtain a Fock space representation for spherically symmetric Lévy processes in non-compact semi-simple Lie groups.

In the case of compact Lie groups whose Lie algebra is also compact (in a certain technical sense), Albeverio and Høegh-Krohn [3] have introduced a non-factorisable representation, called the *energy representation* which is intimately related to Brownian motion on the group. It would be nice to have some greater insight into this from a Fock space point of view. Some interesting connections between current algebra representations and quantum Lévy processes are explored in [1]. A great deal of information about representations of current and other infinite dimensional groups can be found in the monograph [46], and Fock space certainly plays a major role here.

Current groups of the general type $C(M, G)$, where M is a manifold, are of interest to physics - where M represents space-time and G is a gauge group. If you take $M = S^1$ then you get *loop groups*, which have been intensively studied by mathematicians. The link with probability is through the Brownian bridge and this leads to the wonderful world of logarithmic Sobolev inequalities and spectral gaps (see [44] for a very readable introduction to these concepts).

References

1. L.Accardi, U.Franz, M.Skeide, Renormalised squares of white noise and other non-Gaussian noise as Lévy processes on real Lie algebras, *Commun. Math.Phys.* **228**, 123-50 (2002)

2. S.Albeverio, H.Gottschalk, J-L.Wu, Convoluted generalised white noise, Schwinger functions, and their analytic continuation to Wightman functions, *Rev. Math. Phys.* **8**, 763-817 (1996)

3. S.Albeverio, R. Høegh-Krohn The energy representation of Sobolev-Lie groups, *Comp. Math.* **36**, 37-52 (1978)

4. E.Alos, O.Mazet, D.Nualart, Stochastic calculus with respect to fractional Brownian motion with Hurst parameter less than $\frac{1}{2}$, *Stoch.Proc.App.* **86**, 121-39 (1999)

5. D.Applebaum, M.Brooks, Infinite series of quantum spectral stochastic integrals, *J.Operator Theory* **36**, 295-316 (1996)

6. D.Applebaum, S.Cohen, Lévy processes, pseudo-differential operators and Dirichlet forms in the Heisenberg group, preprint (2002) - see http://www.scm.ntu.ac.uk/dba

7. D.Applebaum, A.Estrade, Isotropic Levy processes on Riemannian manifolds, *Ann. Prob.* **28**, 166-84 (2000)

8. D.Applebaum, H.Kunita, Lévy flows on manifolds and Lévy processes on Lie groups, *J.Math Kyoto Univ* **33**, 1103-23 (1993)

9. D.Applebaum, Compound Poisson processes and Lévy processes in groups and symmetric spaces, *J.Theor. Prob.* **13**, 383-425 (2000)

10. D.Applebaum, Lévy processes in stochastic differential geometry, in *Lévy Processes:Theory and Applications* ed. O.Barndorff-Nielsen, T.Mikosch, S.Resnick (Birkhäuser Boston Basel Berlin), 111-39 (2001)

11. D.Applebaum, Operator-valued stochastic differential equations arising from unitary group representations, *J. Theor. Prob.* **14**, 61-76 (2001)

12. D.Applebaum, Brownian motion and Lévy processes in locally compact groups, preprint (2003) - see http://www.scm.ntu.ac.uk/dba

13. D.Applebaum, *Lévy Processes and Stochastic Calculus*, Cambridge University Press, to appear (2004)

14. H.Araki, Factorisable representations of current algebra, *Publ. Res. Inst. Math. Sci. Kyoto Univ.* **5**, 361-422 (1970)

15. O.E.Barndorff-Nielsen, T.Mikosch, S.Resnick (eds), *Lévy Processes: Theory and Applications*, Birkhäuser, Basel (2001)

16. O.E.Barndorff-Nielsen, N.Shephard, Non-Gaussian Ornstein-Uhlenbeck-based models and some of their uses in financial economics, *J.R.Statis.Soc. B*, **63**, 167-241 (2001)

17. A.Bendikov, L.Saloff-Coste, On the sample paths of Brownian motions on compact infinite dimensional groups, *Ann. Prob.* **31**, 1464-94 (2003)

18. C.Berg, G.Forst, *Potential Theory on Locally Compact Abelian Groups*, Springer-Verlag Berlin-Heidelberg New York (1975)

19. J.Bertoin, *Lévy Processes*, Cambridge University Press (1996)

20. J.Bertoin, Some properties of Burgers turbulence with white or stable noise initial data, in *Lévy Processes:Theory and Applications* ed. O.Barndorff-Nielsen, T.Mikosch, S.Resnick (Birkhäuser Boston Basel Berlin), 267-79 (2001)

21. W.R.Bloom, H.Heyer, *Harmonic Analysis of Probability Measures on Hypergroups*, de Gruyter, Berlin, New York (1995)

22. E.Born, An explicit Lévy-Hinčin formula for convolution semigroups on locally compact groups, *J.Theor. Prob.* **2**, 325-42 (1989)

23. N.Bouleau, Visco-elasticité et processus de Lévy, *Potential Anal.* **11**, 289-302 (1999)

24. P.J.Brockwell, Lévy-driven CARMA processes, *Ann.Inst.Statist.Math.* **53**, 113-24 (2001)

25. R.Carmona, W.C.Masters, B.Simon, Relativistic Schrödinger operators: asymptotic behaviour of the eigenvalues, *J.Funct.Anal.* **91**, 117-42 (1990)

26. R.Carter, G.Segal, I.Macdonald, *Lectures on Lie Groups and Lie Algebras*, London Math.Soc Student Texts **32**, Cambridge University Press (1995)

27. S.Cohen, Some Markov properties of stochastic differential equations with jumps, in *Séminaire de Probabilités XXIX*, ed. J.Azéma, M.Emery, P.A.Meyer, M.Yor, Lecture Notes in Math, **1613**, Springer-Verlag Berlin Heidelberg, 181-94 (1995)

28. P.Diaconis, *Group Representations in Probability and Statistics*, Lecture Notes-Monograph Series Volume 11, Institute of Mathematical Statistics, Hayward, California (1988)

29. E.B.Davies, *One-Parameter Semigroups*, Academic Press (1980)

30. J.Erven, B-J.Falkowski, *Low Order Cohomology and Applications*, Lecture Notes in Mathematics **877**, Springer-Verlag Berlin Heidelberg New York (1981)

31. P.Embrechts, M.Maejima, *Selfsimilar Processes*, Princeton University Press (2002)

32. M.Fannes, J.Quaegebeur, Infinite divisibility and central limit theorems for completely positive mappings, in *Quantum Probability and Applications II*, ed. L.Accardi, W. von Waldenfels, Lecture Notes in Mathematics **1136**, Springer-Verlag Berlin Heidelberg, 177-202 (1985)

33. P. Feinsilver, Processes with independent increments on a Lie group, *Trans Amer Math Soc* **242**, 73-121 (1978)

34. W.Feller, *An Introduction to Probability Theory and its Applications, Volume 1* (second edition), Wiley, New York (1957)

35. W.Feller, *An Introduction to Probability Theory and its Applications, Volume 2* (second edition), Wiley, New York (1971)

36. U.Franz, M.Schürmann, Lévy processes on quantum hypergroups, in *Infinite Dimensional Harmonic Analysis - Kyoto 1999*, ed. H.Heyer, T.Hirai, N.Obata, 93-115 D. and M.Gräbner (2000)

37. T.Fujiwara, H.Kunita, Stochastic differential equations of jump type and Lévy processes in diffeomorphisms group, *J.Math.Kyoto Univ* **25**, 71-106 (1985)

38. M.Fukushima, Y.Oshima, M.Takeda, *Dirichlet Forms and Symmetric Markov Processes*, Walter de Gruyter, Berlin, New York (1994)

39. H.Geman, D.B.Madan, M.Yor, Time changes for Lévy Processes, *Math. Finance* **11**, 79-86 (2001)

40. B.V.Gnedenko, A.N.Kolmogorov, *Limit Distributions for Sums of Independent Random Variables* (second edition), Addison-Wesley, Reading Mass. (1968)

41. A.Guichardet, *Symmetric Hilbert Spaces and Related Topics*, Lecture Notes in Mathematics **261**, Springer-Verlag Berlin Heidelberg New York (1972)

42. H.Heyer, L'analyse de Fourier non-commutative et applications à la théorie des probabilités, *Ann. Inst. Henri Poincaré (Prob. Stat.)* **4**, 143-68 (1968)

43. H.Heyer, *Probability Measures on Locally Compact Groups*, Springer-Verlag, Berlin-Heidelberg (1977)

44. E.Hsu, Analysis on path and loop spaces, in *Probability Theory and Applications*, IAS/Park City Math Series, Vol 6 (E.Hsu, S.R.S.Varadhan eds), 279-347 AMS (1999)

45. G.A.Hunt, Semigroups of measures on Lie groups, *Trans. Amer. Math. Soc.* **81**, 264-93 (1956)

46. R.S.Ismagilov, *Representations of Infinite Dimensional Groups*, Translations of Mathematical Monographs Vol.152, AMS (1996)

47. N.Ikeda, S.Watanabe *Stochastic Differential Equations and Diffusion Processes* (second edition), North Holland-Kodansha, Amsterdam New York (1989)

48. N.Jacob, *Pseudo-differential Operators and Markov Processes*, Akademie-Verlag, Mathematical Research vol **94**, Berlin (1996)

49. N.Jacob, *Pseudo-differential Operators and Markov Processes: 1. Fourier Analysis and Semigroups*, World Scientific (2001)

50. N.Jacob, *Pseudo-differential Operators and Markov Processes: 2. Generators and their Potential Theory*, World Scientific (2002)

51. J.Jacod, A.N.Shiryaev, *Limit Theorems for Stochastic Processes*, Springer-Verlag, Berlin Heidelberg (1987), second edition (2003)

52. F.B.Knight, *Essentials of Brownian Motion and Diffusion*, AMS (1981) (reprinted (1989))

53. H.Kunita, *Stochastic Flows and Stochastic Differential Equations*, Cambridge University Press (1990)

54. H.Kunita, Stochastic differential equations based on Lévy processes and stochastic flows of diffeomorphisms, to appear in *Real and Stochastic Analysis* M.M. Rao (ed.) (2004)

55. J-F.Le Gall, Y.Le Jan, Branching processes in Lévy processes: the exploration process, *Ann. Prob.* **26**, 213-52 (1998)

56. P.Lévy, *Processus Stochastiques et Mouvement Brownien*, (second edition) Gauthier-Villars, Paris (1954) (first edition 1948)

57. M.Liao, *Lévy Processes in Lie Groups*, Cambridge University Press, (2004).

58. M. Liao, Lévy processes and Fourier analysis on compact Lie groups, *Ann. Prob.*, to appear.

59. W.Linde, *Probability in Banach Spaces - Stable and Infinitely Divisible Distributions*, Wiley-Interscience (1986)

60. P.Malliavin, *Stochastic Analysis*, Springer-Verlag Berlin Heidelberg (1997)

61. E.Nelson, *Dynamical Theories of Brownian Motion*, Princeton University Press (1967)

62. G.Pap, Construction of processes with stationary independent increments in Lie groups, *Arch.Math.* **69**, 146-55 (1997)

63. K.R.Parthasarathy, K.Schmidt, *Positive Definite Kernels, Continuous Tensor Products and Central Limit Theorems of Probability Theory*, Lecture Notes in Mathematics **272**, Springer-Verlag Berlin (1972)
64. K.R.Parthasarthy, *Probability Measures on Metric Spaces*, Academic Press, New York (1967)
65. K.R.Parthasarthy, *An Introduction to Quantum Stochastic Calculus*, Birkhäuser Verlag Basel (1992)
66. P.Protter, *Stochastic Integration and Differential Equations*, Springer-Verlag, Berlin Heidelberg (1992)
67. M.Reed, B.Simon *Methods of Modern Mathematical Physics, Volume 1 : Functional Analysis* (revised and enlarged edition), Academic Press (1980)
68. D.Revuz, M.Yor *Continuous Martingales and Brownian Motion*, Springer-Verlag, Berlin Heidelberg (1990)
69. L.C.G.Rogers, D.Williams, *Diffusions, Markov Processes and Martingales, Volume 1: Foundations*, J.Wiley and Sons Ltd (1979,1994), Cambridge University Press (2000)
70. L.C.G.Rogers, D.Williams, *Diffusions, Markov Processes and Martingales, Volume 2: Itô Calculus*, J.Wiley and Sons Ltd (1994), Cambridge University Press (2000)
71. J.Rosiński, W.A.Woyczyński, On Itô stochastic integration with respect to p-stable motion: inner clock, integrability of sample paths, double and multiple integrals, *Ann. Prob.*,**14**, 271-86 (1986)
72. W.Rudin, *Fourier Analysis on Groups*, (1st edition 1962), Wiley Classics (1990)
73. G.Samorodnitsky, M.S.Taqqu, *Stable non-Gaussian Random Processes*, Chapman and Hall, New York (1994)
74. K-I.Sato, *Lévy Processes and Infinite Divisibility*, Cambridge University Press (1999)
75. W.Schoutens, *Lévy Processes in Finance: Pricing Financial Derivatives*, Wiley (2003)
76. E.Siebert, Fourier analysis and limit theorems for convolution semigroups on a locally compact group, *Advances in Math.* **39**, 111-54 (1981)
77. B.Simon, *Representations of Finite and Compact Groups*, Graduate Studies in Math, Vol. **10**, Amer. Math. Soc. (1996)
78. J.M.Steele, *Stochastic Calculus and Financial Applications*, Springer-Verlag New York Inc. (2001)
79. M.E.Taylor, *Partial Differential Equations II, Qualitative Studies of Linear Equations*, Springer-Verlag New York Inc. (1996)
80. S.J.Wolfe, On a continuous analogue of the stochastic difference equation $X_n = \rho X_{n-1} + B_n$, *Stoch.Proc.Appl.* **12**, 301-12 (1982)

Locally compact quantum groups

Johan Kustermans[*]

KU Leuven
Departement Wiskunde
Celestijnenlaan 200B
3001 Heverlee, Belgium
johan.kustermans@wis.kuleuven.ac.be

---
[*] Supported by the National Science Foundation - Flanders

Introduction

These lecture notes are intended as an introduction to the theory of locally compact quantum groups that are studied in the framework of operator algebras, i.e. C*-algebras and von Neumann algebras. The presentation revolves around the definition of a locally compact quantum group as given in [KuV00a] and [KuV03].

Historically the first aim in constructing axiomatizations of 'quantized' locally compact groups was the extension of the Pontryagin duality to non-abelian groups. Because in general the dual of a non-abelian group will not be a group any more, one searched for a larger category which included both groups and group duals. After pioneering work by Tannaka, Krein, Kac and Takesaki, among others, this problem was completely solved independently by M. Enock and J.-M. Schwartz (see [ES92] for a survey) and by Kac and Vainerman ([KaV73], [KaV74]) in the seventies. The object they defined is called a Kac algebra.

In [Wor87b] S. L. Woronowicz constructed a C*-algebra with comultiplication, quantum $SU(2)$, which had so many group-like properties that it was justified to call it a 'quantum group'. But this example did not fit in the category of Kac algebras. So it became clear that the category of Kac algebras was too small to include all quantum groups, and that it should be enlarged.

The first success in this direction was obtained by Woronowicz who succeeded to define the compact quantum groups ([Wor87a],[Wor98]) in a simple way and who proved, most importantly, the existence and uniqueness of a Haar state.

The next success provided us with another approach. In [BS] S. Baaj and G. Skandalis made a study of multiplicative unitaries, which can be considered as an abstract study of the Kac-Takesaki operator of a locally compact group. With an irreducible and regular multiplicative unitary they associate two C*-algebras, which are each others dual, with a comultiplication and a densely defined antipode. In this way they obtain both the compact quantum groups and, in a certain sense, the Kac algebras. At the same time a multitude of different aspects of multiplicative unitaries are investigated in this paper, rendering it an invaluable source of information.

Still one wanted to give a more intrinsic definition of a locally compact quantum group, with a C*-algebra (or von Neumann algebra) with comultiplication as a starting point. An essential idea in this direction was put forward by Kirchberg in [Kir], who proposed to allow the antipode of a Kac algebra to be deformed by a 'scaling group' which should be a one-parameter group of automorphisms of the underlying von Neumann algebra. Then T. Masuda and Y. Nakagami formulated the definition of a Woronowicz algebra in [MN], generalizing Kac algebras by introducing this scaling group. They were able to construct the dual within the same category, and their theory included the known examples, the Kac algebras and the compact quantum groups in a certain sense. However there is an objection to their theory and that is the complexity of the axioms : a Woronowicz algebra is a quintuple consisting of a von Neumann algebra, a comultiplication, a Haar weight, a unitary antipode and a scaling group, satisfying a lot of relations. For these reasons, this definition was not satisfactory. A C*-algebraic version of this definition is discussed in [MNW].

Finally a relatively simple definition was proposed by Vaes and the author in [KuV00a] and [KuV03] but is has to be said that a lot of ideas of the work of Kac, Vainerman, Enock, Schwartz, Baaj and Skandalis (and the polar decomposition due to Kirchberg) play an important role in [KuV00a] and [KuV03].

In these lecture notes we will follow [KuV03] in which the definition of a locally compact quantum group is formulated in the framework of von Neumann algebras, because this provides us with the simplest definition. Since the operator algebra theory involved is rather complicated a large part of these lecture notes are devoted to a careful explanation of all the concepts involved. We therefore opted to be precise, rather than to be exhaustive.

The first section gives a quick introduction to the basic theory of C*-algebras. The second section looks at locally compact quantum groups in an effort to motivate the later definitions of quantum groups. The third section discusses compact quantum groups in the C*-algebra framework due to Woronowicz. Here the C*-algebra framework is preferable because of the presence of an existence theorem for the quantum analogue of the Haar measure for compact quantum groups. In the fourth section we introduce the necessary tools from the theory of von Neumann algebras.

The definition of a locally compact quantum group and its main consequences are discussed in the fifth section. Two examples of locally compact quantum groups are studied in the sixth section. In the appendix we collected a number of concepts for the convenience of the reader that is not that well acquainted with the theory of unbounded operators in Hilbert spaces.

In the last part of this introduction we fix some notations and conventions. We want to stress that the notations and conventions used in this text might differ from the ones used in other texts of these lecture notes.

If X is a set we will denote the identity mapping on X by ι_X or even ι if it is clear which set is under consideration. The domain of a function f is denoted by $D(f)$. Let Y be another set and S, T two functions $S : D(S) \subseteq X \to Y$ and $T : D(T) \subseteq X \to Y$. We say that $S \subseteq T$ if and only if $D(S) \subseteq D(T)$ and $S(x) = T(x)$ for all $x \in D(S)$.

The set of all natural numbers, not including 0, is denoted by $\mathbb{N}$. Also, $\mathbb{N}_0 = \mathbb{N} \cup \{0\}$. For a vector space V and a subset $U \subseteq V$, the linear span of U in V is denoted by $\langle U \rangle$.

We will always work with complex Hilbert spaces. Let $\mathcal{H}$ be a Hilbert space. The inner product on $\mathcal{H}$ will be denoted by $\langle . , . \rangle$. In this text, all such inner products will be linear in the first and antilinear in the second argument. Notice that this is contrary to most other texts in these lecture notes! A linear operator T in $\mathcal{H}$ is a map $T : D(T) \subseteq \mathcal{H} \to \mathcal{H}$ for which the domain

$D(T)$ is a linear subspace of $\mathcal{H}$ and the action of T on $D(T)$ is linear. If $D(T) = \mathcal{H}$, we call T a linear operator on $\mathcal{H}$. You can find extra information on linear operators in Hilbert spaces in the appendix.

If S, T are linear operators in $\mathcal{H}$ and $\lambda \in \mathbb{C}$ the linear operators $S + T$, λS and ST in $\mathcal{H}$ are defined such that $D(S + T) = D(S) \cap D(T)$, $D(\lambda S) = D(S)$, $D(ST) = \{\, v \in D(T) \mid T(v) \in D(S) \,\}$ and $(S + T)(v) = S(v) + T(v)$ for all $v \in D(S + T)$, $(\lambda S)(v) = \lambda S(v)$ for all $v \in D(S)$ and $(ST)(v) = S(T(v))$ for all $v \in D(ST)$.

A *-algebra A is an (associative) complex algebra together with an anti-linear map $.^* : A \to A$ satisfying $(ab)^* = b^* a^*$ and $(a^*)^* = a$ for all $a, b \in A$. The map $.^*$ is called the *-operation (star-operation) of the *-algebra. If A, B are *-algebras, we call a map $\pi : A \to B$ a *-homomorphism from A to B if π is linear, multiplicative and satisfies $\pi(a^*) = \pi(a)^*$ for all $a \in A$.

1 Elementary C*-algebra theory

This section serves as a quick introduction to basic C*-algebra theory needed to understand and work with the definition of a locally compact quantum group that we will give later on. There are several good books on the subject available, eg. [Mur], [KR1] and [KR2]. The order in which results are stated does not respect the chronology of the build up of the theory.

Definition 1.1. *A C*-algebra A is a *-algebra equipped with a norm $\|\cdot\|$ for which A is complete and such that (i) $\|ab\| \le \|a\| \, \|b\|$ and (ii) $\|a^* a\| = \|a\|^2$ for all $a, b \in A$.*

It follows from (i) and (ii) that $\|a^*\| = \|a\|$ for all $a \in A$. We do in general not assume that A has a unit element for the multiplication, but if it does have a unit element and $A \ne \{0\}$, we call A unital and denote the unit element by 1. The space of continuous linear functionals on A will be denoted by A^*.

A *C*-subalgebra* of A is a *-subalgebra of A that is closed for the norm topology. By restricting all algebraic operations and the norm to such a C*-subalgebra one obtains a new C*-algebra.

If $\mathcal{A}$ is a *-algebra and $\|.\|$ is a norm on $\mathcal{A}$ satisfying property (i) and (ii) in the definition above, we say that $\|.\|$ is a *C*-norm* on $\mathcal{A}$. Let A denote the Banach space that is the completion of $\mathcal{A}$ with respect to this norm. By extending the product and *-operations of $\mathcal{A}$ by continuity to a product and *-operation on A, we turn A into a genuine C*-algebra.

Example 1.2. (1) Consider a locally compact Hausdorff space X and let $C_0(X)$ be the space of complex valued continuous functions on X that vanish at infinity. By definition, a function $f : X \to \mathbb{C}$ vanishes at infinity if for every $\varepsilon > 0$ there exists a compact subset $K \subseteq X$ so that $|f(x)| \le \varepsilon$ for all $x \in X \setminus K$.

This set of functions $C_0(X)$ is a commutative C*-algebra for the pointwise algebraic operations and the sup-norm: if $f, g \in C_0(X)$ and $\lambda \in \mathbb{C}$, then

$$(f + g)(x) = f(x) + g(x) , \qquad (\lambda f)(x) = \lambda f(x) ,$$
$$(fg)(x) = f(x)g(x) , \qquad f^*(x) = \overline{f(x)}$$

for all $x \in X$ and

$$\|f\| = \sup \{ |f(x)| \mid x \in X \} .$$

Notice that $C_0(X)$ has a unit if and only if X is compact, in which case the unit is given by the constant function that takes the value 1 on the whole of X.

There is also another natural commutative C*-algebra associated to X, namely the space $C_b(X)$ of all bounded continuous functions on X. One turns $C_b(X)$ into a *-algebra and defines a norm on $C_b(X)$ by the same formulas as above. The resulting *-algebra always has a unit given by the constant function that takes the value 1 on the whole of X. Of course, if X is compact, then $C_0(X) = C_b(X) = C(X)$, the space of all continuous functions on X.

(2) Let $\mathcal{H}$ be a Hilbert space. Recall that, in *this* text, all such inner products will be linear in the first and antilinear in the second argument! The algebra of bounded linear operators on $\mathcal{H}$ will be denoted by $B(\mathcal{H})$. It is a unital C*-algebra for the usual operator adjoint and operator norm, i.e. (i) $\langle T^*v, w \rangle = \langle v, Tw \rangle$ and (ii) $\|T\| = \sup\{ \|Tu\| \mid u \in \mathcal{H}, \|u\| \leq 1 \}$ for all $T \in B(\mathcal{H})$ and $v, w \in \mathcal{H}$.

The algebra $B_0(\mathcal{H})$ of compact operators on $\mathcal{H}$ is a C*-subalgebra of $B(\mathcal{H})$, it is even a two-sided ideal inside $B(\mathcal{H})$ (in the literature one also uses the notation $\mathcal{K}(\mathcal{H})$ for $B_0(\mathcal{H})$). Recall that a linear operator T on $\mathcal{H}$ is called compact if for every bounded subset $D \subseteq \mathcal{H}$, the closure of $T(D)$ is compact. If $v, w \in \mathcal{H}$ we define $\theta_{v,w} \in B_0(\mathcal{H})$ by $\theta_{v,w}(u) = \langle u, w \rangle\, v$ for all $u \in \mathcal{H}$. The linear span of such operators $\theta_{v,w}$ is norm dense in $B_0(\mathcal{H})$.

The algebra $B_0(\mathcal{H})$ has an identity if and only if the Hilbert space is finite-dimensional. In this case, $B_0(\mathcal{H}) = B(\mathcal{H}) \cong M_n(\mathbb{C})$, the *-algebra of n by n complex matrices, where n is the dimension of the Hilbert space $\mathcal{H}$.

Since C*-algebras are *-algebras we can talk about *-*homomorphisms* and *-*isomorphisms* between C*-algebras. The next proposition guarantees that these are the natural morphisms between C*-algebras. As a consequence, if A,B are C*-algebras we say that $A \cong B$ if and only if A is *-isomorphic to B.

Proposition 1.3. *Consider two C*-algebras A, B and a *-homomorphism $\pi : A \to B$. Then, π is contractive, i.e. $\|\pi(x)\| \leq \|x\|$ for all $x \in A$. Moreover, $\pi(A)$ is a C*-subalgebra of B. If π is injective, then π is isometric, i.e. $\|\pi(x)\| = \|x\|$ for all $x \in A$.*

This implies that if A is a *-algebra, there exists at most one norm on A for which A is C*-algebra (there might be no such norm at all). Notice that the closedness of $\pi(A)$ in B is a non-trivial claim!

Let $\mathcal{H}$ be a Hilbert space. If B is a *-subalgebra of $B(\mathcal{H})$, we call B *non-degenerate* if the subspace $B\mathcal{H} := \langle bv \mid b \in B, v \in \mathcal{H} \rangle$ is dense in $\mathcal{H}$.

If A is a C*-algebra, a *-*representation* π of A on $\mathcal{H}$ is by definition a *-homomorphism $\pi : A \rightarrow B(\mathcal{H})$. We call π non-degenerate if $\pi(A)$ is non-degenerate in $B(\mathcal{H})$. We call π faithful if and only if it is injective.

The next two theorems show that the examples given above are in fact typical examples of C*-algebras.

Theorem 1.4 (Gelfand).   *Consider a commutative C*-algebra A. Then, there exists a locally compact space X, unique up to a homeomorphism, such that $A \cong C_0(X)$.*

Theorem 1.5 (Gelfand-Naimark).   *Consider a C*-algebra A. Then, there exists a Hilbert space $\mathcal{H}$ and an injective non-degenerate *-representation $\pi : A \rightarrow B(\mathcal{H})$.*

Remark 1.6. Consider a *-algebra $\mathcal{A}$ and define the map $\|.\|_* : \mathcal{A} \rightarrow [0, \infty]$ by

$$\|a\|_* = \sup\{\, \|\theta(a)\| \mid \mathcal{K} \text{ a Hilbert space}, \theta : \mathcal{A} \rightarrow B(\mathcal{K}) \text{ a *-homomorphism}\,\}$$

for all $a \in \mathcal{A}$. Suppose that if $a \in \mathcal{A}$, then (1) $\|a\|_* < \infty$ and (2) $\|a\|_* = 0 \Leftrightarrow a = 0$. Then $\|.\|_*$ is a C*-norm and the completion of $\mathcal{A}$ with respect to $\|.\|_*$ is called the *enveloping C*-algebra of $\mathcal{A}$.*

Being a *-algebra we can identify special kinds of elements inside a C*-algebra A. Recall the following terminology for $a \in A$.

1. a is *self-adjoint* $\Leftrightarrow a^* = a$.
2. a is *normal* $\Leftrightarrow a^*a = aa^*$.
3. a is an (orthogonal) *projection* $\Leftrightarrow a = a^* = a^2$.
4. If A is unital, the element a is *unitary* $\Leftrightarrow a^*a = aa^* = 1$.

In a C*-algebra the notion of positivity is a powerful one and can be stated in several different ways, one of which is the following one:

Definition 1.7. *Consider a C*-algebra A. An element $a \in A$ is called positive $\Leftrightarrow$ there exists $b \in A$ such that $a = b^*b$. The set of positive elements of A is denoted by A^+.*

The set of positive elements A^+ is closed in the norm topology, closed under addition and scalar multiplication by positive numbers and $A^+ \cap (-A^+) = \{0\}$. We define a partial order relation $\leq$ on the real linear space of self-adjoint elements A_h of A as follows. If $a, b \in A_h$, then $a \leq b \Leftrightarrow b - a \in A^+$.

In our examples above positivity takes on the familiar form:

Example 1.8. (1) If X is a locally compact Hausdorff space and $f \in C_0(X)$, then $f \geq 0 \Leftrightarrow f(x) \geq 0$ for all $x \in X$.

(2) If $\mathcal{H}$ is a Hilbert space and $T \in B(\mathcal{H})$, then $T \geq 0 \Leftrightarrow \langle Tv, v \rangle \geq 0$ for all $v \in \mathcal{H}$.

It is important to note that positivity does not depend on the C*-algebra in which the element is being considered: if B is a C*-subalgebra of A and $a \in B$, then a is positive in $A \Leftrightarrow a$ is positive in B.

Having a notion of positivity for elements in a C*-algebra we can introduce the notion of positivity for linear functionals.

Definition 1.9. *If A is a C*-algebra, a linear functional ω on A is called positive if $\omega(x) \geq 0$ for all $x \in A^+$. The set of positive linear functionals on A is denoted by A^*_+.*

A positive linear functional ω on A is automatically continuous (and if A is unital, $\|\omega\| = \omega(1)$). A *state* on A is by definition a positive linear functional on A of norm one. The set A^*_+ is closed for the norm topology and is closed under addition and scalar multiplication by positive numbers. Any continuous linear functional is a linear combination of positive linear functionals.

Example 1.10. (1) Let X be a locally compact Hausdorff space. By the theorem of Riesz we have a bijection from the vector space of regular (necessarily finite) complex measures on X and $C_0(X)^*$ which associates to every regular complex measure μ on X the linear functional $\omega_\mu \in C_0(X)^*$ given by $\omega_\mu(f) = \int_X f \, d\mu$ for all $f \in C_0(X)$. Under this bijection the set of finite positive measures on X corresponds to $C_0(X)^*_+$.

(2) Consider a Hilbert space $\mathcal{H}$. If $v, w \in \mathcal{H}$ we define $\omega_{v,w} \in B(\mathcal{H})^*$ by $\omega_{v,w}(x) = \langle xv, w \rangle$ for all $x \in B(\mathcal{H})$. Notice that $\|\omega_{v,w}\| = \|v\| \|w\|$. The linear functional $\omega_{v,v}$ is positive.

Just as a positive measure defines an L^2-space, any positive linear functional on a C*-algebra has a natural L^2-space associated to it.

Definition 1.11 (Gelfand-Naimark-Segal). *Consider a C*-algebra A and a positive linear functional ω on A. A triple $(\mathcal{H}, \pi, \xi)$ is called a cyclic GNS-construction for ω if (i) $\mathcal{H}$ is a Hilbert space, (ii) π is a *-representation of A on $\mathcal{H}$ and (iii) ξ is a vector in $\mathcal{H}$ so that $\omega(a) = \langle \pi(a)\xi, \xi \rangle$ for all $a \in A$ and $\pi(A)\xi$ is dense in $\mathcal{H}$.*

Such a GNS-construction always exist (and is not so difficult to construct) and is easily seen to be unique up to a unitary transformation. The representation π will be referred to as a *GNS-representation* of ω and the vector ξ will be referred to as a *cyclic vector* of the GNS-construction. If $a \in A$ one should think of $\pi(a)\xi$ as the equivalence class corresponding to a in the L^2-space $\mathcal{H}$.

A powerful tool for C*-algebras is the possibility of 'defining continuous functions of normal elements' in a C*-algebra. We will give a precise statement of this fact in the next proposition but in order to do so we need the notion of a spectrum of an element in a unital C*-algebra A. If $a \in A$, the spectrum $\sigma(a) \subseteq \mathbb{C}$ is defined as

$$\sigma(a) = \{\, \lambda \in \mathbb{C} \mid a - \lambda 1 \text{ is not invertible in } A \,\}$$

The spectrum $\sigma(a)$ is a non-empty compact subset of $\mathbb{C}$ contained in the closed disc of radius $\|a\|$.

Proposition 1.12 (Functional Calculus). *Consider a unital C*-algebra A and a normal element a in A. There exists a unique unital *-homomorphism $\pi : C(\sigma(a)) \to A$ such that $\pi(\iota_{\sigma(a)}) = a$. We call π the (continuous) functional calculus of a. For any continuous function $f \in C(\sigma(a))$ we set $f(a) = \pi(f)$.*

Notice that if p is a complex polynomial in two variables, then $p(z, \bar{z})(a) = p(a, a^*)$ by the algebraic properties of π. By Stone-Weierstrass the polynomial functions $p(z, \bar{z})$ are norm dense in $C(\sigma(a))$. If one combines this with the fact that π is continuous, the uniqueness of π is obvious.

One can also easily see that if $\sum_{n=0}^{\infty} r_n z^n$ is a complex power series so that $\sum_{n=0}^{\infty} |r_n| \|a\|^n < \infty$ and $f \in C(\sigma(a))$ is defined by $f(z) = \sum_{n=0}^{\infty} r_n z^n$ for all $z \in \sigma(a)$, then

$$f(a) = \sum_{n=0}^{\infty} r_n a^n = \lim_{k \to \infty} \sum_{n=0}^{k} r_n a^n \,,$$

where the last limit is to be understood in the norm-topology. So we get for instance

$$e^a = \sum_{n=0}^{\infty} \frac{a^n}{n!} \,.$$

Notice that we can only construct this continuous functional calculus for normal elements. If a is any element in a unital Banach algebra, one can also define a functional calculus for complex analytic functions defined on an open subset of $\mathbb{C}$ that contains the spectrum $\sigma(a)$. This functional calculus is called the Riesz functional calculus of a. If a is a normal element in a unital C*-algebra, the continuous functional calculus of a extends the Riesz functional calculus of a. But we will not make any use of the Riesz functional calculus in these notes.

If B is a C*-subalgebra of A so that B contains the unit element of A and a is a normal element of B, then $\sigma_A(a) = \sigma_B(b)$ and the functional calculus of a with respect to A and the functional calculus of a with respect to B agree.

If A is not unital, one defines a functional calculus for normal elements by extending A to a unital C*-algebra in which A sits as a closed two sided ideal and using the functional calculus above. There are several ways of extending A to a unital algebra just as there are several ways to compactify a non-compact Hausdorff space. Let us first look at the simplest one that in the commutative case agrees with the one-point compactification.

Definition 1.13. *Consider a non-unital C*-algebra A. The unital C*-algebra $\tilde{A}$ is defined as follows. As a vector space, $\tilde{A} = A \oplus \mathbb{C}$. The multiplication, *-operation and norm on $\tilde{A}$ are defined by the formulas*

$$(a, \lambda)(b, \nu) = (ab + \lambda b + \nu a, \lambda \nu), \qquad (a, \lambda)^* = (a^*, \bar{\lambda})$$

and

$$\|(a, \lambda)\| = \sup\{\|ac + \lambda c\| \mid c \in A, \|c\| \le 1\}$$

for all $a, b \in A$ and $\lambda, \nu \in \mathbb{C}$.

Notice that $(0, 1)$ is the unit element of $\tilde{A}$. The C*-algebra A is embedded in $\tilde{A}$ via the map $A \to \tilde{A} : a \to (a, 0)$ and as such A is a closed two sided essential ideal of $\tilde{A}$. Notice that these two facts determine the formulae for the product and *-operation on $\tilde{A}$. Recall that an ideal I in an algebra B is called essential if for all $b \in B$, we have $(\forall x \in I : xb = 0) \Rightarrow b = 0$ and $(\forall x \in I : bx = 0) \Rightarrow b = 0$.

If A is unital one can still define the *-algebra $\tilde{A}$ this way but one needs another formula for the norm to obtain a C*-algebra. But we will not make use of this fact.

If A is a non-unital algebra and $a \in A$ one defines the spectrum $\sigma(a)$ of a as $\sigma(a) = \sigma_{\tilde{A}}(a) \subseteq \mathbb{C}$. Notice that $0 \in \sigma(a)$. If $f \in C(\sigma(a))$ and $f(0) = 0$ we define the element $f(a) \in A$ as $f(a) = \pi(f)$, where π is the functional calculus of a with respect to $\tilde{A}$. The condition $f(0) = 0$ guarantees that $f(a)$ belongs to A (as opposed to $\tilde{A} \setminus A$).

Among all normal elements the self-adjoint and positive elements can be identified by their spectrum:

Proposition 1.14. *Consider a C*-algebra A and a normal element $a \in A$. Then*

1. *a is self-adjoint $\Leftrightarrow \sigma(a) \subseteq \mathbb{R}$,*
2. *a is positive $\Leftrightarrow \sigma(a) \subseteq \mathbb{R}^+$.*

The functional calculus is also used to define powers of positive elements.

Definition 1.15. *Consider a C*-algebra A and $a \in A^+$. If $\alpha \in \mathbb{R}^+$ we define $a^\alpha = f(a) \in A^+$, where $f \in C(\sigma(a))$ is defined by $f(t) = t^\alpha$ for all $t \in \sigma(a)$.*

The fact that the functional calculus of a is a *-homomorphism implies several familiar formulas for its powers, e.g. $a^{\alpha+\beta} = a^\alpha a^\beta$ for all $\alpha, \beta \in \mathbb{R}^+$. If moreover A is unital and a is invertible, this same method can be used to define any complex power of a.

Although a C*-algebra does not have to be unital, we can always approximate a unit.

Proposition 1.16. *Consider a C*-algebra A. There exists a net $(u_i)_{i \in I}$ in A^+ such that*

1. *$\|u_i\| \leq 1$ for all $i \in I$,*
2. *$u_i \leq u_j$ if $i, j \in I$ and $i \leq j$,*
3. *For all $a \in A$, the nets $(u_i a)_{i \in I}$ and $(a u_i)_{i \in I}$ converge to a in the norm topology.*

A net $(u_i)_{i \in I}$ in A satisfying conditions (1),(2),(3) of the above proposition is called an *approximate unit* for A. If A is separable (that is, A contains a countable dense subset), then A has an approximate unit that is a sequence.

Tensor Products of C*-algebras

Let A, B be C*-algebras. The algebraic tensor product $A \odot B$ is a *-algebra, where the product and *-operation are determined by $(a \otimes b)(c \otimes d) = (ac) \otimes (bd)$ and $(a \otimes b)^* = a^* \otimes b^*$ for all $a, b, c, d \in A$. We can therefore look at C*-norms on $A \odot B$ (see the comments after Definition 1.1).

There exist in general different C*-norms α on the algebraic tensor product $A \odot B$ and all of them are automatically compatible with the original norms on A and B (that is, $\alpha(a \otimes b) = \|a\| \|b\|$ for all $a \in A$, $b \in B$) but there is a smallest and biggest C*-norm.

Out of all these norms, we will now single out and describe the smallest one. For this purpose, take a faithful *-representation π of A on a Hilbert space $\mathcal{H}$ and a faithful *-representation θ of B on a Hilbert space $\mathcal{K}$.

Recall that $B(\mathcal{H}) \odot B(\mathcal{K})$ is naturally embedded in $B(\mathcal{H} \otimes \mathcal{K})$ in the following way. Given two elements $x \in B(\mathcal{H})$, $y \in B(\mathcal{K})$, the element $x \otimes y \in B(\mathcal{H} \otimes \mathcal{K})$ is defined so that $(x \otimes y)(v \otimes w) = (xv) \otimes (yw)$ for all $v \in \mathcal{H}$, $w \in \mathcal{K}$. Moreover, $\|x \otimes y\| = \|x\| \|y\|$. Therefore we have a faithful *-representation $\pi \odot \theta : A \odot B \to B(\mathcal{H} \otimes \mathcal{K})$ given by $(\pi \odot \theta)(a \otimes b) = \pi(a) \otimes \theta(b)$ for all $a \in A$ and $b \in B$.

The mapping $A \odot B \to \mathbb{R}^+ : z \mapsto \|(\pi \odot \theta)(z)\|$ is a C*-norm on $A \odot B$ and the completion of $A \odot B$ with respect to this norm is denoted by $A \otimes B$ (or $A \otimes_{\min} B$ if ones wants to distinguish it from other C*-norms) and called the *spatial* (or *minimal*) tensor product of the C*-algebras A and B. One can prove that this norm does not depend on the choice of the faithful *-representations π and θ.

Example 1.17. (1) If X and Y are two locally compact Hausdorff spaces there exists a natural injective *-homomorphism $C_0(X) \odot C_0(Y) \hookrightarrow C_0(X \otimes Y)$ which identifies a simple tensor $f \otimes g$ in the algebraic tensor product $C_0(X) \odot C_0(Y)$ with a function on $X \times Y$ given by $(f \otimes g)(x, y) = f(x) g(y)$ for all $x \in X$, $y \in Y$. This way the algebraic tensor product $C_0(X) \odot C_0(Y)$ becomes a dense *-subalgebra of $C_0(X \times Y)$. As such $C_0(X) \odot C_0(Y)$ inherits the norm of $C_0(X \times Y)$ and this norm is the spatial C*-norm (which is in fact the only C*-norm on $C_0(X) \odot C_0(Y)$). As a consequence, $C_0(X) \otimes C_0(Y) = C_0(X \times Y)$.

(2) Let $\mathcal{H}$ and $\mathcal{K}$ be Hilbert spaces, A a C*-subalgebra of $B(\mathcal{H})$ and B a C*-subalgebra of $B(\mathcal{K})$. As discussed above can $A \odot B$ be considered as a *-subalgebra of $B(\mathcal{H} \otimes \mathcal{K})$ and is $A \otimes B$ nothing else than the norm closure of $A \odot B$ in $B(\mathcal{H} \otimes \mathcal{K})$. In this setting $B_0(\mathcal{H}) \otimes B_0(\mathcal{K}) = B_0(\mathcal{H} \otimes \mathcal{K})$ but if $\mathcal{H}$ and $\mathcal{K}$ are infinite dimensional, $B(\mathcal{H}) \otimes B(\mathcal{K}) \subsetneqq B(\mathcal{H} \otimes \mathcal{K})$.

We will need the following results concerning bounded linear functionals.

Proposition 1.18. *Let $\omega \in A^*$ and $\eta \in B^*$. There exists a unique element $\omega \otimes \eta \in (A \otimes B)^*$ satisfying $(\omega \otimes \eta)(a \otimes b) = \omega(a)\,\eta(b)$ for all $a \in A$, $b \in B$. Moreover, $\|\omega \otimes \eta\| = \|\omega\|\,\|\eta\|$. If ω and η are positive, then $\omega \otimes \eta$ is positive.*

The minimal tensor product enjoys moreover the useful property that the set $\{\,\omega \otimes \eta \mid \omega \in A^*, \eta \in B^*\,\}$ separates the elements of $A \otimes B$, a property not shared by the other C*-algebraic tensor products of A and B.

Proposition 1.19. *Let $\omega \in A^*$. Then there exists a unique bounded linear mapping $\omega \otimes \iota : A \otimes B \to B$ such that $(\omega \otimes \iota)(a \otimes b) = \omega(a)\,b$ for all $a \in A$, $b \in B$. Such a mapping $\omega \otimes \iota$ is referred to as a slice-map. Moreover, $(\omega \otimes \eta)(x) = \eta((\omega \otimes \iota)(x))$ for all $x \in A \otimes B$ and $\eta \in B^*$.*

Of course, a similar results holds for mappings of the form $\iota \otimes \omega$.

Proposition 1.20. *Consider C*-algebras A_1, A_2, B_1, B_2 and *-homomorphisms $\pi : A_1 \to A_2$ and $\theta : B_1 \to B_2$. There exists a unique *-homomorphism $\pi \otimes \theta : A_1 \otimes A_2 \to B_1 \otimes B_2$ so that $(\pi \otimes \theta)(a \otimes b) = \pi(a) \otimes \theta(b)$ for all $a \in A_1$ and $b \in B_1$. If π and θ are injective, then $\pi \otimes \theta$ is injective.*

We defined the tensor products between two C*-algebras, but the same principles can be adapted to define the minimal tensor product of any finite number of C*-algebras. Then the obvious associativity results hold and we will use them without further mention. For instance, if A_1, A_2, A_3 are C*-algebras, there exists unique isomorphisms of C*-algebras $\pi_1 : (A_1 \otimes A_2) \otimes A_3 \to A_1 \otimes A_2 \otimes A_3$ and $\pi_2 : A_1 \otimes (A_2 \otimes A_3) \to A_1 \otimes A_2 \otimes A_3$ so that $\pi_1\big((a_1 \otimes a_2) \otimes a_3\big) = a_1 \otimes a_2 \otimes a_3$ and $\pi_2\big(a_1 \otimes (a_2 \otimes a_3)\big) = a_1 \otimes a_2 \otimes a_3$ for all $a_1 \in A_1$, $a_2 \in A_2$ and $a_3 \in A_3$. For the rest of these lecture notes we will therefore identify $A_1 \otimes A_2 \otimes A_3$, $(A_1 \otimes A_2) \otimes A_3$ and $A_1 \otimes (A_2 \otimes A_3)$.

There also exists a unique *-isomorphism $\chi : A_1 \otimes A_2 \to A_2 \otimes A_1$ so that $\chi(a_1 \otimes a_2) = a_2 \otimes a_1$ for all $a_1 \in A_1$ and $a_2 \in A_2$. We call χ the *flip-map* on $A_1 \otimes A_2$.

The multiplier C*-algebra

If A is a non-unital C*-algebra we saw in Definition 1.13 how one can extend A to a unital algebra. In the next part we introduce another way of extending A that in the commutative case corresponds to the Stone-Čech compactification of a locally compact Hausdorff space.

Definition 1.21. *Let A be a C^*-algebra. A multiplier x of A is a pair $x = (r, \ell)$ of maps $r, \ell : A \to A$ satisfying $r(a)\, b = a\,\ell(b)$ for all $a, b \in A$. The maps r, ℓ are automatically linear, bounded and satisfy the module properties*

$$r(ab) = a\, r(b) \qquad\qquad \ell(ab) = \ell(a)\, b$$

for all $a, b \in A$. We set $a\,x = r(a)$ and $x\,a = \ell(a)$ for all $a \in A$. The set of multipliers of A is denoted by $M(A)$.

The boundedness of l and r follows from the Closed Graph Theorem 7.1. The defining relation for the multiplier x is nothing but an associativity-relation $(ax)b = a(xb)$ for all $a, b \in A$. A multiplier x is obviously determined if one knows how elements of A are multiplied to the left and the right of x and this is the way one should think of multipliers (not as pairs of mappings).

The set of multipliers $M(A)$ is made into a C*-algebra by defining the sum and the product in such a way that the following obvious formulas hold. If $x, y \in M(A)$ and $a \in A$,

$$\begin{aligned} (x + y)a &= xa + ya & a(x + y) &= ax + ay \\ (xy)a &= x(ya) & a(xy) &= (ax)y \end{aligned} \quad,$$

and similarly for the scalar multiplication. The *-operation is defined such that

$$x^*a = (a^*x)^* \qquad\qquad ax^* = (xa^*)^*.$$

The norm is given by

$$\|x\| = \sup\{\, \|xb\| \mid b \in A, \|b\| \le 1 \,\} = \sup\{\, \|bx\| \mid b \in A, \|b\| \le 1 \,\}.$$

The multiplier algebra $M(A)$ always contains a unit element (the pair consisting of the identity mappings on A) and if we define for all $c \in A$, the linear maps $L_c, R_c : A \to A$ by $L_c(a) = ca$ and $R_c(a) = ac$ for all $a \in A$, we obtain an injective *-isomorphism $A \to M(A) : c \mapsto (L_c, R_c)$. From now on we will identify A with its image in $M(A)$. As such, A is an essential two-sided ideal in $M(A)$. If A is unital, then $M(A) = A$.

In practice we will work with concrete realizations of the multiplier algebra $M(A)$ as follows. Most of the time one can find a natural C*-algebra B so that A is an essential two-sided ideal of B and their exists a unital *-isomorphism $\pi : M(A) \to B$ such that $\pi(L_a, R_a) = a$ for all $a \in A$. In this case we will always identify $M(A)$ with B and work with the explicit C*-algebra B instead of the abstract definition of $M(A)$. Let us give two examples.

Example 1.22. i) If X is a locally compact Hausdorff space then $M(C_0(X)) = C_b(X)$.

ii) If $\mathcal{H}$ is a Hilbert space, then $M(B_0(\mathcal{H}))$ equals $B(\mathcal{H})$.

We already mentioned that *-homomorphism are the natural morphisms between C*-algebras but this is not completely accurate. In the next definition we introduce another natural notion of a morphism which is sometimes (like in the context of quantum groups) better suited to the needs of non-unital algebras.

Definition 1.23. *Consider two C^*-algebras A and B and a *-homomorphism $\pi : A \to M(B)$. It is called non-degenerate if $\pi(A)B := \langle \pi(a)b \mid a \in A, b \in B \rangle$ is dense in B.*

Notice that for unital A and B, the non-degeneracy of π is equivalent to $\pi(1) = 1$. We need to extend non-degenerate *-homomorphisms to the multiplier algebra $M(A)$:

Proposition 1.24. *Consider two C^*-algebras A, B and a non-degenerate *-homomorphism $\pi : A \to M(B)$. Then there exists a unique *-homomorphism $\bar\pi : M(A) \to M(B)$ extending π. This extension is unital. For all $x \in M(A)$, we set $\pi(x) = \bar\pi(x)$.*

The extension is determined by

$$\bar\pi(x)\,(\pi(a)b) = \pi(xa)\,b \qquad\qquad \text{and} \qquad\qquad (b\pi(a))\,\bar\pi(x) = b\,\pi(ax)$$

whenever $a \in A$, $x \in M(A)$ and $b \in B$. The uniqueness of this extension is a consequence of the non-degeneracy of π.

The above property makes it possible to compose non-degenerate *-homomorphisms. Simply, if $\pi : A \to M(B)$ and $\theta : B \to M(C)$ are non-degenerate *-homomorphisms, we set $\theta\,\pi = \bar\theta \circ \pi$.

Example 1.25. Consider two locally compact Hausdorff spaces X and Y. If $j : X \to Y$ is a continuous map, we can define a *-homomorphism $\pi_j : C_0(Y) \to C_b(X) : f \to f \circ j$. Then $\pi_j : C_0(Y) \to M(C_0(X))$ is non-degenerate and $\pi_j(f) = f \circ j$ for all $f \in C_b(Y)$. Let $\mathrm{Mor}(C_0(Y), C_0(X))$ be the set of non-degenerate *-homomorphisms from $C_0(Y)$ to $M(C_0(X))$. It is not so hard to prove that the mapping $C(X, Y) \to \mathrm{Mor}(C_0(Y), C_0(X)) : j \to \pi_j$ is a bijection. This way we get a contravariant functor from the category of locally compact Hausdorff spaces (with continuous functions as morphisms) to the category of C*-algebras (with non-degenerate *-homomorphisms as morphisms).

The *-homomorphism π_j maps $C_0(Y)$ into $C_0(X)$ if and only if j is proper, i.e. $j^{-1}(K)$ is compact for every compact subset $K \subseteq Y$. Check all the claims above by appealing to Urysohn's lemma. Notice also that j is injective if $\pi_j(C_0(Y)) \supseteq C_0(X)$.

2 Locally compact quantum groups in the C*-algebra setting

Let G be a locally compact group, i.e. G is a group and possesses a locally compact Hausdorff topology for which the mappings $G \times G \to G : (s,t) \to st$ and $G \to G : s \to s^{-1}$ are continuous.

By the discussion in the previous section we can associate to G a **commutative** C*-algebra $A := C_0(G)$. Notice however that we nowhere use the group operation to define this C*-algebra. In order to lift the group operation to the level of the C*-algebra A we mimic the procedure that is followed in the theory of Hopf-algebras to introduce a comultiplication $\Delta : A \to M(A \otimes A) = C_b(G \times G) : f \to \Delta(f)$ by setting $\Delta(f)(s,t) = f(st)$ for all $f \in A$ and $s, t \in G$. Since it is the adjoint of the continuous map $G \times G \to G : (s,t) \to st$, the *-homomorphism Δ is non-degenerate, cfr. Example 1.25. This example also gives us a formula for the extension of Δ to the multiplier algebra.

Using Proposition 1.20 we get a *-homomorphism $\Delta \otimes \iota : C_0(G \times G) = A \otimes A \to C_b(G \times G) \otimes C_b(G)$. Similar to the discussion in Example 1.17 (1) we get that $C_b(G \times G) \otimes C_b(G)$ is naturally embedded into $C_b(G \times G \times G) = M(A \otimes A \otimes A)$ and that under this embedding, $(\Delta \otimes \iota)(f)(s,t,u) = f(st,u)$ for all $f \in A \otimes A$ and $s, t, u \in G$. This is immediately seen if f is a simple tensor and is then extended to the whole of $A \otimes A$ by continuity of $\Delta \otimes \iota$. Note that this also implies that $\Delta \otimes \iota$ is non-degenerate.

Similarly we see that $\iota \otimes \Delta : A \otimes A \to M(A \otimes A \otimes A)$ is given by $(\iota \otimes \Delta)(f)(s,t,u) = f(s,tu)$ for all $s, t, u \in G$.

As a consequence, the associativity of the group operation implies (and is in fact equivalent) to the well known *coassociativity* formula

$$(\Delta \otimes \iota)\Delta = (\iota \otimes \Delta)\Delta \tag{2.1}$$

which makes sense because $\iota \otimes \Delta$ and $\Delta \otimes \iota$ are non-degenerate and can be extended to $C_b(G \times G)$.

Next one might wonder how the existence of the unit element of G and the existence and continuity of the inverse operation on G can be translated to the level of the pair (A, Δ). The obvious way is to define the natural *counit* $\varepsilon : A \to \mathbb{C}$ and *antipode* or *coinverse* $S : A \to A$ by the formulas

1. $\varepsilon(f) = f(e)$ for all $f \in A$, whence ε is a non-zero *-homomorphism,
2. $S(f)(s) = f(s^{-1})$ for all $f \in A$ and $s \in G$, whence S is a *-automorphism.

The group axioms for the unit and inverse elements are then equivalent to the formulas

$$(\varepsilon \otimes \iota)\Delta(f) = (\iota \otimes \varepsilon)\Delta(f) = f \text{ and } m(S \otimes \iota)\Delta(f) = m(\iota \otimes S)\Delta(f) = \varepsilon(f)\,1 \tag{2.2}$$

for all $f \in A$. Here, $m : A \otimes A \to A$ is the non-degenerate *-homomorphism such that $m(a \otimes b) = ab$ for all $a, b \in A$. Or explicitly, $m(f)(s) = f(s, s)$ for all $f \in C_0(G \times G)$ and $s \in G$. Notice that Eq. (2.2) makes sense because of the non-degeneracy of the maps involved (see Proposition 1.24).

Thanks to Gelfand's Theorem 1.4 and the covariant functor of Example 1.25 we can easily go the other way around in the following way. Suppose that $(A, \Delta, \varepsilon, S)$ is a quadruple consisting of (1) a commutative C*-algebra A, (2) a non-degenerate *-homomorphism $\Delta : A \to M(A \otimes A)$, (3) a non-zero *-homomorphism $\varepsilon : A \to \mathbb{C}$ and a *-automorphism $S : A \to A$ satisfying Eqs. (2.1) and (2.2). Then, there exists a locally compact group G so that the quadruple $(A, \Delta, \varepsilon, S)$ is isomorphic to the concrete quadruple constructed from G as explained above.

The basic idea behind the theory of locally compact quantum groups is very simple: is it possible to develop a rich theory if we replace the **commutative** C*-algebra A by a **non-commutative** C*-algebra A. Luckily for us the answer is no, at least not in this very naive way. Let us indicate some of the problems involved in such a naive generalization.

There will be no problem in generalizing the notion of a *comultiplication* Δ. This will still be a non-degenerate *-homomorphism $\Delta : A \to M(A \otimes A)$ that is coassociative , i.e. $(\Delta \otimes \iota)\Delta = (\iota \otimes \Delta)\Delta$. Problems arise if we try to generalize the counit and coinverse.

(1) The map m used in the equalities on the right hand side of (2.2) is ill-defined in the non-commutative setting. It is still defined on the algebraic tensor product $m : A \odot A \to A$ by $m(a \otimes b) = ab$ for all $a, b \in A$, but is not continuous and cannot be extended to the completed tensor product $A \otimes A$.

(2) There are examples of (even unital) C*-algebras with a comultiplication that deserve to be considered 'locally compact quantum groups' but for which the corresponding antipode is unbounded (also, in general, it does not preserve the *-operation and it is anti-multiplicative). On the plus side, it turns out that this ill-behavior can be controlled very well and the antipode still plays an important role.

(3) The counit can be unbounded. It plays a minor role in the theory of quantum groups but turns out to be useful from time to time.

In short, the behavior of the antipode and the counit seems to be too erratic for them to be part of the definition of a locally compact quantum group but they still have a role to play in the theory.

Exercise 2.1. Consider the Hilbert space $\mathcal{H} = \ell^2(\mathbb{N})$ and the C*-algebra $B_0(\mathcal{H})$. We saw in Example 1.17 (2) that the minimal tensor product $B_0(\mathcal{H}) \otimes B_0(\mathcal{H})$ is naturally identified with $B_0(\mathcal{H} \otimes \mathcal{H})$ (as a side remark, there is only one C*-norm on $B_0(\mathcal{H}) \odot B_0(\mathcal{H})$). Consider the linear map $m : B_0(\mathcal{H}) \odot B_0(\mathcal{H}) \to B_0(\mathcal{H})$ defined by $m(a \otimes b) = ab$ for all $a, b \in B_0(\mathcal{H})$. Show that there exists a sequence $(x_n)_{n=1}^\infty$ in $B_0(\mathcal{H}) \odot B_0(\mathcal{H})$ such that $(x_n)_{n=1}^\infty \to 0$ in the norm topology but for which $(\|m(x_n)\|)_{n=1}^\infty \to \infty$.

So we are led to the following vague problem: *Consider a C*-algebra A and a non-degenerate *-homomorphism $\Delta : A \to M(A \otimes A)$ such that*

$$(\iota \otimes \Delta)\Delta = (\Delta \otimes \iota)\Delta$$

What kind of extra conditions does the pair (A, Δ) have to satisfy in order for it to be called a locally compact quantum group?

These extra axioms have to be chosen in such a way that

1. The axioms are as simple as possible and they should be not 'too' hard to verify in explicit examples.
2. From such a definition of a locally compact quantum group, it should be possible to derive a rich theory that has similarities with the classical theory of locally compact groups but should produce 'quantum' phenomena not seen in the classical case.
3. Certain examples that have been constructed over the years should satisfy these axioms.

The search for the right set of axioms has lasted for about 30 years and a acceptable general definition has been presented in [KuV00a] (more precisely, there are at least two persons who like to think so). The definition in [KuV00a] is formulated in the C*-algebra setting but in these lecture notes we will concentrate on the (equivalent) definition in the von Neumann algebra setting discussed in [KuV03] (see Definition 5.1).

A short discussion of the history and contributions of mathematicians involved has been given in the introduction of these lecture notes. In special cases like the compact and discrete one, there exist simpler sets of axioms and they have in fact been discovered much earlier. These will be discussed in section 3.

How does one get hold of these extra axioms? The basic idea is, again, pretty simple. In the definition of a locally compact group, one tries to replace the existence of unit element an inverse operation by other equivalent conditions which can be translated to conditions on the level of the associated commutative C*-algebra and successfully survive the transformation to the non-commutative setting.

Let us illustrate this. Consider a locally compact group G and set $A = C_0(G)$. Define the continuous mapping

$$j : G \times G \to G \times G : (s, t) \mapsto (st, s) .$$

Then j is a homeomorphism and $j^{-1}(s, t) = (s^{-1}t, s)$ for all $s, t \in G$. Hence, we can define a *-isomorphism $\pi : A \otimes A \to A \otimes A$ by $\pi(h) = h \circ j$ for every $h \in A \otimes A$. If $f, g \in A$, then $\pi(f \otimes g) = \Delta(f)(g \otimes 1)$, implying that

1. $\langle \Delta(a)(b \otimes 1) \mid a, b \in A \rangle$ is a dense subspace of $A \otimes A$.

2. The linear map $T : A \odot A \rightarrow A \otimes A$ defined by $T(a \otimes b) = \Delta(a)(b \otimes 1)$ is injective.

Of course, we get similar results involving expressions of the form $\Delta(a)(1 \otimes b)$. Notice that these kind of conditions still make sense if A is non-commutative. Imposing these kind of extra conditions is not sufficient in the general case (it is even not sufficient in the general commutative case) but we will see in the next section that they are sufficient in the compact and discrete case. It turns out that the final general solution will hinge on assuming the existence of generalizations of Haar measures (see Definition 5.1).

3 Compact quantum groups

Arguably the most satisfactory set of axioms of a locally compact quantum group can be given in the compact case, i.e. when the underlying C*-algebra has a unit. In this case the existence of the Haar measure can be derived from a simple set of axioms and a generalization of the Peter-Weyl theory (with some modifications) can be developed. The definition and its associated theory are mainly due to S.L. Woronowicz.

3.1 The theoretical setting

Let us first take a look at the classical case:

Exercise 3.1. Consider a compact associative semi-group G, i.e. G carries a compact Hausdorff topology and the product $G \times G \rightarrow G : (s,t) \rightarrow st$ is continuous and associative. Set $A = C(G)$ and define the unital *-homomorphism $\Delta : A \rightarrow A \otimes A = C(G \times G)$ by $\Delta(f)(s,t) = f(st)$ for all $f \in C(G)$ and $s, t \in G$. Assume that the spaces $\langle \Delta(a)(b \otimes 1) \mid a, b \in A \rangle$ and $\langle \Delta(a)(1 \otimes b) \mid a, b \in A \rangle$ are dense in $A \otimes A$. Show that G is a compact group:

(1) Define the continuous map $j : G \times G \rightarrow G \times G : (s,t) \rightarrow (st, s)$. Then j is injective (why?) so G satisfies the left cancellation property: if $s, t, u \in G$ and $st = su$, then $t = u$. Similarly G satisfies the right cancellation property.
(2) Take s in G and define H to be the closure of $\{ s^n \mid n \in \mathbb{N} \}$. We call J a closed ideal of H if J is a non-empty closed subset of H such that $gJ \subseteq J$ for all $g \in H$. Prove that there is a smallest closed ideal I in H. Then $I \subseteq gI$ for all $g \in H$. Conclude that G has a unit element that belongs to I. Show also that s has an inverse in H.
(3) Show that $G \rightarrow G : s \rightarrow s^{-1}$ is continuous (remember the map j).

As a consequence the classical case suggests that the following definition might be the correct generalization to the quantum setting.

Definition 3.2. *Consider a unital C*-algebra A and a unital *-homomorphism $\Delta : A \to A \otimes A$ such that $(\Delta \otimes \iota)\Delta = (\iota \otimes \Delta)\Delta$ and*

$$\langle \Delta(a)(b \otimes 1) \mid a, b \in A \rangle \qquad \text{and} \qquad \langle \Delta(a)(1 \otimes b) \mid a, b \in A \rangle$$

are dense in $A \otimes A$. Then (A, Δ) is called a compact quantum group.

Of course, the fact that in the commutative case this definition is equivalent to the fact that (A, Δ) arises from a compact group as described in the previous section is no guarantee that the definition above is the right one. It is only a minimal condition that any definition should satisfy. However, Woronowicz has showed in [Wor87a] and [Wor98] that this is indeed the correct definition by developing the associated theory. In the rest of this section we will give an overview of the most important aspects of this theory. For the rest of this section we fix a compact quantum group (A, Δ).

The invariant state

The essential first step in the development of the theory is the proof of the existence of the quantum version of the Haar measure. Let us quickly look at the classical case to find out what kind of object this should be.

Example 3.3. Suppose that $A = C(G)$ where G is a compact group. The Haar measure μ on G is the unique regular Borel measure on G so that $\mu(G) = 1$ and

$$\int_G f(st) \, d\mu(s) = \int_G f(ts) \, d\mu(s) = \int_G f(s) \, ds$$

for all $t \in G$. As in Example 1.10 (1), μ is translated to the positive linear functional $\varphi : A \to \mathbb{C}$ given by $\varphi(f) = \int_G f(s) \, d\mu(s)$ for all $f \in C(G)$. Then $(\varphi \otimes \iota)(h)(t) = \int h(s, t) \, d\mu(s)$ and $(\iota \otimes \varphi)(h)(t) = \int h(t, s) \, d\mu(s)$ for all $h \in C(G \times G)$ and $t \in G$. This is immediate for elements of the form $h_1 \otimes h_2$, where $h_1, h_2 \in C(G)$ and follows for any element in $C(G \times G)$ by linearity and continuity since $C(G) \odot C(G)$ is dense in $C(G \times G)$. It follows that the invariance of μ is equivalent to

$$(\varphi \otimes \iota)\Delta(f) = (\iota \otimes \varphi)\Delta(f) = \varphi(f)\,1$$

for all $f \in C(G)$.

Hence, in the quantum world, the 'quantum Haar measure' takes on the following form.

Theorem 3.4. *There exists a unique state φ on A so that*

$$(\iota \otimes \varphi)\Delta(a) = (\varphi \otimes \iota)\Delta(a) = \varphi(a)\,1$$

for all $a \in A$. We call φ the Haar state of (A, Δ).

The first proof of the existence of φ was given by Woronowicz in [Wor87a] albeit in the case that A is separable. The general case was dealt with by Van Daele in [VD95]. The uniqueness is in fact a triviality. If $\omega \in A^*$ and $(\omega \otimes \iota)\Delta(a) = \omega(a)\,1$ for all $a \in A$, then

$$\omega(a) = \varphi\big((\omega \otimes \iota)\Delta(a)\big) = \omega\big((\iota \otimes \varphi)\Delta(a)\big) = \varphi(a)\,\omega(1)\ .$$

Unlike in the classical case, the Haar state does not have to be faithful! A positive linear functional ω on A is called faithful if for $x \in A^+$, we have $\omega(x) = 0 \Rightarrow x = 0$. This can already happen if (A, Δ) is the universal dual of discrete group (see the first example in the next subsection) .

Corepresentation theory

Another important aspect of group theory is the study of group representations on Hilbert spaces. The representation theory of compact groups works extremely well because of the Peter-Weyl theory. This theory can be generalized to the quantum setting but in order to do so we first need a good notion of a 'strongly continuous unitary quantum group representation'. In order to define these objects, we first need some extra notation.

Let $\mathcal{K}$ be a Hilbert space. The mapping $\Delta \otimes \iota : A \otimes B_0(\mathcal{K}) \to A \otimes A \otimes B_0(\mathcal{K})$ is a non-degenerate *-homomorphism and as such extends uniquely to a unital *-homomorphism between the multiplier algebras $M(A \otimes B_0(\mathcal{K})) \to M(A \otimes A \otimes B_0(\mathcal{K}))$.

We have also non-degenerate *-homomorphisms $\pi_{13}, \pi_{23} : A \otimes B_0(\mathcal{K}) \to A \otimes A \otimes B_0(\mathcal{K})$ determined by $\pi_{13}(a \otimes t) = a \otimes 1 \otimes t$ and $\pi_{23}(a \otimes t) = 1 \otimes a \otimes t$ for all $a \in A$, $t \in B_0(K)$. The existence of π_{23} is obvious and $\pi_{13} = (\chi \otimes \iota)\pi_{23}$, where $\chi : A \otimes A \to A \otimes A$ is the flip map. These maps also extend to the multiplier algebra and for every $x \in M(A \otimes B_0(\mathcal{K}))$, we define $x_{13}, x_{23} \in M(A \otimes A \otimes B_0(\mathcal{K}))$ as $x_{13} = \pi_{13}(x)$ and $x_{23} = \pi_{23}(x)$.

Definition 3.5. *Consider a Hilbert space $\mathcal{K}$. If $U \in M(A \otimes B_0(\mathcal{K}))$ is an invertible element and $(\Delta \otimes \iota)(U) = U_{13}\,U_{23}$, we call U a corepresentation of (A, Δ). If U is moreover a unitary element in $M(A \otimes B_0(\mathcal{K}))$, we call U a unitary corepresentation of (A, Δ).*

One uses the terminology 'corepresentation' for these objects to avoid confusion with the familiar notion of *-representations of the C*-algebra A.

Example 3.6. Suppose that $A = C(G)$ where G is a compact group and let $\mathcal{K}$ be a Hilbert space . Using the fact that there is only one C*-norm on $A \odot B_0(\mathcal{K})$ it is not difficult to check that we can identify the C*-algebra $C(G) \otimes B_0(\mathcal{K})$ with the C*-algebra $C(G, B_0(\mathcal{K}))$ of continuous function from G into $B_0(\mathcal{K})$ in such a way that for $f \in C(G)$, $x \in B_0(\mathcal{K})$ and $s \in G$, we have $(f \otimes x)(s) = f(s)\,x$. The algebraic operations on $C(G, B_0(K))$ are defined pointwise (same formulas as in Example 1.2) and the norm is given by $\|g\| = \sup\{ \|g(s)\| \mid s \in G \}$ for all $g \in C(G, B_0(K))$.

Similarly one can look at the C*-algebra $C_b^s(G, B(\mathcal{K}))$ of all bounded functions from G into $B(\mathcal{K})$ that are continuous with respect to the strong*-topology on $B(\mathcal{K})$ (see the comments after Definition 4.8). Also here, the algebraic operations are defined pointwise and is the norm the sup-norm defined using the same formula as above. Observe that $C_b^s(G, B(\mathcal{K}))$ contains $C(G, B_0(\mathcal{K}))$ as an essential ideal. Using the fact that $M(B_0(\mathcal{K})) = B(\mathcal{K})$ it is not so hard to show that $M(A \otimes B_0(\mathcal{K})) = C_b^s(G, B(\mathcal{K}))$.

If $U \in M(A \otimes B_0(\mathcal{K}))$, then $(\Delta \otimes \iota)(U)$ belongs to $M(A \otimes A \otimes B_0(\mathcal{K})) = M(C(G \times G) \otimes B_0(\mathcal{K})) = C_b^s(G \times G, B(\mathcal{K}))$ and

$$(\Delta \otimes \iota)(U)(s, t) = U(st) \qquad (*)$$

for all $s, t \in G$. This is immediate if U is an elementary tensor in $A \odot B_0(\mathcal{K})$. The equality is then extended by linearity and norm-continuity to any element in $U \in C(G, B_0(\mathcal{K}))$. Since the formula in the right hand side of $(*)$ clearly defines a unital *-homomorphism from $M(A \otimes B_0(\mathcal{K}))$ into $M(A \otimes A \otimes B_0(\mathcal{K}))$, it follows that the equality $(*)$ holds in general.

Similarly, one proves that $U_{13}(s, t) = U(s)$ and $U_{23}(s, t) = U(t)$ for all $s, t \in G$. As a consequence the multipicativity of U is equivalent with the equality $(\Delta \otimes \iota)(U) = U_{13} U_{23}$.

The theory of corepresentations of compact quantum groups (but the same is true for the theory of corepresentations of general locally compact quantum groups) has straightforward generalizations of notions like intertwiners, invariant subspaces, irreducibility and direct sum of corepresentations.

Definition 3.7. *Consider Hilbert spaces* $\mathcal{K}, \mathcal{L}$ *and corepresentations* U *of* (A, Δ) *on* $\mathcal{K}$, V *of* (A, Δ) *on* $\mathcal{L}$. *We call* T *an intertwiner from* U *to* $V \Leftrightarrow$ $T \in B(\mathcal{K}, \mathcal{L})$ *and* $(\omega \otimes \iota)(V) T = T (\omega \otimes \iota)(U)$ *for all* $\omega \in A^*$. *The set of intertwiners from* U *to* V *is denoted by* Mor(U, V). *We say that* U *is equivalent to* V, *notation* $U \cong V$, *if* Mor(U, V) *contains a bijective operator (by the Closed Graph Theorem, such an element is necessarily a homeomorphism). We say that* U *is unitarily equivalent to* V *if* Mor(U, V) *contains a unitary operator.*

If U and V are unitary, then equivalence is the same as unitary equivalence (this follows from the polar decomposition of any bijective intertwiner). The next generalization of a classical result implies that we can restrict our attention most of the time to unitary corepresentations.

Proposition 3.8. *Consider a Hilbert space* $\mathcal{K}$ *and a corepresentation* U *of* (A, Δ) *on* $\mathcal{K}$ *and define* $T = (\varphi \otimes \iota)(U^*U) \in B(\mathcal{K})^+$. *Then,* T *in invertible and the operator* $V := (1 \otimes T^{\frac{1}{2}})U(1 \otimes T^{-\frac{1}{2}})$ *is a unitary corepresentation of* (A, Δ) *such that* $V \cong W$.

Invariant subspaces are also easily defined:

Definition 3.9. *Consider a corepresentation U of (A, Δ) on a Hilbert space $\mathcal{K}$. We call $\mathcal{L}$ an invariant subspace of U if and only if $\mathcal{L}$ is a closed subspace of $\mathcal{K}$ such that $(\omega \otimes \iota)(U)\mathcal{L} \subseteq \mathcal{L}$ for all $\omega \in A^*$. We call U irreducible $\Leftrightarrow \{0\}$ and $\mathcal{K}$ are the only invariant subspaces of U.*

Suppose that U is unitary and $\mathcal{L}$ is an invariant subspace of U. Then, the orthogonal complement $\mathcal{L}^\perp$ is also invariant with respect to U and we can restrict U to $\mathcal{L}$ in the following way.

If $x \in B_0(\mathcal{L})$ we define $\tilde{x} \in B_0(\mathcal{K})$ so that $\tilde{x}\!\restriction_\mathcal{L} = x$ and $\tilde{x}\!\restriction_{\mathcal{L}^\perp} = 0$. Thus, we get an injective *-homomorphism $\pi : B_0(\mathcal{L}) \to B_0(\mathcal{K}) : x \mapsto \tilde{x}$ implying that $\iota_A \otimes \pi$ embeds $A \otimes B_0(\mathcal{L})$ into $A \otimes B_0(\mathcal{K})$. The image of this embedding is in fact nothing else than the closure of the *-algebra $\langle a \otimes \tilde{x} \mid a \in A, x \in B_0(\mathcal{L}) \rangle$. Then, there exists a unique unitary corepresentation $U_\mathcal{L}$ of (A, Δ) on $\mathcal{L}$ such that

$$(\iota \otimes \pi)(U_\mathcal{L} z) = U(\iota \otimes \pi)(z) \qquad \text{and} \qquad (\iota \otimes \pi)(z\, U_\mathcal{L}) = (\iota \otimes \pi)(z)\, U$$

for all $z \in B_0(\mathcal{L})$. If $\mathcal{H}$ is a Hilbert space and $\theta : A \to B(\mathcal{H})$ is a unital *-homomorphism, then $(\theta \otimes \iota)(U_\mathcal{L}) = (\theta \otimes \iota)(U)\!\restriction_{\mathcal{H} \otimes \mathcal{L}}$.

Unitary representations of a compact group are finite dimensional, this remains true in the quantum setting:

Proposition 3.10. *All irreducible corepresentations of (A, Δ) are finite dimensional.*

Let $\mathcal{K}$ be a finite dimensional Hilbert space. Then $A \odot B(\mathcal{K})$ is complete and $A \otimes B_0(\mathcal{K}) = A \odot B(\mathcal{K})$ and this C*-algebra is unital. Thus, $M(A \otimes B_0(\mathcal{K})) = A \otimes B(\mathcal{K})$. If we fix an orthonormal basis $(e_i)_{i=1}^n$ for $\mathcal{K}$, we get an obvious *-isomorphism $\eta : A \otimes B(\mathcal{K}) \to M_n(A)$ given by $\eta(x)_{ij} = (\iota \otimes \omega_{e_i, e_j})(x)$ for all $x \in A \otimes B(\mathcal{K})$ and $i, j \in \{1, \ldots, n\}$. Notice that $x = \sum_{i,j=1}^n \eta(x)_{ij} \otimes \theta_{e_j, e_i}$. If $U \in A \otimes B(\mathcal{K})$, one checks that $(\Delta \otimes \iota)(U) = U_{13} U_{23}$ if and only if $\Delta(\eta(U)_{ij}) = \sum_{k=1}^n \eta(U)_{ik} \otimes \eta(U)_{kj}$ for all $i, j \in \{1, \ldots, n\}$. In this respect one calls a unitary matrix $u \in M_n(A)$ satisfying $\Delta(u_{ij}) = \sum_{k=1}^n u_{ik} \otimes u_{kj}$ for all $i, j \in \{1, \ldots, n\}$, a *unitary matrix corepresentation* of (A, Δ) of dimension n and is it sometimes customary to translate notions like intertwiners, invariant subspaces, irreducibility and direct sums in terms of matrix corepresentations.

Proposition 3.11 (Schur's Lemma). *Consider two irreducible corepresentations U, V of (A, Δ) on Hilbert spaces $\mathcal{K}, \mathcal{L}$ respectively. Then*

1. *If U, V are inequivalent, then $\text{Mor}(U, V) = \{0\}$.*
2. *If U, V are equivalent, there exists a bijective $T \in B(\mathcal{K}, \mathcal{L})$ such that $\text{Mor}(U, V) = \mathbb{C}\, T$.*

Proposition 3.12. *Consider a unitary corepresentation U of (A, Δ) on a Hilbert space $\mathcal{K}$. Then, there exists a family $(\mathcal{K}_i)_{i \in I}$ of mutually orthogonal finite dimensional invariant subspaces of $\mathcal{K}$ such that $\mathcal{K} = \oplus_{i \in I} \mathcal{K}_i$ and each $U_{\mathcal{K}_i}$ is irreducible.*

Consider a family of Hilbert spaces $(\mathcal{K}_i)_{i \in I}$ and define the Hilbert space $\mathcal{K}$ as the direct sum $\mathcal{K} = \oplus_{i \in I} \mathcal{K}_i$. Then we have natural embedding

$$\oplus_{i \in I} A \otimes B_0(\mathcal{K}_i) \hookrightarrow A \otimes B_0(\mathcal{K}) \qquad (*)$$

by a non-degenerate *-homomorphism. As a set, $\oplus_{i \in I} A \otimes B_0(\mathcal{K}_i)$ is the set of all I-tuples $(x_i)_{i \in I}$ such that $x_i \in A \otimes B_0(\mathcal{K}_i)$ for all $i \in I$ and $(\|x_i\|)_{i \in I}$ belongs to $C_0(I)$. The algebraic operations are defined componentwise and for $(x_i)_{i \in I}$ in $\oplus_{i \in I} A \otimes B_0(\mathcal{K}_i)$, the norm is defined as $\| (x_i)_{i \in I} \| = \sup_{i \in I} \|x_i\|$. Notice that the space of elements in $\oplus_{i \in I} A \otimes B_0(\mathcal{K}_i)$ for which only a finite number of components are non-zero, is dense in $\oplus_{i \in I} A \otimes B_0(\mathcal{K}_i)$. If $a \in A$, $i \in I$ and $x \in B_0(\mathcal{K}_i)$ then the embedding (*) mentioned above sends $a \otimes x$ onto $a \otimes \tilde{x}$ where $\tilde{x} \in B_0(\mathcal{K})$ is defined so that $\tilde{x} \restriction_{\mathcal{K}_i} = x$ and $\tilde{x} \restriction_{\mathcal{K}_i^\perp} = 0$.

Since the embedding (*) is non-degenerate, it extends canonically to an injective unital *-homomorphism

$$M\big(\oplus_{i \in I} A \otimes B_0(\mathcal{K}_i) \big) \hookrightarrow M(A \otimes B_0(\mathcal{K})) \ .$$

We have a natural isomorphism

$$M\big(\oplus_{i \in I} A \otimes B_0(\mathcal{K}_i) \big) \cong \prod_{i \in I} M(A \otimes B_0(\mathcal{K}_i)) \ .$$

where $\prod_{i \in I} M(A \otimes B_0(\mathcal{K}_i))$ is the set of I-tuples $(x_i)_{i \in I}$ so that x_i belongs to $M(A \otimes B_0(\mathcal{K}_i))$ for all $i \in I$ and $(\|x_i\|)_{i \in I} \in C_b(I)$. Again, the algebraic operations are defined componentwise and the norm is the sup-norm extending the one defined above. Thus, we have a natural embedding

$$\prod_{i \in I} M(A \otimes B_0(\mathcal{K}_i)) \subseteq M(A \otimes B_0(\mathcal{K})) \ .$$

So if we have for every $i \in I$ a unitary corepresentation U_i of (A, Δ), we define the direct sum $\oplus_{i \in I} U_i \in M(A \otimes B_0(\mathcal{K}))$ as the image of $(U_i)_{i \in I}$ under the above embedding. Then $\oplus_{i \in I} U_i$ is a unitary corepresentation of (A, Δ) on $\oplus_{i \in I} \mathcal{K}_i$.

In the notation of proposition 3.12 we have that $\oplus_{i \in I} U_{\mathcal{K}_i} \cong U$.

Peter-Weyl theory

Although we are working within the framework of C*-algebras, compact quantum groups are in essence algebraic structures as we shall explain below. However, the big advantage of the C*-algebraic approach is the automatic existence of the Haar state.

Theorem 3.13. *Define* $\mathcal{A} \subseteq A$ *as the set* $\mathcal{A}$ *consisting of all elements of the form* $(\iota \otimes \omega)(U)$, *where* U *is a corepresentation of* (A, Δ) *on a finite dimensional Hilbert space* $\mathcal{K}$ *and* $\omega \in B(\mathcal{K})^*$. *Then* $\mathcal{A}$ *is a dense *-subalgebra* A *of* A. *For all* $a \in \mathcal{A}$, *the element* $\Delta(a)$ *belongs to* $\mathcal{A} \odot \mathcal{A}$, *the algebraic tensor product of* $\mathcal{A}$ *with itself. In fact,* $(\mathcal{A}, \Delta \restriction_{\mathcal{A}})$ *is a Hopf*-algebra and* φ *is faithful on* $\mathcal{A}$. *We call* $\mathcal{A}$ *the coefficient *-algebra of* A

We can even be more explicit. Denote the counit of $(A, \Delta\restriction_A)$ by ε and the antipode of $(A, \Delta\restriction_A)$ by S. Note that these linear mappings do not have to be bounded. In any case, S is closable.

Let $(u^\alpha)_{\alpha \in \mathbb{A}}$ be a complete family of mutually inequivalent, irreducible unitary corepresentations of (A, Δ). Complete means that every irreducible corepresentation of (A, Δ) is equivalent to one of the corepresentations in $(u^\alpha)_{\alpha \in \mathbb{A}}$. We will assume that there exists $\alpha_0 \in \mathbb{A}$ so that $u^{\alpha_0} = 1$, considered as a corepresentation on $\mathbb{C}$.

Let $\alpha \in \mathbb{A}$. Denote the carrier space of u^α by $\mathcal{K}_\alpha$ and the dimension of $\mathcal{K}_\alpha$ by n_α. Also fix a basis $(e_i^\alpha)_{i=1}^{n_\alpha}$ for $\mathcal{K}_i$ and define for all $i, j \in \{1, \ldots, n_\alpha\}$ the element $u_{ij}^\alpha = (\iota \otimes \omega_{e_i^\alpha, e_j^\alpha})(u^\alpha) \in A$.

Proposition 3.14. *The family* $\left(u_{i,j}^\alpha \mid \alpha \in \mathbb{A}, i, j = 1, \ldots, n_\alpha \right)$ *is a linear basis for* $\mathbb{A}$ *and if* $\alpha \in \mathbb{A}$ *and* $i, j \in \{1, \ldots, n_\alpha\}$, *then*

$$\Delta(u_{ij}^\alpha) = \sum_{k=1}^{n_\alpha} u_{ik}^\alpha \otimes u_{kj}^\alpha$$

$$S(u_{ij}^\alpha) = \left(u_{ji}^\alpha \right)^*$$

$$\varepsilon(u_{ij}^\alpha) = \delta_{ij}$$

$$\varphi(u_{ij}^\alpha) = 0 \text{ if } \alpha \neq \alpha_0$$

Theorem 3.15. *For each* $\alpha \in \mathbb{A}$ *there exists a unique positive invertible matrix* $Q^\alpha \in M_{n_\alpha}(\mathbb{C})$ *such that for all* $\alpha, \beta \in \mathbb{A}$, $i, r \in \{1, \ldots, n_\alpha\}$ *and* $j, s \in \{1, \ldots, n_\beta\}$,

$$\varphi\big((u_{j,s}^\beta)^* u_{i,r}^\alpha\big) = \delta_{\alpha,\beta}\, \delta_{r,s}\, Q_{ji}^\alpha \ .$$

The left regular corepresentation of a compact quantum group

In order to define the left regular corepresentation of (A, Δ) we assume that A is a unital C*-subalgebra of $B(\mathcal{H})$ for some Hilbert space $\mathcal{H}$ (which we may because of the Gelfand-Naimark Theorem 1.5). Also fix a cyclic GNS-construction $(\mathcal{H}_\varphi, \pi_\varphi, \xi_\varphi)$ for the Haar state (Definition 1.11).

Proposition 3.16. *There exists a unique unitary operator* $V \in B(\mathcal{H} \otimes \mathcal{H}_\varphi)$ *such that*

$$V^*(u \otimes \pi_\varphi(a)\xi_\varphi) = (\iota \otimes \pi_\varphi)(\Delta(a))(u \otimes \xi_\varphi)$$

for all $u \in \mathcal{H}$ *and* $a \in A$. *The operator* V *belongs to* $M(A \otimes B_0(\mathcal{H}_\varphi))$ *and* $(\Delta \otimes \iota)(V) = V_{13} V_{23}$. *We call* V *the left regular corepresentation of* (A, Δ).

Notice that the existence of V follows easily from the left invariance of φ: if $a, b \in A$ and $u, v \in \mathcal{H}$, then

$$\langle (\iota \otimes \pi_\varphi)(\Delta(a))(u \otimes \xi_\varphi), (\iota \otimes \pi_\varphi)(\Delta(b))(v \otimes \xi_\varphi) \rangle = (\omega_{u,v} \otimes \varphi)(\Delta(b^*a))$$
$$= \langle u, v \rangle \, \varphi(b^*a) = \langle u \otimes \pi_\varphi(a)\xi_\varphi, v \otimes \pi_\varphi(b)\xi_\varphi \rangle \ .$$

The fact that $V^*(\mathcal{H} \otimes \mathcal{H}_\varphi) = \mathcal{H} \otimes \mathcal{H}_\varphi$ follows from the density axioms in the definition of a compact quantum group. Check that

1. $(\iota \otimes \omega_{\pi_\varphi(a)\xi_\varphi, \pi_\varphi(b)\xi_\varphi})(V) = (\iota \otimes \varphi)(\Delta(b^*)(1 \otimes a))$ for all $a, b \in A$,

2. $(\iota \otimes \pi_\varphi)\Delta(x) = V^*(1 \otimes \pi_\varphi(x))V$ for all $x \in A$.

Notice that the first equality implies that $\langle (\iota \otimes \omega_{v,w})(V) \mid v, w \in \mathcal{H}_\varphi \rangle$ is dense in A.

Each irreducible corepresentation of (A, Δ) is contained in the left regular corepresentation:

Theorem 3.17. *Consider an irreducible unitary corepresentation U of (A, Δ) on some Hilbert space $\mathcal{K}$. Then, there exists a finite dimensional subspace $\mathcal{L}$ of $\mathcal{H}_\varphi$ such that $\mathcal{L}$ is invariant with respect to the left regular corepresentation and $U \cong V_\mathcal{L} = V\restriction_{\mathcal{H} \otimes \mathcal{L}}$.*

The dual of a compact quantum group

Consider a compact quantum group (A, Δ) with Haar state φ. We denote the coefficient *-algebra of (A, Δ) by $\mathcal{A}$ and set $\Phi = \Delta\restriction_\mathcal{A}$ so that $(\mathcal{A}, \Phi)$ is a Hopf *-algebra. Let us denote the counit of $(\mathcal{A}, \Phi)$ by ε and the antipode of $(\mathcal{A}, \Phi)$ by S. The algebraic dual $\mathcal{A}'$ is a *-algebra for the product and *-operation defined by

$$\omega \eta = (\omega \otimes \eta)\Phi \qquad \text{and} \qquad \omega^*(a) = \overline{\omega(S(a)^*)}$$

for $\omega, \eta \in \mathcal{A}'$ and $a \in \mathcal{A}$. The counit ε is the unit of the algebra $\mathcal{A}'$. One can define a 'comultiplication' $\hat{\Phi} : \mathcal{A} \to (\mathcal{A} \odot \mathcal{A})'$ such that $\hat{\Phi}(\omega)(a \otimes b) = \omega(b\,a)$ for all $\omega \in \mathcal{A}'$ and $a, b \in \mathcal{A}$. We have a natural embedding $\mathcal{A}' \odot \mathcal{A}' \subseteq (\mathcal{A} \odot \mathcal{A})'$ but this is not an equality in general since $\mathcal{A}$ is not assumed to be finite dimensional, and there is no way to give a characterization of $(\mathcal{A} \odot \mathcal{A})'$ in terms of $\mathcal{A}' \odot \mathcal{A}'$. This can be solved, how strange this may sound, by looking at a smaller subalgebra of $\mathcal{A}'$.

For this purpose we single out the subspace

$$\hat{\mathcal{A}} := \{\, \varphi(a\,.\,) \mid a \in \mathcal{A} \,\} \subseteq \mathcal{A}' \,,$$

where, obviously, $\varphi(a\,.\,)$ is the element in $\mathcal{A}'$ defined by $\varphi(a\,.\,)(x) = \varphi(ax)$ for all $x \in \mathcal{A}$.

In general φ is not a trace so that $\varphi(a\,.\,)$ is not always equal to $\varphi(\,.\,a)$. One can however prove the existence of an algebra automorphism $\sigma : \mathcal{A} \to \mathcal{A}$ such that $\varphi(a\,x) = \varphi(x\,\sigma(a))$ for all $a, x \in \mathcal{A}$. As a consequence,

$$\hat{\mathcal{A}} = \{\, \varphi(\,.\,a) \mid a \in \mathcal{A} \,\} \,.$$

Take $a, b \in \mathcal{A}$. There exist $p_1, \ldots, p_n, q_1, \ldots, q_n \in \mathcal{A}$ for which $a \otimes b = \sum_{i=1}^n (p_i \otimes 1)\Phi(q_i)$. Thus, we get for all $x \in \mathcal{A}$,

$$(\varphi(a\,.\,)\,\varphi(b\,.\,))(x) = (\varphi \otimes \varphi)((a \otimes b)\Phi(x))$$

$$= \sum_{i=1}^n (\varphi \otimes \varphi)((p_i \otimes 1)\Phi(q_i\,x)) = \sum_{i=1}^n \varphi(p_i)\,\varphi(q_i\,x) \,.$$

It follows that $\varphi(a.)\varphi(b.)$ belongs to $\hat{A}$. The uniqueness of the Haar state on (A, Φ) implies that $\varphi(S(x)) = \varphi(x)$ for all $x \in A$. As a consequence $\varphi(a.)^* = \varphi(S(a)^*.)$ for all $a \in A$. It follows that $\hat{A}$ is a *-subalgebra of A'. It is important to mention that the unit ε of A' does not have to belong to $\hat{A}$ in general.

An analogous calculation as above implies that $\hat{A}$ is an essential two sided ideal of A'. Although $\hat{A}$ is not a C*-algebra, one can define the multiplier algebra $M(\hat{A})$ of $\hat{A}$ in the same way as in Definition 1.21. It is not very hard to show that $A' = M(\hat{A})$.

The pair $(A \odot A, (\iota \odot \chi \odot \iota)(\Phi \odot \Phi))$ is a Hopf *-algebra with Haar state $\varphi \otimes \varphi$ (here, χ is the flip map op $A \odot A$). It is obvious that $(A \odot A)^{\hat{}} = \hat{A} \odot \hat{A}$, which by the discussion above implies that $M(\hat{A} \odot \hat{A}) = (A \odot A)'$. Thus we get a comultiplication $\hat{\Phi} : \hat{A} \to M(\hat{A} \odot \hat{A})$.

The pair $(\hat{A}, \hat{\Phi})$ is an example of a *multiplier Hopf*-algebra* as introduced in [VD94]. This means that $\hat{\Phi}$ is coassociative and that the linear mappings $T_1, T_2 : \hat{A} \odot \hat{A} \to M(\hat{A} \odot A)$ defined by

$$T_1(x \otimes y) = \hat{\Phi}(x)(y \otimes 1) \qquad \text{and} \qquad T_2(x \otimes y) = \hat{\Phi}(x)(1 \otimes y)$$

for all $x, y \in \hat{A}$ are bijections from $\hat{A} \odot \hat{A}$ to $\hat{A} \odot \hat{A}$. The formulas $\hat{S}(\omega) = \omega \circ S^{-1}$ and $\hat{\varepsilon}(\omega) = \omega(1)$ for all $\omega \in \hat{A}$ define an antipode $\hat{S}$ and counit $\hat{\varepsilon}$ for $(\hat{A}, \hat{\Phi})$ in the sense of [VD94].

If one defines a linear mapping $\hat{\varphi} : \hat{A} \to \mathbb{C}$ by $\hat{\varphi}(\varphi(a.)) = \varepsilon(a)$ for all $a \in A$, we get a left invariant functional on $(\hat{A}, \hat{\Phi})$ in the sense that

$$(\iota \odot \hat{\varphi})(\hat{\Phi}(x)(y \otimes 1)) = y\, \hat{\varphi}(x)$$

for all $x, y \in \hat{A}$. One can also prove that $\hat{\varphi}(x^*x) \geq 0$ for all $x \in \hat{A}$. Of course, $\hat{\varphi}\hat{S}$ provides a right invariant functional for $(\hat{A}, \hat{\varphi})$ that is not always proportional to $\hat{\varphi}$. Since $(\hat{A}, \hat{\Delta})$ is the dual of a compact quantum group, we consider it to be a discrete quantum group. Also remember that, unlike the classical case, a discrete quantum group does not have to be unimodular.

Since $(\hat{A}, \hat{\Phi})$ possesses a left invariant functional, the pair fits into the framework of [VD98] which basically investigates an algebraic version of locally compact quantum groups. Since $\hat{\varphi}$ is positive ([VD98] also allows for non-positive $\hat{\varphi}$), $(\hat{A}, \hat{\Phi})$ fits into the framework studied in [Kus02]. This paper shows that multiplier Hopf *-algebras that have positive invariant functionals have the same rich structure as general locally compact groups discussed in these lecture notes. They can however be studied in an algebraic context.

One expects that the dual of a compact quantum group is a 'discrete' quantum group. This is reflected by the following proposition. Notice that if G is a discrete group, $C_0(G) = \sum_{g \in G} \mathbb{C} = \sum_{g \in G} M_1(\mathbb{C})$.

Proposition 3.18. *Let $(u^\alpha)_{\alpha \in A}$ be a complete family of mutually inequivalent irreducible unitary corepresentations of (A, Δ) and let n_α denote the dimension of u^α. Then $\hat{A} \cong \sum_{\alpha \in A} M_{n_\alpha}(\mathbb{C})$.*

Here, $\sum_{\alpha \in A} M_{n_\alpha}(\mathbb{C})$ is an algebraic direct sum where we consider only A-tuples with only a finite number of non-zero components. Can you prove this proposition? So it follows that $\hat{A}$ embeds into the C*-algebra $\oplus_{\alpha \in A} M_{n_\alpha}(\mathbb{C})$. Also note that each element in $\hat{A}$ extends uniquely to an element in the topological dual A^* and as such has a norm as a continuous linear functional. However, this does not define a C*-norm on $\hat{A}$!

So we get discrete quantum groups as duals of compact quantum groups and this is the road followed in [PoW]. An intrinsic definition for discrete quantum groups has been introduced in [ER] and [VD96], albeit in an algebraic setting. The C*-algebraic version of the definition in [VD96] goes as follows.

Definition 3.19. *Consider a C*-algebra B so that there exists a family of natural numbers $(n_i)_{i \in I}$ such that $B \cong \oplus_{i \in I} M_{n_i}(\mathbb{C})$. Let $\Delta : B \to M(B \otimes B)$ be a non-degenerate *-homomorphism such that*

1. *$(\Delta \otimes \iota)\Delta = (\iota \otimes \Delta)\Delta$.*
2. *The vector spaces $\Delta(B)(1 \otimes B)$ and $\Delta(B)(B \otimes 1)$ are dense subspaces of $B \otimes B$.*
3. *The linear mappings $T_1 : B \odot B \to B \otimes B$ and $T_2 : B \odot B \to B \otimes B$ defined by*

$$T_1(x \otimes y) = \Delta(x)(y \otimes 1) \qquad and \qquad T_2(x \otimes y) = \Delta(x)(1 \otimes y)$$

for all $x, y \in B$, are injective.

Then, (B, Δ) is called a discrete quantum group.

In this definition, the C*-algebraic direct sum $\oplus_{i \in I} M_{n_i}(\mathbb{C})$ is defined in a similar way as the one introduced after Proposition 3.12.
Notice that we do not assume the existence of Haar measures, these can be constructed but do not have to be the canonical traces on B (which is the case in the classical setting where the Haar measure is just the counting measure). It is also important to note that T_1 and T_2 are not continuous in general.

3.2 Examples of compact quantum groups

'The' dual of a discrete group

Fix a discrete group G and let $K(G)$ be the vector space of complex valued functions on G with finite support. If $s \in G$, we define the function δ_s on G as the one that takes the value 1 in s and that is 0 elsewhere. Of course, $(\delta_s \mid s \in G)$ is a linear basis for $K(G)$. We consider $K(G)$ as the convolution *-algebra of G. This means that the product $\star$ and *-operation $.^\circ$ are given by the formulas

1. $(f \star g)(t) = \sum_{s \in G} f(s)g(s^{-1}t)$,
2. $f^\circ(t) = \overline{f(t^{-1})}$

for all $f, g \in K(G)$ and $t \in G$. Thus $\delta_s \star \delta_t = \delta_{st}$ and $\delta_s^\circ = \delta_{s^{-1}}$ for all $s, t \in G$. Notice that $K(G)$ is unital with unit element δ_e and that $K(G)$ is commutative if and only if the group G is commutative. If we define the *-homomorphism $\Phi : K(G) \to K(G) \odot K(G)$ such that $\Phi(\delta_s) = \delta_s \odot \delta_s$ for all $s \in G$, then $(K(G), \Phi)$ is a Hopf *-algebra that is *cocommutative*, i.e. $\chi \Phi = \Phi$. It possesses an invariant linear functional $h : K(G) \to \mathbb{C}$ determined by $h(\delta_s) = \delta_{s,e}$ for all $s \in G$.

Define the linear mapping $T : K(G) \odot K(G) \to K(G) \odot K(G)$ such that $T(f \otimes g) = \Phi(g)(f \otimes 1)$ for all $f, g \in K(G)$. Since $(K(G), \Phi)$ is a Hopf *-algebra, T is a bijection. This is not so hard to prove because one can write down an explicit formula for the inverse of T: if S denotes the antipode of $(K(G), \Phi)$, then $T^{-1}(f \otimes g) = (S^{-1} \otimes 1)(\Phi(g))(f \otimes 1)$ for all $f, g \in K(G)$. It follows that $K(G) \odot K(G) = \langle \Phi(g)(f \otimes 1) \mid f, g \in K(G) \rangle$. In a similar way, $K(G) \odot K(G) = \langle \Phi(g)(1 \otimes g) \mid f, g \in K(G) \rangle$.

There are in general different ways to put a C*-norm on the convolution algebra $K(G)$. Let us discuss the two most natural ones.

(1) Let $\|.\|_*$ be the norm introduced in Remark 1.6 with $\mathcal{A} = K(G)$. Consider a Hilbert space $\mathcal{K}$ and a *-representation $\theta : K(G) \to B(\mathcal{K})$. If $s \in G$, then $\delta_s^\circ \star \delta_s = \delta_e$ implies that $\theta(\delta_s)^* \theta(\delta_s) = \theta(\delta_e)$ is a projection, thus $\|\theta(\delta_s)\| \leq 1$. Consequently the triangle inequality implies that $\|\theta(f)\| \leq \sum_{s \in G} |f(s)|$ and thus $\|f\|_* < \infty$ for all $f \in K(G)$.

The *-representation of the next paragraph guarantees that that $\|f\|_* = 0 \Leftrightarrow f = 0$, so that we can apply remark 1.6 and define the C*-algebra $C^*(G)$ as the enveloping C*-algebra of $K(G)$. The C*-algebra $C^*(G)$ is coined the *group C*-algebra* of G. The definition of the norm $\|.\|_*$ implies that the *-homomorphism $\Phi : K(G) \to K(G) \odot K(G) \subseteq C^*(G) \otimes C^*(G)$ extends to a *-homomorphism $\Delta : C^*(G) \to C^*(G) \otimes C^*(G)$ and one easily sees that $(C^*(G), \Delta)$ is a compact quantum group.

(2) Let $\ell^2(G)$ be the space of square summable functions on G (remember that the Haar measure of G is the counting measure so that $\ell^2(G) = L^2(G)$). Define the linear map $\pi : K(G) \to B(\ell^2(G))$ such that $\pi(f) g = f \star g$ for all $f \in K(G)$ and $g \in \ell^2(G)$. Note that $(\pi(\delta_s) g)(t) = g(s^{-1}t)$ for all $g \in \ell^2(G)$ and $s, t \in G$, implying that $\pi(\delta_s)$ is unitary and that $\pi(f)$ indeed defines a bounded operator on $\ell^2(G)$. The *-homomorphism is easily seen to be faithful. One defines $C_r^*(G)$ as the closure of $\pi(K(G))$ in $B(\ell^2(G))$, $C_r^*(G)$ is coined the *reduced group C*-algebra* of G.

Define the unitary $W \in B(\ell^2(G \times G))$ by $(Wk)(s,t) = k(s, s^{-1}t)$ for all $k \in \ell^2(G \times G)$ and $s, t \in G$. One checks that $(\pi \odot \pi)(\Phi(f)) = W(\pi(f) \otimes 1)W^*$ for all $f \in K(G)$. So we can define a unital *-homomorphism $\Delta_r : C_r^*(G) \to C_r^*(G) \otimes C_r^*(G)$ such that $\Delta_r(x) = W(x \otimes 1)W^*$ for all $x \in C_r^*(G)$. Thus, $\Delta_r \pi = (\pi \odot \pi)\Phi$. Again, $(C_r^*(G), \Delta_r)$ is a compact quantum group and its Haar state φ_r is given by the vector functional $\varphi_r = \omega_{\delta_e, \delta_e}$.

This Haar state φ_r is faithful. In order to see this, let us define for every $s \in G$ the unitary operator $R_s \in B(\ell^2(G))$ by $R_s(f)(t) = f(ts^{-1})$ for all

$f \in \ell^2(G)$ and $t \in G$. Then $R_s\pi(f) = \pi(f)R_s$ for all $f \in K(G)$, implying that $x\, R_s = R_s\, x$ for all $x \in C_r^*(G)$. So if $x \in C_r^*(G)$ and $\varphi_r(x^*x) = 0$, then $x\, \delta_e = 0$. Hence, for every $s \in G$, $0 = R_s x \delta_e = x R_s \delta_e = x\delta_s$. We conclude that $x = 0$.

The definition of the norm $\|.\|_*$ implies immediately the existence of a unital surjective *-homomorphism $\pi_r : C^*(G) \to C_r^*(G)$ that extends π. Thus $(\pi_r \otimes \pi_r)\Delta = \Delta_r \pi_r$. Also, the Haar state φ of $(C^*(G), \Delta)$ is given by $\varphi = \varphi_r \pi_r$ since it extends h. As a consequence we see that φ is faithful if and only if π_r is faithful. The faithfulness of π_r is equivalent to the amenability of G and there are numerous examples of discrete groups that are not amenable, for instance the free group on 2 generators $\mathbb{F}_2$.

Quantum SU(2)

In the framework of locally compact quantum groups, the compact quantum group $SU_q(2)$ was introduced by Woronowicz. He also unravelled the correpresentation theory of this quantum group by analyzing the 'quantum Lie algebra' of this quantum group (see [Wor87b] and [Wor87a]).

Recall that $SU(2)$ is the group of unitary complex matrices in $M_2(\mathbb{C})$ of determinant 1. Alternatively,

$$SU(2) = \left\{ \begin{pmatrix} a & -\bar{c} \\ c & \bar{a} \end{pmatrix} \mid a, c \in \mathbb{C} \text{ s.t. } |a|^2 + |c|^2 = 1 \right\}.$$

Define the functions $\alpha, \gamma \in C(SU(2))$ by

$$\alpha \begin{pmatrix} a & -\bar{c} \\ c & \bar{a} \end{pmatrix} = a \qquad \text{and} \qquad \gamma \begin{pmatrix} a & -\bar{c} \\ c & \bar{a} \end{pmatrix} = c$$

for all $a, c \in \mathbb{C}$ such that $|a|^2 + |c|^2 = 1$. Define $\mathcal{A}$ as the dense commutative unital *-subalgebra of $C(SU(2))$ generated by α and γ. Thus, α and γ are normal, $\alpha\gamma = \gamma\alpha$ and $\alpha^*\alpha + \gamma^*\gamma = 1$.

Define the *-homomorphism $\Phi : \mathcal{A} \to \mathcal{A} \odot \mathcal{A} \subseteq C(SU(2) \times SU(2))$ by $\Phi(f)(s,t) = f(st)$ for all $f \in \mathcal{A}$ and $s, t \in SU(2)$. Note that $(\mathcal{A}, \Phi)$ is a Hopf *-algebra such that

$$\Phi(\alpha) = \alpha \otimes \alpha - \gamma^* \otimes \gamma \qquad \text{and} \qquad \Phi(\gamma) = \gamma \otimes \alpha + \alpha^* \otimes \gamma.$$

Now we are going to deform this Hopf *-algebra $\mathcal{A}$. For this purpose we fix a non-zero number $q \in (-1, 1)$.

Define $\mathcal{A}$ as the universal unital *-algebra generated by two elements α, γ and relations

$$\alpha^* \alpha + \gamma^* \gamma = 1 \qquad\qquad \alpha \alpha^* + q^2 \gamma \gamma^* = 1$$

$$\gamma \gamma^* = \gamma^* \gamma \qquad q \gamma \alpha = \alpha \gamma \qquad q \gamma^* \alpha = \alpha \gamma^* .$$

The universality property of the *-algebra $\mathcal{A}$ implies the existence of a unique *-homomorphism $\Phi : \mathcal{A} \to \mathcal{A} \odot \mathcal{A}$ such that

$$\Phi(\alpha) = \alpha \otimes \alpha - q \gamma^* \otimes \gamma \qquad \text{and} \qquad \Phi(\gamma) = \gamma \otimes \alpha + \alpha^* \otimes \gamma .$$

Then, $(\mathcal{A}, \Phi)$ is a Hopf *-algebra (one can easily produce a counit and an antipode). Hence, $\mathcal{A} \odot \mathcal{A} = \langle \Phi(a)(b \otimes 1) \mid a, b \in \mathcal{A} \rangle = \langle \Phi(a)(b \otimes 1) \mid a, b \in \mathcal{A} \rangle$, as explained in the previous example.

But we want to get hold of a C*-algebra. In order to do so we follow the same procedure as in the previous example. Define the Hilbert space $H = \ell^2(\mathbb{N}_0) \otimes \ell^2(\mathbb{Z})$. Let $(e_n)_{n=0}^{\infty}$ be the standard orthonormal basis of $\ell^2(\mathbb{N}_0)$ and let $(f_k)_{k=-\infty}^{\infty}$ be the standard orthonormal basis of $\ell^2(\mathbb{Z})$. Set $e_{-1} = 0$. Then we can define two bounded linear operators S and T on H such that

$$S(e_n \otimes f_k) = \sqrt{1 - q^{2n}} \, e_{n-1} \otimes e_k \qquad \text{and} \qquad T(e_n \otimes f_k) = q^n e_n \otimes e_{k+1}$$

for all $n \in \mathbb{N}_0$ and $k \in \mathbb{Z}$. Then, S and T satisfy the same relations as α and γ do. Therefore the universality of $\mathcal{A}$ implies the existence of a unique *-representation $\pi : \mathcal{A} \to B(H)$ such that $\pi(\alpha) = S$ and $\pi(\gamma) = T$.

Note first that the relations above imply easily that $\mathcal{A}$ is the linear span of elements of the form $\alpha^r (\gamma^*)^k \gamma^l$ and $(\alpha^*)^{r'} (\gamma^*)^k \gamma^{l'}$, where $r, r', k, k', l, l' \in \mathbb{N}_0$. It is a little bit tedious but not too hard to check that the elements $S^r (T^*)^k T^l$ and $(S^*)^{r'} (T^*)^{k'} T^{l'}$, where $r, k, k', l, l' \in \mathbb{N}_0$, $r' \in \mathbb{N}$, are linearly independent. This implies that π is injective and that the elements $\alpha^r (\gamma^*)^k \gamma^l$ and $(\alpha^*)^{r'} (\gamma^*)^{k'} \gamma^{l'}$, where $r, k, k', l, l' \in \mathbb{N}_0$, $r' \in \mathbb{N}$ form a basis for $\mathcal{A}$.

As before we want to define a C*-algebra using Remark 1.6. If $\mathcal{K}$ is a Hilbert space and $\theta : \mathcal{A} \to B(\mathcal{K})$ is a unital *-homomorphism, then obviously, $\theta(\alpha)^* \theta(\alpha) + \theta(\gamma)^* \theta(\gamma) = 1$ implying that $\|\theta(\alpha)\| \leq 1$ and $\|\theta(\gamma)\| \leq 1$. This implies easily that the number $\|x\|_*$ is finite for any $x \in \mathcal{A}$. Note that the injectivity of π guarantees that $\|.\|_*$ is a norm and not a semi-norm. Therefore we can apply Remark 1.6 and define the C*-algebra A to be the enveloping C*-algebra of $\mathcal{A}$.

By the definition of the norm $\|.\|_*$ there exists a unique unital *-homomorphism $\Delta : A \to A \otimes A$ that extends Φ. The properties of $(\mathcal{A}, \Phi)$ immediately imply that (A, Δ) is a compact quantum group that we call quantum $SU(2)$ and denote by $SU_q(2) = (A, \Delta)$. A formula for the Haar state of $SU_q(2)$ can be found in [Wor87a].

There exists a unique *-homomorphism $\pi_r : A \to B(H)$ that extends π. In principle, one could use π_r to define a reduced version of quantum $SU(2)$ but it can be shown that π_r is injective.

The irreducible corepresentations of (A, Δ) have been computed in
[Wor87b]. As for $SU(2)$ itself, there exist for every $n \in \mathbb{N}$ a unique (up to uni-
tary equivalence) irreducible unitary corepresentation of $SU_q(2)$ of dimension
n. The coefficient $*$-algebra of $SU_q(2)$ is nothing else but $\mathcal{A}$.

Universal quantum groups

Proposition 3.20. *Let Q be in $M_n(\mathbb{C})$ and assume that it is invertible. Let $\mathcal{A}$
be the universal unital $*$-algebra generated by elements $\{u_{ij} \mid i, j = 1, \ldots, n\}$
subject to the following relations*

$$u^* u = u u^* = 1$$

and

$$u^t \, Q \, \overline{u} \, Q^{-1} = Q \, \overline{u} \, Q^{-1} u^t = 1 \ .$$

*Here $\overline{u}$ is the matrix $(u_{ij}^*)_{ij}$ and u^t is the matrix $(u_{ji})_{ij}$. The $*$-algebra $\mathcal{A}$
becomes a Hopf $*$-algebra with comultiplication Φ determined by $\Phi(u_{ij}) = \sum_k u_{ik} \otimes u_{kj}$.*

Again, one obtains a compact quantum group (A, Δ) by considering $*$-
representations of A over all possible Hilbert spaces as in the previous two
examples. Just as in the case of $SU_q(2)$, elements of $\mathcal{A}$ are always represented
by bounded operators because u is unitary. The quantum group (A, Δ) is
called the *universal compact quantum group* (see [VDWa]). The reason for using this terminology is that any compact quantum group will be a 'quantum
subgroup' of the universal one (for a suitable Q).

A multitude of properties about universal quantum groups have been proven
by T. Banica in [Ban97] (and about related matters in [Ban96]). A description
of the corepresentation theory of universal compact quantum groups is given
in [Ban97, Thm. 1]. This paper also provides a remarkable characterization
of universal compact quantum groups as compact matrix quantum groups
for which the character of the universal corepresentation, divided by 2, is a
circular element with respect to the Haar measure.

The free product (following Voiculescu) of two compact quantum groups turns
out to be again a compact quantum group, where the Haar measure is the free
product of the Haar measures on the original quantum groups, a fact that is
explained in [Wan95].

Tannaka-Krein duality for compact quantum groups

In the case of compact quantum groups one can use the Tannaka-Krein duality
(see [Wor88] and [Wan97]) to construct examples. Instead of trying to define
a C$*$-algebra and a comultiplication one constructs a well-behaved category of
abstract corepresentations. The general Tannaka-Krein duality Theorem for
compact quantum groups then guarantees the existence of a unique compact
quantum group such that the category of concrete corepresentations of this

quantum group agrees with the constructed category of abstract corepresentations. In [Ros] this method is used to associate compact quantum groups to quantized universal enveloping Lie algebras.

4 Weight theory on von Neumann algebras

In the previous section we saw that for compact and discrete quantum groups, the existence of the Haar measure follows from a relatively simple set of axioms. This seems unfortunately not to be true in the general case, the existence of the Haar measure is incorporated in the definition and plays a pivotal role in the development of the theory. In the framework of C*-algebras, the role of measures is played by *weights*. Weight theory works better for von Neumann algebras which are the proper generalizations of measure spaces. That is one of the reasons why we will shift our attention in this and the following section to von Neumann algebras. Later we will go back to the theory of C*-algebras.

Let us first look at the classical case to motivate the definition of weights in general and Haar weights in the next section. Therefore take a locally compact group G and fix a left Haar measure μ on G. Let A be the commutative C*-algebra $C_0(G)$.

Just as finite measures on G easily translate to positive linear functionals on A, one can easily translate the measure μ to an object on the level of the C*-algebra $A = C_0(G)$ that contains all information about μ. Using μ (but this is possible for any measure), one defines the map

$$\varphi_\mu : A^+ \to [0, \infty] : f \mapsto \varphi_\mu(f) = \int_G f \, d\mu .$$

This map satisfies the following properties

1. $\varphi_\mu(f + g) = \varphi_\mu(f) + \varphi_\mu(g)$ for all $f, g \in A^+$,
2. $\varphi_\mu(\lambda f) = \lambda \varphi_\mu(f)$ for all $f \in A^+$ and $\lambda \in \mathbb{R}^+$,
3. $\varphi_\mu(f) < \infty$ if $f \in K(G)^+$,
4. A special case of the lemma of Fatou: Let $f \in A^+$ and $(f_i)_{i=1}^\infty$ a sequence of functions so that f converges pointwise to f. Then, $\varphi_\mu(f) \le \liminf_{i \to \infty} \varphi_\mu(f_i)$.

This will shortly motivate the definition of a weight on a C*-algebra as an analogue of a measure. Remember that $K(G)$ is the *-algebra of all continuous functions on G with compact support.

We will also need to translate the left invariance of μ to a condition on φ_μ but this is not so difficult and is based on the same principle as the one used in the compact case (see Example 3.3). Take a positive linear functional ω on A. The theorem of Riesz guarantees the existence of a regular, finite, positive Borel measure (see [Coh]) ν on G such that $\omega(f) = \int f d\nu$ for all $f \in A$. Then, the theorem of Fubini and the left invariance of μ imply for $f \in K(G)^+$,

$$\varphi_\mu((\omega \otimes \iota)\Delta(f)) = \int (\omega \otimes \iota)(\Delta(f))(s)\,d\mu(s) = \int\int f(ts)d\nu(t)\,d\mu(s)$$

$$= \int\int f(ts)d\mu(s)\,d\nu(t) = \int\int f(s)d\mu(s)\,d\nu(t) = \varphi_\mu(f)\,\omega(1) \ ,$$

an equation that still makes sense in a general C*-algebra framework.

4.1 Weights on C*-algebras

Fix a C*-algebra A. Recall that we denote the set of positive elements in A by A^+ and the set of positive linear functionals on A by A^*_+.

Let us start of by formalizing the notion of a weight on A as motivated by the discussion in the introduction above.

Definition 4.1. *Consider a function* $\varphi : A^+ \to [0,\infty]$ *such that*

1. $\varphi(a + b) = \varphi(a) + \varphi(b)$ *for all* $a, b \in A^+$,
2. $\varphi(\lambda a) = \lambda \varphi(a)$ *for all* $a \in A^+$ *and* $\lambda \in \mathbb{R}^+$,

Then we call φ *a weight on A. We say that φ is densely defined if the set* $\{ a \in A^+ \mid \varphi(a) < \infty \}$ *is norm dense in A^+. We call the weight φ proper if φ is densely defined and φ is lower semi-continuous with respect to the norm topology.*

We use the convention that $0 \cdot \infty = 0$. Note that this implies that $\varphi(0) = 0$. Also note that the first condition implies that $\varphi(a) \leq \varphi(b)$ if $a, b \in A^+$ and $a \leq b$.

In the framework of C*-algebras that are not von Neumann algebras (!) one usually works with proper weights. In the case of von Neumann algebras, one strengthens the continuity condition and weakens the density condition by using another topology but we will come to this shortly.

The introduction above explains this definition but there is one striking difference. There is no condition involving the analogue of continuous functions with compact support; the density condition just assumes that there are enough integrable elements. The reason for this is simple: there is no useful analogue for continuous functions with compact support. In fact, this explains to a certain extent the technical difficulties in the theory of locally compact quantum groups.

Let us recall the notion of lower semi-continuity. Let X be a topological space and $f : X \to [0,\infty]$ a function. The easiest characterization of *lower semi-continuity* is the following one: f is lower semi-continuous $\Leftrightarrow$ for all $\lambda \in \mathbb{R}^+$, the set $\{ x \in X \mid f(x) \leq \lambda \}$ is closed in X.

But lower semi-continuity can also be characterized in terms of nets: f is lower semi-continuous $\Leftrightarrow$ for all $x \in X$ and every net $(x_i)_{i \in I}$ in X such that $(x_i)_{i \in I} \to x$, we have $f(x) \leq \liminf_{i \in I} f(x_i)$. This characterization has the advantage that it makes sense locally, in a point.

This also implies that a lower semi-continuous weight φ on A satisfies a kind of dominated convergence theorem: if $a \in A^+$ and $(a_i)_{i \in I}$ is a net in A^+ such that $(a_i)_{i \in I} \to a$ for the norm topology and $a_i \leq a$ for all $i \in I$, then $\big(\varphi(a_i)\big)_{i \in I} \to \varphi(a)$.

Let us distinguish some special elements with respect to a weight.

Definition 4.2. *Let φ be a weight on A. Then we define the following sets:*

1. $\mathcal{M}_\varphi^+ = \{\, a \in A^+ \mid \varphi(a) < \infty \,\}$,
2. $\mathcal{M}_\varphi = $ *the linear span of* $\mathcal{M}_\varphi^+$ *in* A,
3. $\mathcal{N}_\varphi = \{\, a \in A \mid \varphi(a^*a) < \infty \,\}$.

The set $\mathcal{M}_\varphi^+$ is a *hereditary cone* in A^+ (norm dense in A^+ if φ is densely defined). One calls it a 'cone' since this set is closed under addition and scalar multiplication with positive elements. One calls it 'hereditary' because if $a \in A^+$, $b \in \mathcal{M}_\varphi^+$ and $a \leq b$, the element a also belongs to $\mathcal{M}_\varphi^+$.

The set $\mathcal{M}_\varphi$ is a *-subalgebra of A (norm dense in A if φ is densely defined). One can show that $\mathcal{M}_\varphi^+ = \mathcal{M}_\varphi \cap A^+$ and that $\mathcal{M}_\varphi = \mathcal{N}_\varphi^* \mathcal{N}_\varphi := \langle b^*a \mid a, b \in \mathcal{N}_\varphi \rangle$. There exists a unique linear map $F : \mathcal{M}_\varphi \to \mathbb{C}$ such that $F(a) = \varphi(a)$ for all $a \in \mathcal{M}_\varphi^+$. If $x \in \mathcal{M}_\varphi$, one defines $\varphi(x) := F(x)$.

The set $\mathcal{N}_\varphi$ is a left ideal in A (dense in A if φ is densely defined).

Notice that $\mathcal{M}_\varphi$ fulfills the role of the set of integrable elements, while $\mathcal{N}_\varphi$ fulfills the role of square integrable elements. Concerning $\mathcal{N}_\varphi$ it should however be mentioned that our choice to work with a^*a instead of aa^* is a matter of taste.

Just as for positive linear functionals, one can introduce for any weight a GNS-construction (see Definition 1.11). The main difference lies in the fact that for weights that are not continuous positive linear functionals, the cyclic vector is non-existent.

Definition 4.3. *Consider a weight φ on A together with a Hilbert space $\mathcal{H}_\varphi$, a *-homomorphism $\pi_\varphi : A \to \mathcal{H}_\varphi$ and a linear map $\Lambda_\varphi : \mathcal{N}_\varphi \to \mathcal{H}_\varphi$ such that*

1. $\Lambda_\varphi(\mathcal{N}_\varphi)$ *is dense in* $\mathcal{H}_\varphi$,
2. $\langle \Lambda_\varphi(a), \Lambda_\varphi(b) \rangle = \varphi(b^*a)$ *for all* $a, b \in \mathcal{N}_\varphi$,
3. $\pi_\varphi(x) \Lambda_\varphi(a) = \Lambda_\varphi(xa)$ *for all* $x \in A$ *and* $x \in \mathcal{N}_\varphi$.

Then, we call the triple $(\mathcal{H}_\varphi, \pi_\varphi, \Lambda_\varphi)$ a GNS-construction for φ.

Such a GNS-construction can be easily constructed and is unique up to unitary equivalence. It is clear that $\mathcal{H}_\varphi$ is just the analogue of $L^2(G, \mu)$ and that $\pi_\varphi(x)$ is just the (left) multiplication operator with x. The map Λ_φ is used to distinguish between a as an element of A and a as an element of $\mathcal{H}_\varphi$.

If φ is lower semi-continuous, the *-representation π_φ is non-degenerate and the mapping $\Lambda_\varphi : \mathcal{N}_\varphi \to \mathcal{H}_\varphi$ is closed with respect to the norm topologies on A and $\mathcal{H}_\varphi$. The notion of a closed linear map can be found in the appendix.

If $\varphi \in A_*^+$ and $(\mathcal{H}_\varphi, \pi_\varphi, \xi_\varphi)$ is a cyclic GNS-construction for φ (see Definition 1.11), we get a GNS-construction $(\mathcal{H}_\varphi, \pi_\varphi, \Lambda_\varphi)$ for φ according to the definition above by defining $\Lambda_\varphi : A \to \mathcal{H}_\varphi : a \to \Lambda_\varphi(a) = \pi_\varphi(a)\xi_\varphi$.

We will be mainly (but not exclusively) be interested in weights that are faithful:

Definition 4.4. *We call a weight φ on A faithful if for every $a \in A^+$, we have $\varphi(a) = 0 \Leftrightarrow a = 0$.*

Note that φ is faithful $\Leftrightarrow \Lambda_\varphi$ is injective. If φ is faithful and densely defined, then π_φ is faithful. The faithfulness of φ_μ in the introduction is equivalent to the fact the support of μ equals G.

Loosely speaking, any lower semi-continuous weight can be approximated by positive linear functionals. Let us make this precise:

Proposition 4.5. *Let φ be a lower semi-continuous weight on A. Define $\mathcal{F} = \{\omega \in A_*^+ \mid \omega \leq \varphi\}$. Then $\varphi(a) = \sup\{\omega(a) \mid \omega \in \mathcal{F}\}$ for all $a \in A^+$.*

If $\omega \in A_*^+$, the inequality $\omega \leq \varphi$ means that $\omega(a) \leq \varphi(a)$ for all $a \in A^+$. There exists a stronger form of this property:

Proposition 4.6. *Let φ be a lower semi-continuous weight on A. Define $\mathcal{G} = \{t\omega \mid \omega \in A_*^+, \omega \leq \varphi, 0 < t < 1\}$. Then $\mathcal{G}$ is upwardly directed and $\varphi(a) = \sup\{\omega(a) \mid \omega \in \mathcal{G}\} = \lim_{\omega \in \mathcal{G}} \omega(a)$ for all $a \in A^+$.*

The statement that $\mathcal{G}$ is upwardly directed means that for all $\omega, \eta \in \mathcal{G}$ there exists $\theta \in \mathcal{G}$ such that $\omega \leq \theta$ and $\eta \leq \theta$. As a consequence, $\mathcal{G}$ can be used as an index set for a net, as we have done above. Note that by linearity this proposition also implies that $\varphi(a) = \lim_{\omega \in \mathcal{G}} \omega(a)$ for all $a \in \mathcal{M}_\varphi$.

Up till now, this discussion concerning weights revolved around concepts that clearly have their roots in the classical framework. One might hope that the notion of (possibly faithful) weights allows for a completely satisfactory non-commutative integration theory. This is indeed the case if one works in the framework of von Neumann algebra's but in the framework of C*-algebra's one needs sometimes to impose extra conditions to get a useful weight theory. In a weakened form these extra conditions are automatically satisfied in the von Neumann algebra framework. To formulate these extra conditions we have to enter into the world of one-parameter groups of *-automorphisms on C*-algebras and von Neumann algebras.

But let us look at the simplest non-trivial case. Fix $n \in \mathbb{N}$ and set $A = M_n(\mathbb{C})$, the C*-algebra of complex n by n matrices. Define the positive linear functional $\tau : A \to \mathbb{C}$ as the trace on A, i.e. $\tau(x) = \sum_{i=1}^n x_{ii}$ for all $x \in A$. Although A is not commutative, any 2 elements from A always commute under τ: $\tau(xy) = \tau(yx)$ for all $x, y \in A$.

This is not the case for any positive linear functional on A, but with the necessary modifications there still is a certain degree of commutation under the functional. We will make this statement precise in the following. Therefore fix a positive, invertible matrix δ in $M_n(A)$ and define the faithful, positive (why?) functional φ of A by $\varphi(x) = \tau(x\,\delta)$ for all $x \in A$.

Exercise 4.7. Prove that every faithful, positive linear functional on A is of this form.

One easily checks that for $x, y \in A$,

$$\varphi(xy) = \varphi\big(y\,(\delta\,x\,\delta^{-1})\big) . \tag{4.1}$$

In Proposition 1.12 we have seen that if f is a complex valued function on the spectrum $\sigma(\delta)$, one defines a new operator $f(\delta)$. Of course, in this case this takes on a more familiar form as explained below. There exists a unitary matrix $u \in A$ and positive numbers $\lambda_1, \ldots, \lambda_n > 0$ such that

$$\delta = u^* \begin{pmatrix} \lambda_1 & 0 & \cdots & 0 \\ 0 & \lambda_2 & \ddots & \vdots \\ \vdots & \ddots & \ddots & 0 \\ 0 & \cdots & 0 & \lambda_n \end{pmatrix} u .$$

Thus $\sigma(\delta) = \{\lambda_1, \ldots, \lambda_n\}$ and if we have a function $f : \{\lambda_1, \ldots, \lambda_n\} \to \mathbb{C}$, then

$$f(\delta) = u^* \begin{pmatrix} f(\lambda_1) & 0 & \cdots & 0 \\ 0 & f(\lambda_2) & \ddots & \vdots \\ \vdots & \ddots & \ddots & 0 \\ 0 & \cdots & 0 & f(\lambda_n) \end{pmatrix} u .$$

Recall that if $z \in \mathbb{C}$, the complex power $\delta^z \in A$ is defined as $\delta^z = f(\delta)$ where $f : \sigma(\delta) \to \mathbb{C} : \lambda \mapsto \lambda^z$. Thus, the following properties hold:

1. If $r \in \mathbb{R}$, then δ^r is an invertible, positive operator in A and δ^{ir} a unitary operator in A.

2. For all $y, z \in \mathbb{C}$, we have $\delta^{y+z} = \delta^y\,\delta^z$ and $(\delta^z)^* = \delta^{\bar{z}}$.

Using δ we now can define a *one-parameter group* σ of *-automorphisms on A, $\sigma : \mathbb{R} \mapsto \mathrm{Aut}\,(A) : t \mapsto \sigma_t$ where $\sigma_t(x) = \delta^{it} x\,\delta^{-it}$ for all $x \in A$ and $t \in \mathbb{R}$. Note that $\sigma_{s+t} = \sigma_s\,\sigma_t$ for all $s, t \in \mathbb{R}$.

It makes perfect sense to extend the definition of σ_t to complex parameters $z \in \mathbb{C}$ and define an algebra automorphism σ_z on A by $\sigma_z(x) := \delta^{iz} x\,\delta^{-iz}$ for all $x \in A$. Note that σ_z is an algebra homomorphism but not always a *-homomorphism. Now,

1. $\sigma_{y+z} = \sigma_y\,\sigma_z$ for all $y, z \in \mathbb{C}$,
2. $\sigma_z(x)^* = \sigma_{\bar{z}}(x^*)$ for all $z \in \mathbb{C}$ and $x \in A$.

Later on we will get into situations where we have been given a one-parameter group $\sigma : \mathbb{R} \to \text{Aut}(A)$ which can not be defined in terms of such an operator δ. Therefore we would like to find a characterization of the algebra automorphism σ_z in terms of the *-automorphism $(\sigma_t)_{t \in \mathbb{R}}$ without referring to δ. Here the theory of analytic functions enters the story.

In this context we call a function $f : \mathbb{C} \to A$ analytic if for every $i, j \in \{1, \ldots, n\}$, the function $\mathbb{C} \to \mathbb{C} : z \mapsto f(z)_{ij}$ is analytic. For instance, the function $\mathbb{C} \to A : z \to \delta^z$ is analytic.

The identity theorem for analytic functions from complex analysis implies that σ_z can be defined in terms of the family $(\sigma_t)_{t \in \mathbb{R}}$ as follows:

Consider $a \in A$. There exists a unique analytic function $f : \mathbb{C} \to A$ such that $f(t) = \sigma_t(a)$ for all $t \in \mathbb{R}$. Moreover, $\sigma_z(a) = f(z)$ for all $z \in \mathbb{C}$.

Equation (4.1) can be rewritten as

$$\varphi(xy) = \varphi(y \, \sigma_{-i}(x)) \qquad \text{for all } x, y \in A \tag{4.2}$$

and it is also easily checked that

$$\varphi(\sigma_t(x)) = \varphi(x) \qquad \text{for all } x \in A . \tag{4.3}$$

One can show that $\sigma : \mathbb{R} \to \text{Aut}(A)$ is completely determined by these two conditions and σ is called the *modular automorphism group* of φ.

One should however not be fooled by the lack of complexity above. We are working in a situation that is much simpler than the general case because

1. φ is a continuous functional, not an unbounded weight.
2. For every $z \in \mathbb{C} \setminus \mathbb{R}$ the map σ_z is defined on the whole of A and obviously continuous.
3. Because there exists a faithful trace τ on A, we can write φ in terms of such an operator δ.

4.2 Von Neumann algebras

An important class of C*-algebras is formed by the class of von Neumann algebras. If C*-algebras are considered to be the quantizations of locally compact Hausdorff spaces, von Neumann algebras are the non-commutative analogues of measure spaces.

Definition 4.8. *Let $\mathcal{H}$ be a Hilbert space and M a unital *-subalgebra of $B(\mathcal{H})$ that is closed with respect to the weak topology on $B(\mathcal{H})$. Then M is called a von Neumann algebra (on $\mathcal{H}$).*

Recall that the *weak* (also called, *weak operator*), *strong* and *strong** topology on $B(\mathcal{H})$ are determined by the following convergence properties.

Let $(T_i)_{i \in I}$ be a net in $B(\mathcal{H})$ and T an element in $B(H)$. Then

i) $(T_i)_{i \in I}$ converges weakly to $T \Leftrightarrow (\langle T_i \xi, \eta \rangle)_{i \in I} \to \langle T\xi, \eta \rangle$ for all $\xi, \eta \in \mathcal{H}$,

ii) $(T_i)_{i \in I}$ converges strongly to $T \Leftrightarrow (T_i \xi)_{i \in I} \to T\xi$ for all $\xi \in \mathcal{H}$,

iii) $(T_i)_{i \in I}$ converges strongly* to $T \Leftrightarrow (T_i)_{i \in I}$ converges strongly to T and $(T_i^*)_{i \in I}$ converges strongly to T^*.

Example 4.9. i) Consider a σ-finite measure space $(X, \mathcal{M}, \mu)$. The *-algebra of (equivalence classes of) essentially bounded measurable functions on X is denoted by $L^\infty(X)$, the space of (equivalence classes of) square integrable functions on X by $L^2(X)$. Let $\pi : L^\infty(X) \to B(L^2(X))$ denote the *-homomorphism that associates to every element in $L^\infty(X)$ the natural multiplication operator. Then $\pi(L^\infty(X))$ is a von Neumann algebra on $L^2(X)$ and π is an isometry.

ii) If $\mathcal{H}$ is a Hilbert space, $B(\mathcal{H})$ is a von Neumann algebra that is the weak closure of $B_0(\mathcal{H})$.

If $\mathcal{H}$ is a Hilbert space and $A \subseteq B(\mathcal{H})$, we define the set $A' \subseteq B(\mathcal{H})$ as

$$A' = \{ x \in B(\mathcal{H}) \mid \forall y \in A : xy = yx \},$$

the set A' is called the *commutant* of A in $B(\mathcal{H})$.

In a lot of cases, von Neumann algebras are obtained by taking the weak closure of a non-degenerate *-algebra of bounded operators on some Hilbert space. The bicommutant theorem asserts that this topological closure can be obtained by algebraic means through the use of commutants.

Theorem 4.10. *Consider a Hilbert space $\mathcal{H}$ and a non-degenerate *-subalgebra A of $B(\mathcal{H})$. Then the weak closure of A in $B(\mathcal{H})$ equals the bicommutant A''. In particular, if A is a von Neumann algebra on $\mathcal{H}$, we have that $A'' = A$.*

This theorem and the fact that any element in a C*-algebra can be written as a linear combination of unitary elements of this C*-algebra, leads to the following affiliation relation for von Neumann algebras . Let M be a von Neumann algebra acting on a Hilbert space $\mathcal{H}$. A densely defined closed linear operator T on $\mathcal{H}$ is said to be affiliated with M in the von Neumann algebraic sense if $u^*Tu = T$ for every unitary $u \in M'$.

Any von Neumann algebra is obviously a C*-algebra so it carries the norm topology, but in the next paragraph we will describe a topology that is far more relevant to von Neumann algebras.

Definition 4.11. *Let M be a von Neumann algebra acting on a Hilbert space $\mathcal{H}$ and ω a linear functional on M that, on the open unit ball of M, is continuous with respect to the weak operator topology. Then we call ω normal. The space of normal linear functionals on M is called the predual of M and is denoted by M_*. Note that M_* is a closed subspace of M^* and, as such, inherits the norm from M^*.*

The cone of positive linear functionals in the predual M_* is denoted by M_*^+. For $\xi, \eta \in \mathcal{H}$, we define $\omega_{\xi,\eta} \in B(\mathcal{H})_*$ by $\omega_{\xi,\eta}(x) = \langle x\xi, \eta \rangle$ for all $x \in B(\mathcal{H})$. The restriction of $\omega_{\xi,\eta}$ to M will also be denoted by $\omega_{\xi,\eta}$; this restriction obviously belongs to M_*. Even stronger, the linear span of $\{\, \omega_{\xi,\eta} \mid \xi, \eta \in \mathcal{H} \,\}$ is norm dense in M_*.

The predual induces one of the preferred topologies on M:

Definition 4.12. *Consider a von Neumann algebra M. The σ-weak topology on M is by definition the initial topology on M induced by the predual M_*.*

Thus, on bounded subsets of M the relative weak topology and relative σ-weak topology agree. As a Banach space, the von Neumann algebra M can be recovered from its predual in the following way. Let $x \in M$ and define $e_x \in (M_*)^*$ by $e_x(\omega) = \omega(x)$ for all $\omega \in M_*$. The mapping

$$M \to (M_*)^* : x \mapsto e_x$$

is an isometric isomorphism.

In the commutative setting, the predual can be easily described in more familiar terms.

Example 4.13. Let us return to example 4.9 i) and denote the space of (equivalence classes of) integrable functions on X by $L^1(X)$. Any element $f \in L^1(X)$ defines a linear functional $\omega_f \in \pi(L^\infty(X))_*$ by $\omega_f(g) = \int f \, g \, d\mu$ for all $g \in L^\infty(X)$. The mapping $L^1(X) \to \pi(L^\infty(X))_* : f \mapsto \omega_f$ is an isometric isomorphism.

In the theory of von Neumann algebras, *-homomorphisms have to satisfy an extra continuity property to be useful. Hence the following definition.

Definition 4.14. *Consider von Neumann algebras M, N and a unital *-homomorphism $\pi : M \to N$. We call π normal if π is σ-weakly continuous, i.e. if $\omega\pi \in M_*$ for every $\omega \in N_*$.*

Notice that π is normal if and only if π is weakly continuous on bounded subsets of M. If π is normal, then $\pi(M)$ is a von Neumann algebra.

There is also another preferred topology on any von Neumann algebra.

Definition 4.15. *If $\omega \in M_*$ we define semi-norms p_ω and p_ω^* by $p_\omega(x) = \omega(x^*x)^{\frac{1}{2}}$ and $p_\omega(x) = \omega(xx^*)^{\frac{1}{2}}$ for all $x \in M$. The σ-strong* topology is by definition the locally convex vector topology on M induced by the family of semi-norms $\{\, p_\omega, p_\omega^* \mid \omega \in M_* \,\}$.*

On bounded subsets of M the relative strong* topology and the relative σ-strong* agree. In most of the statements involving von Neumann algebras one can replace the σ-weak topology by the σ-strong* topology. For instance, if M, N are von Neumann algebras and $\pi : M \to N$ is a unital *-homomorphism, then π is normal $\Leftrightarrow$ π is σ-strongly*-continuous $\Leftrightarrow$ π is strongly*-continuous on bounded subsets of M. We have moreover for any $\omega \in M^*$ that ω belongs to M_* $\Leftrightarrow$ ω is σ-strongly* continuous. If K is a convex subset of M, then K is σ-weakly closed if and only if it is σ-strongly* closed.

Example 4.16. It is about time to introduce some well-known important operator algebras associated to a locally compact group G. Let μ be a left Haar measure G. First recall the definition of the convolution *-algebra $L^1(G)$. The product $\star$ and *-operation $.^\circ$ on $L^1(G)$ are defined as follows. If $f, g \in L^1(G)$, then

(1) $(f \star g)(t) = \int_G f(s) g(s^{-1}t) \, d\mu(s)$ for almost all $t \in G$,

(2) $f^\circ(t) = \delta(t)^{-1} \overline{f(t^{-1})}$ for $t \in G$ (here, δ is the modular function of G).

Remember that the integral converges for almost all $t \in G$, not necessarily for all $t \in G$. If f and g belong to $K(G)$, the integral is everywhere convergent and $f \star g$ and f° belong to $K(G)$. This convolution *-algebra is a Banach *-algebra for the L^1-norm.

We have a faithful *-representation from the convolution algebra $\lambda : L^1(G) \to B(L^2(G))$ such that

$$(\lambda(f) g)(t) = \int f(s) g(s^{-1}t) \, d\mu(s)$$

for all $f \in L^1(G)$, $g \in L^2(G)$ and $t \in G$.

The group von Neumann algebra $\mathcal{L}(G)$ of G is by definition the weak closure of $\lambda(L^1(G))$ in $B(L^2(G))$. The reduced group C*-algebra $C_r^*(G)$ is by definition the norm closure of $\lambda(L^1(G))$ in $B(L^2(G))$.

Finally we use Example 1.6 with $\mathcal{A} = L^1(G)$ to define the group C*-algebra $C^*(G)$ as the enveloping C*-algebra of $L^1(G)$.

The notion of a tensor product of von Neumann algebras is similar to that of the minimal tensor product of C*-algebras.

Definition 4.17. *Let M, N be two von Neumann algebras acting on Hilbert spaces $\mathcal{H}$, $\mathcal{K}$ respectively. The tensor product of the von Neumann algebras M and N is the von Neumann algebra on $\mathcal{H} \otimes \mathcal{K}$, denoted by $M \bar{\otimes} N$ and defined as the weak operator closure of $M \odot N$ in $B(\mathcal{H} \otimes \mathcal{K})$.*

The commutant of such a tensor product is easily described in terms of M and N by the neat (but difficult to prove) formula

$$(M \bar{\otimes} N)' = M' \bar{\otimes} N'.$$

Let $\omega \in M_*$, $\eta \in N_*$. Then $\omega \bar{\otimes} \eta$ is by definition the unique element in $(M \bar{\otimes} N)_*$ extending the algebraic tensor product $\omega \odot \eta$.

Given $x \in M \bar{\otimes} N$, the slice $(\omega \bar{\otimes} \iota)(x) \in N$ is defined so that $\eta\big((\omega \bar{\otimes} \iota)(x)\big) = (\omega \bar{\otimes} \eta)(x)$ for all $\eta \in N_*$. Similarly one defines the slice $(\iota \bar{\otimes} \eta)(x) \in M$.

Consider von Neumann algebras M_1, M_2, N_1, N_2 and normal unital *-homomorphisms $\pi_1 : M_1 \rightarrow N_1$ and $\pi_2 : M_2 \rightarrow N_2$. Then, there exists a unique normal unital *-homomorphism $\pi_1 \bar\otimes \pi_2 : M_1 \bar\otimes M_1 \rightarrow N_1 \bar\otimes N_2$ such that $(\pi_1 \otimes \pi_2)(x_1 \otimes x_2) = \pi_1(x_1) \otimes \pi_2(x_2)$ for all $x_1 \in M_1$ and $x_2 \in M_2$.

As mentioned in the previous section, the relevant notion of continuity and density for weights on von Neumann algebra are different for the ones used on C*-algebras.

Definition 4.18. *Consider a weight φ on a von Neumann algebra M. Then*

1. *We call φ semi-finite if $\mathcal{M}_\varphi^+$ is σ-strongly* dense in M^+.*

2. *We call φ normal if φ is lower semi-continuous with respect to the σ-strong* topology on M^+.*

There are different equivalent conditions for normality, for instance a weight φ on M is normal $\Leftrightarrow$ for every increasing net $(x_i)_{i \in I}$ in M^+ and every $x \in M^+$ for which $(x_i)_{i \in I} \rightarrow x$ in the σ-strong* topology, $(\varphi(x_i))_{i \in I} \rightarrow \varphi(x)$.

The expression 'normal semi-finite faithful weight' is in the literature also abbreviated to 'nsf weight'. We will almost exclusively work with nsf weights.

Let φ be a nsf weight on a von Neumann algebra M and $(\mathcal{H}_\varphi, \pi_\varphi, \Lambda_\varphi)$ is a GNS-construction φ. Then π_φ is an injective normal *-homomorphism and Λ_φ is closed for the σ-strong* topology on M and the norm topology on $\mathcal{H}_\varphi$. Propositions 4.5 and 4.6 remain true if one changes the definition of $\mathcal{F}$ and $\mathcal{G}$ to include only functionals in M_*^+.

4.3 One-parameter groups and their analytic extensions

Let us first formalize the notion of a one-parameter group of *-automorphisms on a *-algebra and the two relevant (to us) continuity properties.

Definition 4.19. *Consider a *-algebra A and $\sigma : \mathbb{R} \rightarrow Aut(A)$ a map from $\mathbb{R}$ into the set of *-automorphisms on A such that $\alpha_{s+t} = \alpha_s \alpha_t$ for all $s, t \in \mathbb{R}$. Then, we call α a one-parameter group of *-automorphisms on A.*

1. *If A is a C*-algebra, we say that α is norm continuous $\Leftrightarrow$ for every $a \in A$, the function $\mathbb{R} \rightarrow A : t \rightarrow \alpha_t(a)$ is continuous with respect to the norm topology on A.*

2. *If A is a von Neumann algebra, we say that α is strongly continuous $\Leftrightarrow$ for every $a \in A$, the function $\mathbb{R} \rightarrow A : t \rightarrow \alpha_t(a)$ is continuous with respect to the strong topology on A.*

Note that is it enough to check continuity in 0 to conclude continuity everywhere. Remember also that if A is a C*-algebra, then $\|\alpha_t(a)\| = \|a\|$ for all $t \in \mathbb{R}$ and $a \in A$. So if A is a von Neumann algebra, the strong continuity in the above definition is equivalent to the σ-strong* continuity.

Now fix a von Neumann algebra M acting on a Hilbert space $\mathcal{H}$ and a strongly continuous one-parameter group of *-automorphisms on M. In the rest of this section we explain the notion of the analytic extensions of σ that generalizes the discussion at the end of subsection 4.1.

We first need the right notion of analyticity but it turns out that a straight-forward generalization of the more familiar notion suffices. Therefore consider a Banach space E, O an open subset of $\mathbb{C}$ and a function $f : O \to E$.

(1) If $z_0 \in O$, we call f differentiable in $z_0 \Leftrightarrow$ the limit $\lim_{z \to z_0} \frac{f(z) - f(z_0)}{z - z_0}$ exists in E for the norm topology. If f is differentiable in z_0, we define $f'(z_0)$ as this limit.

(2) We call f analytic on O if f is differentiable in every point of O.

The analyticity of E-valued functions can be described in terms of analyticity of ordinary complex valued functions by the following remarkable result:

$$f \text{ is analytic on } O \;\Leftrightarrow\; \omega \circ f : O \to \mathbb{C} \text{ is analytic on } O \text{ for every } \omega \in E^*.$$
(4.4)

The $\Rightarrow$-implication is obvious (and $(\omega \circ f)'(z) = \omega(f'(z))$ for all $z \in O$) but the reverse implication needs the *Uniform Boundedness Principle* (see e.g. [Con]). If $E = M$, we can even replace E^* by the predual M_* in the above statement.

Using the equivalence above together with Hahn-Banach, one can easily trans-fer a number of properties that are known for analytic complex valued func-tions to the world of E-valued analytic functions (like infinite differentiability, Cauchy's Theorem, the Identity Theorem and Morera's Theorem; note that only the first and last of these two properties need the non-trivial implication of the equivalence above).

If $z \in \mathbb{C}$ we define the closed horizontal strip $\mathcal{S}(z)$ in the complex plane as

$$\mathcal{S}(z) = \{ \, y \in \mathbb{C} \mid 0 \le \operatorname{Im} y \le \operatorname{Im} z \text{ or } \operatorname{Im} z \le \operatorname{Im} y \le 0 \, \} \,.$$

Definition 4.20. *Consider $z \in \mathbb{C}$. Define the mapping $\alpha_z : D(\alpha_z) \subseteq M \to M$ such that the domain $D(\alpha_z)$ of α_z consists of all elements $x \in M$ for which there exists a function $f : \mathcal{S}(z) \to M$ satisfying*

1. *f is σ-strongly* continuous and norm bounded on $\mathcal{S}(z)$,*
2. *f is analytic on the interior $\mathcal{S}(z)^\circ$ of $\mathcal{S}(z)$,*
3. *$f(t) = \alpha_t(x)$ for all $t \in \mathbb{R}$.*

If $x \in D(\alpha_z)$, the function f described above is unique and we define $\alpha_z(x) = f(z)$. We call α_z the analytic extension of α in z.

The uniqueness of f can be seen by combining the Schwarz' Mirror principle, the Identity Theorem and Hahn-Banach.

Let $x \in M$. One calls x *analytic with respect to* α if $x \in D(\alpha_z)$ for all $z \in \mathbb{C}$, which is equivalent to the existence of an analytic function $f : \mathbb{C} \to M$ such that $f(t) = \alpha_t(x)$ for all $t \in \mathbb{R}$.

These analytic extensions satisfy the following basic properties for $y, z \in \mathbb{C}$.

1. $D(\alpha_z)$ is a subalgebra of M and $\alpha_z : D(\alpha_z) \to M$ is a homomorphism of algebras.

2. $D(\alpha_z)^* = D(\alpha_{\bar{z}})$ and $\alpha_z(x)^* = \alpha_{\bar{z}}(x^*)$ for all $x \in D(\alpha_z)$.

3. $\alpha_y \alpha_t = \alpha_t \alpha_y = \alpha_{y+t}$ for all $t \in \mathbb{R}$.

4. $\alpha_y \alpha_z \subseteq \alpha_{y+z}$ and $D(\alpha_y \alpha_z) = D(\alpha_{y+z}) \cap D(\alpha_z)$. If y and z lie on the same side of the real axis, then $\alpha_y \alpha_z = \alpha_{y+z}$.

5. If $z \in \mathbb{C}$ and $y \in S(z)$, then $D(\alpha_z) \subseteq D(\alpha_y)$.

6. α_z is injective, $\operatorname{Ran} \alpha_z = D(\alpha_{-z})$ and $(\alpha_z)^{-1} = \alpha_{-z}$.

7. α_z is closed for the σ-strong* topology. Recall that the notion of closedness is given in the appendix.

Some of the proofs of these properties rely on the Phragmen-Lindelöf Theorem (see [Rud, Thm. 12.8]):

Proposition 4.21 (Phragmen-Lindelöf). *Consider a function* $f : S(i) \to \mathbb{C}$ *so that*

1. *f is continuous and bounded on $S(i)$,*

2. *f is analytic on $S(i)^\circ$*

and define $M = \sup\left(\{\, |f(t)| \mid t \in \mathbb{R}\,\} \cup \{\, |f(t+i)| \mid t \in \mathbb{R}\,\}\right)$. *Then* $|f(z)| \le M$ *for all* $z \in S(i)$.

It is very easy to construct elements in the domain of α_z once we have a good notion of integrating M-valued functions because then we can 'smear' elements with respect to α to obtain elements in the domain of α_z.

Proposition 4.22. *Consider* $x \in M$ *and* $n \in \mathbb{N}$. *Define* $x(n) \in M$ *as*

$$x(n) = \frac{n}{\sqrt{\pi}} \int \exp(-n^2 t^2)\, \alpha_t(x)\, dt \ .$$

Then $x(n)$ *is analytic with respect to* α *and*

$$\alpha_z(x(n)) = \frac{n}{\sqrt{\pi}} \int \exp(-n^2(t-z)^2)\, \alpha_t(x)\, dt \ .$$

for all $z \in \mathbb{C}$. *If* $x \in D(\alpha_z)$, *then* $\alpha_z(x(n)) = (\alpha_z(x))(n)$.

The appendix discusses a notion of integration of vector-valued functions that will suffice for our purposes. The integrals above can be understood in the σ-weak sense, i.e. $\omega(x(n)) = \frac{n}{\sqrt{\pi}} \int \exp(-n^2 t^2)\, \omega(\alpha_t(x))\, dt$ for all

$\omega \in M_*$. These equalities also hold in the strong sense, i.e. $x(n)v = \frac{n}{\sqrt{\pi}} \int \exp(-n^2 t^2) \alpha_t(x) v \, dt$ for all $v \in \mathcal{H}$.

This proposition also implies that $\langle a(n) \mid a \in D(\alpha_z), n \in \mathbb{N} \rangle$ is a core for α_z (see the appendix for the notion of a core).

The definition of the analytic extensions (and the statement of its main properties) is given with respect to the σ-strong* topology. But we can everywhere replace the σ-strong* topology by the σ-weak topology and still obtain the same analytic extension.

The process of smearing elements with respect to a one-parameter group of *-automorphisms (or with respect to a finite number of commuting one-parameter groups of automorphisms via a multiple integral) is one of the most useful techniques to create well-behaved elements in the theory of locally compact quantum groups.

By definition, α_i is determined by the family of *-automorphisms $(\alpha_t)_{t \in \mathbb{R}}$. But also the converse is true. In the statement of the next proposition we could have used any number, different from 0, on the imaginary axis.

Proposition 4.23. *Consider two strongly continuous one-parameter groups of *-automorphisms α, β on M. Then, $\alpha = \beta \Leftrightarrow \alpha_i = \beta_i$.*

If we work with a C*-algebra A and a norm continuous one-parameter group α of *-automorphisms on A, then the analytic extension α_z is defined in the same way as above but one replaces the σ-strong* topology by the norm topology.

4.4 The KMS-properties of normal semi-finite faithful weights on von Neumann algebras

Tomita-Takesaki Theory

Consider a von Neumann algebra M and a nsf weight φ on M. The next theorem contains one of the most important results of the theory of von Neumann algebras and which follows from the celebrated *Tomita-Takesaki* theory. Since 2002 a modern and comprehensive account of this theory can be found in [Tak02a]. It is important to mention that there does not exist a C*-algebraic version of the theorem below.

Theorem 4.24. *There exists a unique strongly continuous one-parameter group σ of *-automorphisms on M so that $\varphi \sigma_t = \varphi$ for all $t \in \mathbb{R}$ and so that for all $x, y \in \mathcal{N}_\varphi \cap \mathcal{N}_\varphi^*$ there is a bounded continuous complex function f on the strip $\{z \in \mathbb{C} \mid 0 \leq \mathrm{Im}\, z \leq 1\}$, analytic on the interior of this strip, and satisfying*

$$f(t) = \varphi(\sigma_t(x)y) \qquad and \qquad f(t+i) = \varphi(y\sigma_t(x))$$

for all $t \in \mathbb{R}$. We call σ the modular automorphism group for φ and one uses the notation $\sigma^{\varphi} = \sigma$.

The properties in the statement of the above theorem are referred as the KMS-properties of the weight φ (KMS stands for *Kubo, Martin, Schwinger*). This theorem implies the next useful results for nsf weights.

Proposition 4.25. *i) Consider $a \in D(\sigma_{-i})$ and $x \in M_{\varphi}$. Then $a\,x$ and $x\,\sigma_{-i}(a)$ belong to M_{φ} and $\varphi(a\,x) = \varphi(x\,\sigma_{-i}(a))$.*

ii) Let $a \in D(\sigma_{\frac{i}{2}})$. Then $\varphi(a^*a) = \varphi(\sigma_{\frac{i}{2}}(a)\sigma_{\frac{i}{2}}(a)^*)$.

On the level of a GNS space $(\mathcal{H}_{\varphi}, \pi_{\varphi}, \Lambda_{\varphi})$ of φ, the following operators are of great importance. The notion of a positive operator can be found in Definition 7.3.

Definition 4.26. *i) There exists a unique injective positive operator ∇ on $\mathcal{H}_{\varphi}$ such that $\nabla^{it}\Lambda_{\varphi}(a) = \Lambda_{\varphi}(\sigma_t^{\varphi}(a))$ for all $a \in N_{\varphi}$. We call ∇ the modular operator of φ and use the notation $\nabla_{\varphi} = \nabla$.*

ii) There exists a unique anti-unitary operator J on $\mathcal{H}_{\varphi}$ such that $J\,\Lambda_{\varphi}(a) = \Lambda_{\varphi}(\sigma_{\frac{i}{2}}^{\varphi}(a)^*)$ for all $a \in N_{\varphi} \cap D(\sigma_{\frac{i}{2}}^{\varphi})$. We call J the modular conjugation of φ and use the notation $J_{\varphi} = J$.

The operator ∇_{φ} induces σ^{φ} in the GNS-space, J_{φ} induces an anti-*-isomorphism from $\pi_{\varphi}(A)$ to the commutant $\pi_{\varphi}(A)'$:

1. $\pi_{\varphi}(\sigma_t^{\varphi}(x)) = \nabla_{\varphi}^{it}\,\pi_{\varphi}(x)\,\nabla_{\varphi}^{-it}$ for all $t \in \mathbb{R}$, $x \in M$.
2. $J_{\varphi}\pi_{\varphi}(M)J_{\varphi} = \pi_{\varphi}(M)'$

Let us also mention the following result. If $x \in N_{\varphi}$ and $a \in D(\sigma_{\frac{i}{2}}^{\varphi})$, then $x\,a \in N_{\varphi}$ and
$$\Lambda_{\varphi}(x\,a) = J_{\varphi}\,\pi_{\varphi}(\sigma_{\frac{i}{2}}^{\varphi}(a))^*\,J_{\varphi}\,\Lambda_{\varphi}(x) \ .$$

It should be said that in reality the theory is build up the other way around. Starting from a nsf weight, one introduces the closure T of the map $\Lambda_{\varphi}(N_{\varphi} \cap N_{\varphi}^*) \mapsto \Lambda_{\varphi}(N_{\varphi} \cap N_{\varphi}^*) : \Lambda_{\varphi}(x) \rightarrow \Lambda_{\varphi}(x^*)$. Next, one defines the anti-unitary J and the injective positive self-adjoint operator ∇ by taking the polar decomposition $T = J\nabla^{\frac{1}{2}}$ of T. One uses ∇ to define the modular automorphism group σ of φ and proves all the properties mentioned above (including theorem 4.24) by using the theory of *left Hilbert algebras* and the associated Tomita-Takesaki theory.

In this case there is also another characterization of the analytic extensions σ_z^{φ} for $z \in \mathbb{C}$. Let $x \in M$.

1. If $x \in D(\sigma_z^{\varphi})$, then $\pi_{\varphi}(x)\,\nabla_{\varphi}^{iz} \subseteq \nabla_{\varphi}^{iz}\,\pi_{\varphi}(\sigma_z^{\varphi}(x))$.
2. If $x \in M$, then x belongs to $D(\sigma_z^{\varphi})$ if and only if there exists $y \in M$ such that $\pi_{\varphi}(x)\,\nabla_{\varphi}^{iz} \subseteq \nabla_{\varphi}^{iz}\,\pi_{\varphi}(y)$.

The Radon-Nikodym derivative

Consider a von Neumann algebra M on a Hilbert space $\mathcal{H}$ and a nsf weight φ on M with GNS-construction (H, π, Λ) and modular automorphism group σ. We will also consider an injective positive operator δ in $\mathcal{H}$ (see Definition 7.3) affiliated with M so that there exists a positive number $\lambda > 0$ satisfying $\sigma_t(\delta) = \lambda^t \delta$ for all $t \in \mathbb{R}$.

The fact that δ is affiliated to M is equivalent to the fact that $\delta^{it} \in M$ for all $t \in \mathbb{R}$.

It is natural to look for a precise definition of the weight that is formally equal to $\varphi(\delta^{\frac{1}{2}} \cdot \delta^{\frac{1}{2}})$. If $\lambda \neq 1$, the method of defining this weight in [PT] is not applicable anymore.

Instead we will work with a reverse GNS-construction. Define L as the left ideal of element $x \in M$ so that $x \delta^{\frac{1}{2}}$ is bounded. If $x \in L$, we define $x \cdot \delta^{\frac{1}{2}} \in M$ as the unique continuous linear extension of $x\delta^{\frac{1}{2}}$.

Now, define the subspace N_0 of M as

$$N_0 = \{\, x \in L \mid x \cdot \delta^{\frac{1}{2}} \in \mathcal{N}_\varphi \,\}$$

Then N_0 is a σ-strongly* dense left ideal of M and the mapping $N_0 \to H$: $x \mapsto \Lambda\big(x \cdot \delta^{\frac{1}{2}}\big)$ is closable with respect to the σ-strong* topology on M and the norm topology on H. We define Λ_δ to be the closure of this mapping and its domain by N. Then N is a σ-strongly* dense left ideal of M and $\Lambda_\delta(x\,y) = \pi(x)\, \Lambda_\delta(y)$ for all $x \in A$, $y \in N$.

Define the strongly continuous one-parameter group σ' of *-automorphisms on M by $\sigma'_t(x) = \delta^{it} \sigma_t(x)\, \delta^{-it}$ for $t \in \mathbb{R}$ and $x \in A$. Then,

Proposition 4.27. *There exists a unique nsf weight φ_δ on M so that $\mathcal{N}_{\varphi_\delta} = N$ and $\varphi_\delta(y^*x) = \langle \Lambda_\delta(x), \Lambda_\delta(y)\rangle$ for all $x, y \in \mathcal{N}_{\varphi_\delta}$. Furthermore, σ' is the modular automorphism group of φ_δ.*

Now, we stumble on another aspect of weight theory that works much better in the von Neumann algebraic approach than in the C*-algebraic approach; we have an analogue of the Radon-Nikodym theorem:

Theorem 4.28. *Consider another nsf weight ψ on M with modular automorphism groups σ'. Consider also a number $\lambda > 0$. Then the following statements are equivalent.*

1. *$\varphi\, \sigma'_t = \lambda^t\, \varphi$ for all $t \in \mathbb{R}$.*
2. *$\psi\, \sigma_t = \lambda^{-t}\, \psi$ for all $t \in \mathbb{R}$.*
3. *There exists an injective positive operator δ in $\mathcal{H}$ affiliated with M such that $\sigma_t(\delta) = \lambda^t\, \delta$ for $t \in \mathbb{R}$ and $\psi = \varphi_\delta$.*

If these conditions hold, we call δ the Radon-Nikodym derivative of ψ with respect to φ.

The proof of this result can be found in [Va01b]. In fact, this paper deals with the generalization of this theorem to the case were the modular automorphism groups merely commute. In the general case (where they do not necessarily commute), all the information about the relation between φ and ψ is encoded in the *Connes cocycle*. We will not go further into these matters because we do not need them in these notes.

5 The definition of a locally compact quantum group

At this stage we have gathered the necessary information from the general theory of von Neumann algebras to state the general definition of a locally compact quantum group and discuss its main consequences. Since we work most of the time in the framework of von Neumann algebras we will denote the tensor product between von Neumann algebras, normal *-homomorphism and functionals in the predual by $\otimes$ (and not by $\bar{\otimes}$). Proofs of all the results in this section can be found in [KuV00a] and [KuV03].

5.1 The definition and its basic consequences

Definition 5.1. *Consider a von Neumann algebra M and a unital normal *-homomorphism $\Delta : M \to M \otimes M$ such that*

(a) $(\Delta \otimes \iota)\Delta = (\iota \otimes \Delta)\Delta$

(b) There exists two nsf weights φ, ψ on M such that

   1. $\varphi\big((\omega \otimes \iota)\Delta(x)\big) = \varphi(x)\,\omega(1)$ *for all $x \in \mathcal{M}_\varphi^+$ and $\omega \in M_*^+$,*

   2. $\psi\big((\iota \otimes \omega)\Delta(x)\big) = \psi(x)\,\omega(1)$ *for all $x \in \mathcal{M}_\psi^+$ and $\omega \in M_*^+$.*

Then (M, Δ) is called a locally compact quantum group (in the von Neumann algebraic setting).

We call Δ the *comultiplication* of (M, Δ). A weight φ as describe above is called a *left Haar weight* on (M, Δ), property (1) is called the left invariance of φ. A weight ψ as described above is called a *right Haar weight* on (M, Δ), property (2) is called the right invariance of ψ. It can be proven

   1. $\varphi\big((\omega \otimes \iota)\Delta(x)\big) = \varphi(x)\,\omega(1)$

   2. $\psi\big((\iota \otimes \omega)\Delta(x)\big) = \psi(x)\,\omega(1)$

for all $x \in M^+$ and $\omega \in M_*^+$. Although this is a nice generalization of the invariance properties, it is not very important to develop most of the basic theory. It is however vital in developing certain applications of quantum groups.

It might seem strange to call (M, Δ) a **locally compact** quantum group because von Neumann algebras are generalizations of measure spaces. We will however show later on that there is a natural C*-algebra sitting inside M.

For the rest of this section we fix a locally compact quantum group (M, Δ) and a left Haar weight φ on M. At this point we do not fix a right Haar weight because we will shortly produce a natural one. As before, the order of the statements of results does not agree with the chronological order the theory is built up.

Left invariant weights (but also right ones) are unique up to a constant:

Theorem 5.2. *Consider a semi-finite normal weight η on M. If $\eta\big((\omega \otimes \iota)\Delta(x)\big) = \eta(x)\,\omega(1)$ for all $x \in \mathcal{M}_\eta^+$ and $\omega \in M_*^+$, there exists $\lambda \geq 0$ such that $\eta = \lambda\,\varphi$.*

Note that we do not assume faithfulness of the weight η involved.

Remark 5.3. Let us take a look at the classical situation of a locally compact σ-compact group G. Because we want to work with von Neumann algebras, we use (with some abuse of language) the von Neumann algebra $L^\infty(G)$ of (equivalence classes) of essentially bounded measurable functions on G (instead of $C_0(G)$) and use the natural identification $L^\infty(G) \otimes L^\infty(G) = L^\infty(G \times G)$ to define the normal *-homomorphism $\Delta_G : L^\infty(G) \to L^\infty(G) \times L^\infty(G)$ by $\Delta_G(f)(s,t) = f(st)$ for $f \in L^\infty(G)$ and $s, t \in G$. Thus, $(L^\infty(G), \Delta_G)$ is a locally compact quantum group.

Take a left Haar measure μ on G and define the left Haar weight φ_G on $L^\infty(G)$ by $\varphi_G(f) = \int_G f \, d\mu$ for all $f \in L^\infty(G)^+$. In this case $\mathcal{N}_{\varphi_G} = L^2(G) \cap L^\infty(G)$ and we define a map $\Lambda_G : \mathcal{N}_{\varphi_G} \to L^2(G)$ by $\Lambda_G(f) = f$ for all $f \in L^2(G) \cap L^\infty(G)$. Thus, $(L^2(G), \pi_G, \Lambda_G)$ is a GNS-construction for φ_G.

In order to further develop the theory of quantum groups, it is important to introduce the antipode of the quantum group as a quantum analogue of the inverse operation on a group. As explained in section 2 the classical way of defining the antipode is not possible so we have to find an alternative way to do so.

Let S_G be the antipode on $L^\infty(G)$, i.e. $S_G(f)(s) = f(s^{-1})$ for $f \in L^\infty(G)$ and $s \in G$. Consider the linear map $\iota \odot \varphi_G : K(G) \odot K(G) \subseteq K(G \times G) \to K(G)$. If $f \in K(G) \odot K(G)$, it is clear that $(\iota \odot \varphi_G)(f)(s) = \int f(s,t) \, d\mu(t)$ for all $s \in G$. Therefore, let us denote by $\iota \otimes \varphi_G$ the extension of $\iota \odot \varphi_G$ to $K(G \times G)$ that integrates out the second variable.

Choose $g, h \in K(G)$. The left invariance of μ implies for $s \in G$,

$$
\begin{aligned}
S_G\big((\iota \otimes \varphi_G)(\Delta_G(f)(1 \otimes g))\big)(s) &= (\iota \otimes \varphi_G)(\Delta_G(f)(1 \otimes g))(s^{-1}) \\
&= \int_G (\Delta_G(f)(1 \otimes g))(s^{-1}, t) \, d\mu(t) = \int_G f(s^{-1}t)g(t) \, d\mu(t) \\
&= \int_G f(t)g(st) \, d\mu(t) = (\iota \otimes \varphi_G)((1 \otimes f)\Delta_G(g))(s) \, .
\end{aligned}
$$

Thus, $S_G\big((\iota \otimes \varphi_G)(\Delta_G(f)(1 \otimes g))\big) = (\iota \otimes \varphi_G)((1 \otimes f)\Delta_G(g))$ which provides us with a formula for the antipode S_G that only uses φ_G and Δ_G.

Let us now return to our general quantum group (M, Δ). The above formula in the classical case suggests a definition for the antipode in the general case but necessitates the definition of $\iota \otimes \varphi$ as an extension of $\iota \odot \varphi : M \odot M_\varphi \to M$ in our general framework. First define the set

$$\mathcal{M}^+_{\iota \otimes \varphi} = \{\, x \in (M \otimes M)^+ \mid \forall \omega \in M^*_+ : (\omega \otimes \iota)(x) \in \mathcal{M}^+_\varphi \,\}.$$

Then $\mathcal{M}^+_{\iota \otimes \varphi}$ is a hereditary cone in $(M \otimes M)^+$ and one can show (and it is not very difficult) the existence of a unique map $\iota \otimes \varphi : \mathcal{M}^+_{\iota \otimes \varphi} \to M^+$ such that

$$\omega\big((\iota \otimes \varphi)(x)\big) = \varphi\big((\omega \otimes \iota)(x)\big)$$

for all $x \in \mathcal{M}^+_{\iota \otimes \varphi}$ and $\omega \in M^+_*$. This map is linear.

Define the *-subalgebra $\mathcal{M}_{\iota \otimes \varphi}$ of $M \otimes M$ as the linear span of $\mathcal{M}^+_{\iota \otimes \varphi}$. Since $\mathcal{M}^+_{\iota \otimes \varphi}$ is a hereditary cone, $\mathcal{M}^+_{\iota \otimes \varphi} = \mathcal{M}_{\iota \otimes \varphi} \cap (M \otimes M)^+$. There exists a unique linear map $F : \mathcal{M}_{\iota \otimes \varphi} \to M$ so that $(\iota \otimes \varphi)(a) = F(a)$ for all $a \in \mathcal{M}^+_{\iota \otimes \varphi}$. We set $(\iota \otimes \varphi)(x) = F(x)$ for all $x \in \mathcal{M}_{\iota \otimes \varphi}$. Notice that $M \odot \mathcal{M}_\varphi \subseteq \mathcal{M}_{\iota \otimes \varphi}$ and $(\iota \otimes \varphi)(x \otimes y) = x \varphi(y)$ for all $x \in M$ and $y \in \mathcal{M}_\varphi$.

One also defines the left ideal $\mathcal{N}_{\iota \otimes \varphi}$ in $M \otimes M$ as the set of all elements $x \in M \otimes M$ for which $x^*x \in \mathcal{M}^+_{\iota \otimes \varphi}$. As for weights, $\mathcal{M}_{\iota \otimes \varphi} = \mathcal{N}^*_{\iota \otimes \varphi} \mathcal{N}^*_{\iota \otimes \varphi}$ (linear span!). Note that $M \odot \mathcal{N}_\varphi \subseteq \mathcal{N}_{\iota \otimes \varphi}$.

The above discussion is valid for any normal weight on M.

If $a, b \in \mathcal{N}_\varphi$, the left invariance of φ implies immediately that $\Delta(a)$ and $\Delta(b)$ belong to $\mathcal{N}_{\iota \otimes \varphi}$ and thus, that $(1 \otimes b^*)\Delta(a)$ and $\Delta(b)^*(1 \otimes a)$ belong to $\mathcal{M}_{\iota \otimes \varphi}$ so that we can apply $\iota \otimes \varphi$ to these elements. Therefore the following vital theorem in the theory of quantum groups makes sense.

Theorem 5.4. *There exists a unique σ-strongly* closed, linear operator $S : D(S) \subseteq M \to M$ such that the linear space*

$$\langle\, (\iota \otimes \varphi)(\Delta(b^*)(1 \otimes a)) \mid a, b \in \mathcal{N}_\varphi \,\rangle$$

is a core for S with respect to the σ-strong* topology and

$$S\big((\iota \otimes \varphi)(\Delta(b^*)(1 \otimes a)) \big) = (\iota \otimes \varphi)((1 \otimes b^*)\Delta(a))$$

for all $a, b \in \mathcal{N}_\varphi$. We call S the antipode of (M, Δ). Moreover, the domain and image of S are σ-strongly* dense in M.

The linear operator S is unbounded in general. The domain $D(S)$ is a subalgebra of M and S is an injective algebra anti-homomorphism, i.e. $S(xy) = S(y)\,S(x)$ for all $x, y \in D(S)$. If $x \in D(S)$, then $S(x)^* \in D(S)$ and $S(S(x)^*)^* = x$. Consequently, $D(S)^* = D(S^{-1})$ and $S(x)^* = S^{-1}(x^*)$ for all $x \in D(S)$. It may happen that $S \neq S^{-1}$, as is the case in the example of $SU_q(2)$ discussed in section 3.

Just as for densely defined, closed, linear operators in Hilbert spaces, we can produce a polar decomposition of the antipode (this idea is due to Kirchberg, see [Kir]). As a matter of fact, the polar decomposition of S is obtained through the polar decomposition of the relevant Hilbert space operator.

Proposition 5.5. *There exists a unique $*$-anti-automorphism $R : M \to M$ and a unique strongly continuous one parameter group of $*$-automorphisms $\tau : \mathbb{R} \to Aut(M)$ so that*

$$S = R\tau_{-\frac{i}{2}}, \qquad R^2 = \iota \qquad and \qquad \tau_t R = R\tau_t \ \ for \ all \ t \in \mathbb{R}.$$

We call R the unitary antipode and τ the scaling group of (M, Δ). Note that $S^2 = \tau_{-i}$.

Let us denote the modular automorphism group of φ by σ. Then the following important commutation relations involving the comultiplication Δ and τ, σ and R hold.

Proposition 5.6. *If $t \in \mathbb{R}$,*

1. $\Delta\tau_t = (\tau_t \otimes \tau_t)\Delta$
2. $\Delta\sigma_t = (\tau_t \otimes \sigma_t)\Delta$
3. $\Delta R = \chi(R \otimes R)\Delta$.

Here $\chi : M \otimes M \to M \otimes M$ is the flip $*$-automorphism. The last equality of this proposition justifies the following definition.

Definition 5.7. *We define the right Haar weight ψ on (M, Δ) by $\psi = \varphi R$.*

If one looks at this definition one might wonder why we needed the existence of the right Haar weight as a condition in the definition of a locally compact quantum group. In reality the existence of the right Haar weight is used in the proof of the existence of the antipode and its polar decomposition (but also in the proof of the unitarity of W introduced in Definition 5.12).

Denote the modular automorphism group of ψ by σ'. Thus, $\sigma'_t = R\sigma_{-t}R$ for all $t \in \mathbb{R}$. All the one-parameter groups σ, σ' and τ mutually commute, e.g. $\sigma_s \tau_t = \tau_t \sigma_s$ for all $s, t \in \mathbb{R}$. Let us extend the list of commutation relations.

Proposition 5.8. *If $t \in \mathbb{R}$, then*

1. $\Delta\sigma'_t = (\sigma'_t \otimes \tau_{-t})\Delta$.
2. $\Delta\tau_t = (\sigma_t \otimes \sigma'_{-t})\Delta$

By definition of a modular automorphism, $\varphi\sigma_t = \varphi$ and $\psi\sigma'_t = \psi$ for all $t \in \mathbb{R}$. The next result deals with the remaining invariance properties.

Proposition 5.9. *There exists a unique number* $\nu > 0$ *so that for all* $t \in \mathbb{R}$ *and* $x \in M^+$,

$$\varphi(\tau_t(x)) = \nu^{-t}\,\varphi(x) \qquad\qquad \varphi(\sigma'_t(x)) = \nu^t\,\varphi(x)$$
$$\psi(\tau_t(x)) = \nu^{-t}\,\psi(x) \qquad\qquad \psi(\sigma_t(x)) = \nu^{-t}\,\psi(x)\ .$$

We call ν *de scaling constant of* (M, Δ).

It has taken a long time, but recently the examples of the $az + b$ and $ax + b$ quantum group (see [WZ] and [VD01]) have shown that it can happen that $\nu \neq 1$ (but not in the case of compact and discrete quantum groups).

In the classical case, the modular function connects the left and right Haar measure. Also this result has an analogue in the quantum world. Let us denote the Hilbert space on which M acts by $\mathcal{H}$.

Proposition 5.10. *There exists a unique injective positive operator* δ *in* $\mathcal{H}$ *such that* δ *is affiliated to* M *and*

1. $\sigma_t(\delta) = \nu^t\,\delta$ *for all* $t \in \mathbb{R}$,
2. $\psi = \varphi_\delta$.

We call δ *the modular element of* (M, Δ).

Recall that the definition of ψ_δ is given in Proposition 4.27; formally, $\psi_\delta = \psi(\delta^{\frac{1}{2}}\,.\,\delta^{\frac{1}{2}})$. In the classical case, the modular function is a group homomorphism. Also this property has its generalization.

Proposition 5.11. *If* $t \in \mathbb{R}$, *then* $\Delta(\delta^{it}) = \delta^{it} \otimes \delta^{it}$.

In the last part of this section we introduce the multiplicative unitary of the quantum group. Although we introduce it at the end of this section that explains the basic theory derived from the definition, the multiplicative unitary is in reality constructed in the beginning of the build up of the theory.

In order to define the multiplicative unitary we need the tensor product weight $\varphi \otimes \varphi$ on $M \otimes M$. Set $\mathcal{F} = \{\, \omega \in M_*^+ \mid \forall x \in M : \omega(x) \le \varphi(x)\,\}$. We define the normal weight $\varphi \otimes \varphi$ on $M \otimes M$ by

$$(\varphi \otimes \varphi)(x) = \sup_{\omega,\eta \in \mathcal{F}}\ (\omega \otimes \eta)(x)$$

for all $x \in (M \otimes M)^+$. It follows from the von Neumann algebraic versions of Propositions 4.5 and 4.6 that this is indeed a weight, that $\mathcal{M}_\varphi \odot \mathcal{M}_\varphi \subseteq \mathcal{M}_{\varphi \otimes \varphi}$ and that $(\varphi \otimes \varphi)(a \otimes b) = \varphi(a)\,\varphi(b)$ for all $a, b \in \mathcal{M}_\varphi$. One can show hat $\varphi \otimes \varphi$ is a nsf weight on $M \otimes M$.

Take a GNS-construction $(\mathcal{H}_\varphi, \pi_\varphi, \Lambda_\varphi)$ for φ. There exists a natural GNS-construction for $\varphi \otimes \varphi$ that we describe in the following. One can show that the linear map

$$\Lambda_\varphi \odot \Lambda_\varphi : \mathcal{N}_\varphi \odot \mathcal{N}_\varphi \to \mathcal{H}_\varphi \otimes \mathcal{H}_\varphi$$

is closable with respect to the σ-strong* topology on $M \otimes M$ and the norm topology on $\mathcal{H}_\varphi \otimes \mathcal{H}_\varphi$. Denote its closure with respect to this topologies by $\Lambda_\varphi \otimes \Lambda_\varphi$. Then, $(\mathcal{H}_\varphi \otimes \mathcal{H}_\varphi, \pi_\varphi \otimes \pi_\varphi, \Lambda_\varphi \otimes \Lambda_\varphi)$ turns out to be a GNS-construction for $\varphi \otimes \varphi$. This tensor product construction can be performed for any two nsf weights on any two von Neumann algebras.

The left invariance of φ implies that for $a, b, c, d \in \mathcal{N}_\varphi$, the elements $\Delta(b)(a \otimes 1)$ and $\Delta(d)(c \otimes 1)$ belong to $D(\mathcal{N}_{\varphi \otimes \varphi})$ and

$$\langle (\Lambda_\varphi \otimes \Lambda_\varphi)(\Delta(b)(a \otimes 1)), (\Lambda_\varphi \otimes \Lambda_\varphi)(\Delta(d)(c \otimes 1)) \rangle$$
$$= \langle \Lambda_\varphi(a) \otimes \Lambda_\varphi(b), \Lambda_\varphi(c) \otimes \Lambda_\varphi(d) \rangle .$$

This justifies the following definition

Definition 5.12. *There exists a unique bounded linear operator W on $\mathcal{H}_\varphi \otimes \mathcal{H}_\varphi$ such that*

$$W^*(\Lambda_\varphi(a) \otimes \Lambda_\varphi(b)) = (\Lambda_\varphi \otimes \Lambda_\varphi)(\Delta(b)(a \otimes 1)) .$$

for all $a, b \in \mathcal{N}_\varphi$. The operator W^* is isometric.

The first important step in the build up of the theory is the proof of the next proposition (that requires the existence of a right Haar weight!)

Proposition 5.13. *The operator W is a unitary operator on $\mathcal{H}_\varphi \otimes \mathcal{H}_\varphi$.*

Remark 5.14. Let (A, Δ) be a compact quantum group so that A is a unital C*-subalgebra of $B(\mathcal{H})$ for some Hilbert space $\mathcal{H}$. Let h be the Haar state of (A, Δ) with cyclic GNS-construction $(\mathcal{H}_h, \pi_h, \xi_h)$.
Define M as the weak closure of $\pi_h(A)$ in $B(\mathcal{H}_h)$, so M is a von Neumann algebra acting on $\mathcal{H}_h$. Let $V \in M(A \otimes B_0(\mathcal{H}_h))$ be the left regular corepresentation of (A, Δ) introduced in Proposition 3.16 and consider the unitary $(\pi_h \otimes \iota)(V) \in B(\mathcal{H}_h \otimes \mathcal{H}_h)$. The remarks after Proposition 3.16 imply that

$$(\pi_h \otimes \pi_h)\Delta(a) = (\pi_h \otimes \iota)(V)^*(1 \otimes \pi_h(a))(\pi_h \otimes \iota)(V)$$

for all $a \in A$. Thus, if we define a normal *-homomorphism $\Delta : M \to M \otimes M$ by

$$\Delta(x) = (\pi_h \otimes \iota)(V)^*(1 \otimes x)(\pi_h \otimes \iota)(V)$$

for all $x \in M$, we get $\Delta \pi_h = (\pi_h \otimes \pi_h)\Delta$.
Define the state $\varphi \in M_*$ as $\varphi = \omega_{\xi_h, \xi_h}$, implying that $\varphi \pi_h = h$. Then, (M, Δ) is a locally compact quantum group in the sense of Definition 5.1 and φ is a left and right Haar weight on (M, Δ) (the proof of the faithfulness of φ is non-trivial).
If we define a GNS-construction $(\mathcal{H}_h, \iota, \Lambda_\varphi)$ for φ by $\Lambda_\varphi(x) = x \xi_h$ for all $x \in M$, we see that $(\pi_h \otimes \iota)(V) = W$ in this case.

5.2 The dual quantum group

In this subsection we still work with the locally compact quantum group (M, Δ) and the notations of the previous section but we assume that the GNS-construction $(\mathcal{H}_\varphi, \pi_\varphi, \Lambda_\varphi)$ is chosen such that $\pi_\varphi = \iota$ (if this is not the case, we can always work with $\pi_\varphi(M)$ instead of M). The unitary operator introduced in Definition 5.12 plays a central role in the theory and is called the *multiplicative unitary* of (M, Δ). We say that it is 'multiplicative' since W satisfies the Pentagonal equation

$$W_{12}\, W_{13}\, W_{23} = W_{23}\, W_{12} \ ,$$

which follows from the coassociativity of Δ. Here we use the *leg-numbering* notation explained in the appendix.

This unitary contains all the information of the quantum group (M, Δ). To be more precise,

1. M is the σ-strong* closure of the algebra $\{\, (\iota \otimes \omega)(W) \mid \omega \in B(\mathcal{H}_\varphi)_* \,\}$,
2. $\Delta(x) = W^*(1 \otimes x)W$ for all $x \in M$.

Exercise 5.15. Use the Pentagonal equation to verify that the vector space $\{\, (\iota \otimes \omega)(W) \mid \omega \in B(\mathcal{H}_\varphi)_* \,\}$ is indeed an algebra. Also prove the last claim involving the comultiplication.

If $a, b \in \mathcal{N}_\varphi$, it is easy to check (do so!) that

$$(\iota \otimes \omega_{\Lambda_\varphi(a), \Lambda_\varphi(b)})(W^*) = (\iota \otimes \varphi)((1 \otimes b^*)\Delta(a))$$

and

$$(\iota \otimes \omega_{\Lambda_\varphi(a), \Lambda_\varphi(b)})(W) = (\iota \otimes \varphi)(\Delta(b^*)(1 \otimes a)) \ .$$

Comparing this with Theorem 5.4 one sees that this implies that for all $\omega \in B(\mathcal{H}_\varphi)_*$ the element $(\iota \otimes \omega)(W)$ belongs to $D(S)$ and $S((\iota \otimes \omega)(W)) = (\iota \otimes \omega)(W^*)$. Prove this.

Remark 5.16. The multiplicative unitary W is also used to define the dual quantum group. To motivate this definition we look first at the classical case. Thus, we suppose that $M = L^\infty(G)$ where G is a locally compact σ-compact group with left Haar measure μ. Recall the definition of the group von Neumann algebra $\mathcal{L}(G)$ of G in Example 4.16.

In this case Definition 5.12 implies that $W_G := W \in B(L^2(G \times G))$ is given by

$$(W_G^* f)(s, t) = f(s, st) \qquad \text{and} \qquad (W_G f)(s, t) = f(s, s^{-1}t)$$

for all $f \in L^2(G \times G)$, $s, t \in G$. Note that in this case the unitarity of W_G follows easily.

If $f_1, f_2 \in L^2(G)$, we have for $g \in L^2(G)$ and $s \in G$,

$$\big((\omega_{f_1,f_2} \otimes \iota)(W_G) g\big)(s) = \int \overline{f_2(s)} f_1(s) g(s^{-1}t) \, d\mu(t) = (\lambda(\bar{f}_2 f_1) g)(s), \quad (5.1)$$

implying that $\{ (\omega \otimes \iota)(W_G) \mid \omega \in B(L^2(G))_* \} = \lambda(L^1(G))$.

This discussion motivates the use of the von Neumann algebra in the next theorem. In this theorem Σ denotes the flip operator on $\mathcal{H}_\varphi \otimes \mathcal{H}_\varphi$.

Theorem 5.17. *Define $\hat{M}$ as the σ-strong* closure of the subalgebra $\{ (\omega \otimes \iota)(W) \mid \omega \in B(\mathcal{H}_\varphi)_* \}$ in $B(\mathcal{H}_\varphi)$. Then $\hat{M}$ is a von Neumann subalgebra of $B(\mathcal{H}_\varphi)$ and there exists a unique normal, injective *-homomorphism $\hat{\Delta} : \hat{M} \to \hat{M} \otimes \hat{M}$ so that $\hat{\Delta}(x) = \Sigma W(x \otimes 1)W^* \Sigma$ for all $x \in \hat{M}$. Then, $(\hat{M}, \hat{\Delta})$ is a locally compact quantum group that is called the Pontryagin dual of (M, Δ).*

There are authors that leave the flip Σ out of the definition of $\hat{\Delta}$. We leave it in so that the dual weight defined in proposition 5.22 is a left and not a right Haar weight on $(\hat{M}, \hat{\Delta})$ is.

Proposition 5.18. *The multiplicative unitary W belongs to $M \otimes \hat{M}$ and*

$$(\Delta \otimes \iota)(W) = W_{13}W_{23} \qquad and \qquad (\iota \otimes \hat{\Delta})(W) = W_{13}W_{12} \ .$$

The two equalities are direct consequences of the Pentagonal equation. In the rest of this section we identify the basic structural elements of $(\hat{M}, \hat{\Delta})$.

Recall that the predual M_* is the 'L^1-space' of M. It is a Banach space and we can turn it into a Banach algebra by the following (usual) product $\omega \eta = (\omega \otimes \eta)\Delta$ for all $\omega, \eta \in M_*$. The co-associativity of the comultiplication implies the associativity of this product on M_*.

Define the linear map $\lambda : M_* \to \hat{M} : \omega \to (\omega \otimes \iota)(W)$. One easily checks that λ is an injective algebra homomorphism.

Lemma 5.19. *Consider $x \in \mathcal{N}_\varphi$ and $\omega \in M_*$. Then $(\omega \otimes \iota)\Delta(x) \in \mathcal{N}_\varphi$ and*

$$(\omega \otimes \iota)(W^*) \, \Lambda_\varphi(x) = \Lambda_\varphi\big((\omega \otimes \iota)\Delta(x)\big) \ .$$

Exercise 5.20. Prove this lemma. If $v \in \mathcal{H}_\varphi$, we define the bounded linear mapping $\theta_v : \mathbb{C} \to \mathcal{H}_\varphi : \lambda \mapsto \lambda v$. Then $\omega_{v,w}(x) = \theta_w^* x \, \theta_v$ for all $x \in B(\mathcal{H}_\varphi)$ en $v, w \in \mathcal{H}_\varphi$. Check

$$(\omega_{v,w} \otimes \iota)(y)^* \, (\omega_{v,w} \otimes \iota)(y) \leq \|w\|^2 \, (\omega_{v,v} \otimes \iota)(y^* y)$$

for all $y \in B(\mathcal{H}_\varphi \otimes \mathcal{H}_\varphi)$. Check that if $a \in \mathcal{N}_\varphi$, the element $(\omega_{v,w} \otimes \iota)\Delta(a)$ belongs to $\mathcal{N}_\varphi$. At the same time, estimate the norm $\|\Lambda_\varphi\big((\omega_{v,w} \otimes \iota)\Delta(a)\big)\|$. Check Eq. (5.19) if $\omega = \omega_{\Lambda_\varphi(c), \Lambda_\varphi(d)}$, where $c, d \in \mathcal{N}_\varphi$. Extend it to any element of M_* by appealing to the closedness of Λ_φ.

In the next part we construct the left Haar weight on $(\hat{M}, \hat{\Delta})$. Define

$$\mathcal{I} = \{ \omega \in M_* \mid \exists v \in \mathcal{H}_\varphi, \forall x \in \mathcal{N}_\varphi : \omega(x^*) = \langle v, \Lambda_\varphi(x) \rangle \} \ .$$

We define the linear map $\xi : \mathcal{I} \to \mathcal{H}_\varphi$ as follows. For $\omega \in \mathcal{I}$, the vector v described above is unique and we set $\xi(\omega) := v$.

Exercise 5.21. Prove the following facts.

1. The mapping $\mathcal{I} \to \mathcal{H}_\varphi : \omega \mapsto \xi(\omega)$ is linear and closed with respect to the norm topologies on M_* and $\mathcal{H}_\varphi$.
2. If $\omega \in \mathcal{I}$ and $\eta \in M_*$, then $\eta\omega \in \mathcal{I}$. What is $\xi(\eta\omega)$?
3. Consider $a \in \mathcal{N}_\varphi$ and $b \in \mathcal{N}_\varphi \cap D(\sigma_{\frac{i}{2}})$. Then $\omega_{\Lambda_\varphi(a),\Lambda_\varphi(b)} \in \mathcal{I}$. What is $\xi(\omega_{\Lambda_\varphi(a),\Lambda_\varphi(b)})$? Recall the results of subsection 4.4.

Notice that the last property implies that $\mathcal{I}$ is dense in M_* and $\xi(\mathcal{I})$ is dense in $\mathcal{H}_\varphi$ (both for the norm topology). So we see that $(\mathcal{H}_\varphi, \lambda, \xi)$ satisfies properties similar to that of a GNS-construction of a weight but remember that M_* is not a C*-algebra. However, we have the following

Proposition 5.22. *The linear map* $\lambda(\mathcal{I}) \subseteq \hat{M} \to \mathcal{H}_\varphi : \lambda(\omega) \mapsto \xi(\omega)$ *is closable with respect to the* σ*-strong* topology on* $\hat{M}$ *and the norm topology on* $\mathcal{H}_\varphi$. *Denote the closure of this map by* $\hat{\Lambda}_\varphi : D(\hat{\Lambda}_\varphi) \subseteq \hat{M} \to \mathcal{H}_\varphi$. *There exists a unique nsf weight* $\hat{\varphi}$ *on* $\hat{M}$ *such that* $(\mathcal{H}_\varphi, \iota, \hat{\Lambda}_\varphi)$ *is a GNS-construction for* $\hat{\varphi}$. *We call* $\hat{\varphi}$ *the dual weight of* φ.

We will denote the modular automorphism group of $\hat{\varphi}$ by $\hat{\sigma}$. At the level of $\lambda(\mathcal{I})$ we can easily write down a formula for $\hat{\sigma}$: if $t \in \mathbb{R}$ and $\omega \in M_*$, then $\hat{\sigma}_t(\lambda(\omega)) = \lambda(\omega')$ where $\omega' \in M_*$ is defined by $\omega'(x) = \omega(\delta^{-it}\tau_{-t}(x))$ for all $x \in M$ and $t \in \mathbb{R}$.

Proposition 5.23. *The nsf weight* $\hat{\varphi}$ *is a left Haar weight of* $(\hat{M}, \hat{\Delta})$.

The unitary antipode of $(\hat{M}, \hat{\Delta})$ is denoted by $\hat{R}$, the scaling group of $(\hat{M}, \hat{\Delta})$ by $\hat{\tau}$. These operators are related to R and τ as follows.

1. $\hat{R}(\lambda(\omega)) = \lambda(\omega R)$,
2. $\hat{\tau}_t(\lambda(\omega)) = \lambda(\omega\tau_{-t})$ for all $t \in \mathbb{R}$,

where $\omega \in M_*$. Moreover, ν^{-1} is the scaling constant of $(\hat{M}, \hat{\Delta})$. The minus sign in (2) is a direct consequence of the presence of the flip operator in the definition of $\hat{\Delta}$.

Exercise 5.24. Let us calculate the multiplicative unitary of $(\hat{M}, \hat{\Delta})$:

1. Consider $\omega \in \mathcal{I}$ and $x \in M$. Show that $\omega(\,.\,x) \in \mathcal{I}$ and calculate $\xi(\omega(\,.\,x))$.

2. Let $\omega \in \mathcal{I}$ and $\mu \in M_*$. Calculate $(\mu \otimes \iota)\hat{\Delta}(\lambda(\omega))$ and conclude from (1) that $(\mu \otimes \iota)\hat{\Delta}(\lambda(\omega))$ belongs to $\lambda(\mathcal{I})$. Write down $\hat{\Lambda}_\varphi((\mu \otimes \iota)\hat{\Delta}(\lambda(\omega)))$.

3. Convert these equalities into equalities on the level of $\mathcal{N}_{\hat\varphi}$ and $\mu \in \hat{M}^*$. Go back to lemma 5.19 to write down the multiplicative unitary of $(\hat{M}, \hat{\Delta})$.

Once we know the multiplicative unitary of $(\hat{M}, \hat{\Delta})$, it is just a bookkeeping exercise to generalize the famous Pontryagin Biduality Theorem. Have a go at it! In the formulation of this biduality result, $(\hat{\hat{M}}, \hat{\hat{\Delta}})$ denotes the Pontryagin dual of $(\hat{M}, \hat{\Delta})$, etc. The uniqueness of Haar weights up to a constant comes in handy to conclude that $\mathcal{N}_\varphi = \mathcal{N}_{\hat{\hat\varphi}}$.

Theorem 5.25. *We have that* $(\hat{\hat{M}}, \hat{\hat{\Delta}}) = (M, \Delta)$. *Moreover,* $\hat{\hat{\Lambda}}_\varphi = \Lambda_\varphi$, *whence* $\hat{\hat\varphi} = \varphi$.

We have seen that for a locally compact group G, the convolution algebra $L^1(G)$ possesses a *-operation, but we did not mention this in connection with M_*. In general, the unboundedness of S, implies that there is no *-operation defined on the whole of M_*. For this purpose one introduces a dense subalgebra $M_*^\sharp$ of M_* as:

$$M_*^\sharp = \{\omega \in M_* \mid \exists \theta \in M_* : \bar{\omega}S \subseteq \theta\} .$$

Here, $\bar{\omega} \in M_*$ is defined by $\bar{\omega}(x) = \overline{\omega(x^*)}$ for all $x \in M$. For $\omega \in M_*^\sharp$ the element θ described above is unique and we set $\omega^* := \theta$.
With this *-operation, $M_*^\sharp$ becomes a *-algebra. Note that $M_*^\sharp = \{\omega \in M_* \mid \exists \theta \in M_* : \omega\tau_{\frac{i}{2}} \subseteq \theta\}$. This should convince you that $M_*^\sharp$ is indeed dense in M_* for the norm topology. Also check that $\lambda(\omega^*) = \lambda(\omega)^*$ for all $\omega \in M_*^\sharp$.

5.3 Quantum groups on the Hilbert space level

A lot of the theory of quantum groups is played out on the level of the Hilbert space $\mathcal{H}_\varphi$. In this section we collect some basic formulas. We still work with the locally compact group (M, Δ) and notations of the previous subsections.
Denote the modular operator of φ by ∇ and the modular conjugation of φ by J. Moreover, denote the modular operator of $\hat\varphi$ by $\hat\nabla$ and the modular conjugation of $\hat\varphi$ by $\hat{J}$ (with respect to the GNS-construction $(\mathcal{H}_\varphi, \iota, \Lambda_\varphi)$).
Then the following formulas hold for $x \in M$, $y \in \hat{M}$ and $t \in \mathbb{R}$.

1. $\tau_t(x) = \hat\nabla^{it} x \hat\nabla^{-it}$ and $\hat\tau_t(y) = \nabla^{it} y \nabla^{-it}$,

2. $R(x) = \hat{J} x^* \hat{J}$ and $\hat{R}(y) = J y^* J$,

3. $(\hat{J} \otimes J)W(\hat{J} \otimes J) = W^*$ and $(\hat\nabla^{it} \otimes \nabla^{it})W(\hat\nabla^{-it} \otimes \nabla^{-it}) = W$.

We also have the right Haar weight $\psi = \varphi R$ with a GNS-construction $(\mathcal{H}_\varphi, \iota, \Lambda_\psi)$ defined by $\Lambda_\psi = (\Lambda_\varphi)_\delta$. See the comments before Proposition 4.27, formally $\Lambda_\psi(a) = \Lambda_\varphi(a\delta^{\frac{1}{2}})$. Then, the modular conjugation of ψ with respect to $(\mathcal{H}_\varphi, \iota, \Lambda_\psi)$ is given by $\nu^{\frac{i}{4}} J$.

But it is also possible to connect Λ_φ and Λ_ψ via R and $\hat{J}$: if $x \in \mathcal{N}_\psi$, then $R(x)^* \in \mathcal{N}_\varphi$ and $\Lambda_\psi(x) = \hat{J}\,\Lambda_\varphi(R(x)^*)$. This implies that $\hat{J} J = \nu^{\frac{i}{4}} J \hat{J}$.

There is still another important injective positive operator in $\mathcal{H}_\varphi$ that we will introduce now. Because $\varphi\tau_t = \nu^{-t}\varphi$ for all $t \in \mathbb{R}$, there exists a unique injective, positive operator P in $\mathcal{H}_\varphi$ such that $P^{it}\Lambda_\varphi(x) = \nu^{\frac{t}{2}}\Lambda_\varphi(\tau_t(x))$ for all $x \in \mathcal{N}_\varphi$ and $t \in \mathbb{R}$ (here we use Proposition 5.9). One can show that $P^{it}\hat{\Lambda}_\varphi(y) = \nu^{-\frac{t}{2}}\hat{\Lambda}_\varphi(\hat{\tau}_t(y))$ for all $y \in \mathcal{N}_{\hat{\varphi}}$ and $t \in \mathbb{R}$. Then,

1. $\tau_t(x) = P^{it}x\,P^{-it}$ and $\hat{\tau}_t(y) = P^{it}y\,P^{-it}$ for $t \in \mathbb{R}$, $x \in M$ and $y \in \hat{M}$.
2. $W(P^{it} \otimes P^{it}) = (P^{it} \otimes P^{it})W$.

5.4 The C*-algebra version of a locally compact quantum group

In order to define the C*-algebra version of a quantum group we need some extra information concerning multiplier algebras and tensor products. We always use the minimal tensor product between C*-algebras and denote it by $\otimes_{min}$. The tensor products between von Neumann algebras will still be denoted by $\otimes$.

If $\mathcal{H}$ is a Hilbert space and B a non-degenerate C*-subalgebra of the $B(\mathcal{H})$, the multiplier algebra $M(B)$ is easily described as a unital C*-subalgebra of $B(\mathcal{H})$ as follows:

$$M(B) = \{\, x \in B(\mathcal{H}) \mid \forall b \in B : xb \in B \text{ and } bx \in B \,\}.$$

Given C*-algebra B_1, B_2 there exists a natural injective *-homomorphism

$$\theta : M(B_1) \otimes_{min} M(B_2) \hookrightarrow M(B_1 \otimes_{min} B_2)$$

such that

$$(b_1 \otimes b_2)\,\theta(x_1 \otimes x_2) = (b_1 x_1) \otimes (b_2 x_2) \quad \text{and} \quad \theta(x_1 \otimes x_2)\,(b_1 \otimes b_2) = (x_1 b_1) \otimes (x_2 b_2)$$

for $x_1 \in M(B_1)$, $x_2 \in M(B_2)$ and $b_1 \in B_1$, $b_2 \in B_2$. It is important to mention that the injectivity statement is true because we work with the minimal tensor product. From now on we will use θ to consider $M(B_1) \otimes_{min} M(B_2)$ as a sub-C*-algebra of $M(B_1 \otimes_{min} B_2)$. This also applies to the minimal C*-tensor product of more C*-algebras.

Let us now return to the theory of locally compact quantum groups proper and focus onto our locally compact quantum group (M, Δ) of subsection 5.1. We will still use the notations gathered in the previous three subsections. Using the multiplicative unitary it is easy to associate a C*-algebra to the von Neumann algebraic quantum group (M, Δ).

Theorem 5.26. *We define A_r as the norm closure of the subalgebra $\{(\iota \otimes \omega)(W) \mid \omega \in B(\mathcal{H}_\varphi)_*\}$ in $B(\mathcal{H}_\varphi)$ and Δ_r as the restriction of Δ to A_r. Then A_r is a non-degenerate C^*-subalgebra of $B(\mathcal{H}_\varphi)$ and*

1. *Δ_r is a non-degenerate $*$-homomorphism from A_r to $M(A_r \otimes_{min} A_r)$*

2. *$(\Delta_r \otimes_{min} \iota)\Delta_r = (\iota \otimes_{min} \Delta_r)\Delta_r$*

3. *The linear spaces $\langle \Delta_r(a)(b \otimes 1) \mid a, b \in A_r \rangle$ and $\langle \Delta_r(a)(1 \otimes b) \mid a, b \in A_r \rangle$ are norm dense subspaces of $A_r \otimes_{min} A_r$.*

We call the pair (A_r, Δ_r) the reduced C^*-algebraic quantum group associated to (M, Δ). Note that A_r is σ-strongly* dense in M.

Already the statement that A_r is a C^*-algebra is not immediate since the Pentagonal equation only guarantees that the linear space $\mathcal{B} := \{(\iota \otimes \omega)(W) \mid \omega \in B(\mathcal{H}_\varphi)_*\}$ is a subalgebra of $B(\mathcal{H}_\varphi)$, not necessarily a $*$-subalgebra. If we set $\mathcal{A} = \mathcal{B} \cap \mathcal{B}^*$, we obtain a dense $*$-subalgebra of A_r. On can show that $\mathcal{A} := \{(\iota \otimes \omega)(W) \mid \omega \in \hat{M}_*^\sharp\}$. Notice that $\mathcal{B}$ is isomorphic (as an algebra) to $\hat{M}_*$ and that $\mathcal{A}$ is $*$-isomorphic to $\hat{M}_*^\sharp$.

All objects that we associated to (M, Δ) in the previous section induce corresponding objects on the C^*-algebra A_r by restriction. We set $\varphi_r = \varphi\lceil_{A_r^+}$ and $\psi_r = \psi \lceil_{A_r^+}$ and obtain this way faithful densely defined lower semi-continuous weights on A_r (the fact that these weights are densely defined needs a (simple) argument!). Moreover, these weights satisfy similar invariance conditions and KMS conditions as the nsf weights φ, ψ.

Once again you have to be a little bit careful with these kind of considerations. You have to prove something. Define the map $\sigma^r : \mathbb{R} \to \text{End}(A_r, M) : t \to \sigma_t^r := \sigma_t\lceil_{A_r}$. Then σ^r is a **norm continuous** one-parameter group of $*$-automorphisms **on A_r**. The analytic continuations of σ^r and σ are then related in the following way (for σ we use Definition 4.20 and for σ^r we use the variation of this definition discussed at the end of subsection 4.3). Let $z \in \mathbb{C}$, then $D(\sigma_z^r) = \{a \in A_r \cap D(\sigma_z) \mid \sigma_z^r(a) \in A_r\}$ and σ_z^r is the restriction of σ^z to $D(\sigma_z^r)$.

This same principle applies in fact to the other one-parameter groups σ' and τ'. The unitary antipode R of (M, Δ) can just be restricted to A_r and one obtains a $*$-automorphism on A_r.

The remaining object that we have not discussed yet is the modular element δ of (M, Δ). Classically, the modular function of a locally compact group G is continuous and as such is somehow 'affiliated' to the C^*-algebra $C_0(G)$. There exists an affiliation relation for C^*-algebras (see [Baa80], [BJ], [Lan] and [Wor91b]) that is different from the affiliation relation for von Neumann algebras discussed in subsection 4.2 but we will not give a precise definition. Loosely speaking, an element is affiliated to a C^*-algebra B if it is a 'well behaved (possibly unbounded) multiplier' of B and such an element can be looked upon as the quantum analogue of a (possibly unbounded) continuous function. If $B = C_0(X)$, where X is a locally compact Hausdorff space, the set of elements affiliated to B equals $C(X)$.

In this case δ is 'affiliated' to A_r because $\delta^{it} \in M(A_r)$ for all $t \in \mathbb{R}$. One looks upon δ as an 'unbounded multiplier' of A_r by associating to δ the linear densely defined mapping $\delta_r : D(\delta_r) \subseteq A_r \to A_r$ such that $D(\delta_r) = \{ a \in A_r \mid \delta\, a \in A_r \}$ and $\delta_r(a) = \delta\, a$ for all $a \in D(\delta_r)$.

It is possible to give a definition for reduced locally compact quantum groups in the C*-algebra setting (as was done in [KuV00a]) by requiring the properties in the statement of the above theorem to hold together with the existence of well-behaved faithful left and right Haar weights. Notice that such a definition contains density conditions whereas this is not the case for the von Neumann algebraic version we use in these lecture notes!

By taking a GNS-construction of the left Haar weight and defining the von Neumann algebra as the weak closure of the image of the C*-algebra under the GNS-representation and extending the comultiplication to this von Neumann algebra, one obtains again a von Neumann algebraic quantum group (this generalizes the discussion in Remark 5.14). This procedure and the one discussed above provide us with a bijective correspondence between reduced C*-algebraic quantum groups and von Neumann algebraic quantum groups.

There is also another C*-algebra associated to (M, Δ) that in some cases is different from A_r. Recall that we have the *-algebra $\mathcal{A}$ defined above and we can hope to apply Remark 1.6 once more. If $a \in \mathcal{A}$ then $\|a\|_* < \infty$ because $\mathcal{A}$ is a Banach *-algebra under the *-isomorphism $\mathcal{A} \cong \hat{M}_*^\sharp$. The norm $\|.\|^*$ on $\hat{M}_*^\sharp$ is given by $\|\omega\|^* = \max\{\|\omega\|, \|\omega^*\|\}$ for all $\omega \in \hat{M}_*^\sharp$ where $\|.\|$ denotes the ordinary norm on $\hat{M}_* \subseteq \hat{M}^*$.

The identity representation ensures that $\|.\|_*$ is a norm and not merely a seminorm. Therefore we can define the C*-algebra A_u as the enveloping C*-algebra of $\mathcal{A}$. Note that their exists a unique surjective *-homomorphism $\pi : A_u \to A_r$ so that $\pi(x) = x$ for all $x \in \mathcal{A}$.

It is possible to define on A_u a canonical comultiplication $\Delta_u : A_u \to M(A_u \otimes_{min} A_u)$ so that $(\pi \otimes \pi)\Delta_u = \Delta_r \pi$. One can also define Haar weights $\varphi_u := \varphi_r\, \pi$ and $\psi_u := \psi_r\, \pi$, modular groups for these Haar weights, a scaling group, a unitary antipode and a modular element on A_u. We call (A_u, Δ_u) the *universal C*-algebraic quantum group* associated to (M, Δ). This universal C*-algebraic quantum group has the same rich structure as (M, Δ) and (A_r, Δ_r) but the Haar weights do not have to be faithful (equivalently, π does not have to be faithful). There does however exist a counit $\varepsilon_u : A_u \to \mathbb{C}$ that is a (continuous!) *-homomorphism determined by $\varepsilon_u\big((\iota \otimes \omega)(W)\big) = \omega(1)$ for all $\omega \in M_*^\sharp$ (*) and satisfies the familiar formula

$$(\iota \otimes \varepsilon_u)\Delta_u = (\varepsilon_u \otimes \iota)\Delta_u = \iota \,.$$

Such a counit does not have to exist on A_r. It always can be defined on $\mathcal{B}$ by the formula (*) above but is not always continuous.

Remark 5.27. In the classical case the above discussion takes on a more familiar form. So let G be a locally compact σ-compact group G and use the notation of remark 5.16. The Pontryagin dual of the quantum group $(L^\infty(G), \Delta)$
is the quantum group $(\mathcal{L}(G), \hat{\Delta})$ where $\hat{\Delta}(x) = \Sigma W_G(x \otimes 1) W_G^* \Sigma$ for all
$x \in \mathcal{L}(G)$. For the quantum group $(\mathcal{L}(G), \hat{\Delta})$, we have $\mathcal{A} = \lambda(L^1(G))$. Thus,
the C*-algebra underlying the reduced C*-algebraic quantum group associated
to $(\mathcal{L}(G), \hat{\Delta})$ is the reduced group C*-algebra $C_r^*(G)$. The C*-algebra underlying the universal C*-algebraic quantum group associated to $(\mathcal{L}(G), \hat{\Delta})$ is the
group C*-algebra $C^*(G)$.

6 Examples of locally compact quantum groups

Just as the development of the general definition took quite a while, the
construction of examples of especially non-compact quantum groups has also
been a slow process. But the list of examples is getting quite respectable
by now. There are more or less two kinds of methods to construct quantum
groups at this moment.

Method 1

In the first method we start from a classical group G consisting of matrices
and follow this recipe.

1. Look at the Hopf *-algebra $\mathcal{A}$ of polynomial functions on the group G and
   find some natural generators and relations for $\mathcal{A}$.
2. Deform the relations by some complex number q, consider the Hopf *-
   algebra $\mathcal{A}_q$ generated by generators and these deformed relations and try
   to define a comultiplication $\Phi : \mathcal{A}_q \to \mathcal{A}_q \odot \mathcal{A}_q$ such that $(\mathcal{A}_q, \Phi)$ is a Hopf
   *-algebra.
3. Represent the deformed generators of $\mathcal{A}_q$ by (possibly) unbounded closed
   operators on a Hilbert space $\mathcal{H}$.
4. Define M as the von Neumann algebra on $\mathcal{H}$ generated by these represented generators. Device a method to define a comultiplication $\Delta : M \to$
   $M \otimes M$ that agrees with Φ on the generators. The most difficult aspect
   of constructing the quantum group this way lies in (1) finding a formula
   for the comultiplication and (2) proving its coassociativity.
5. Define left and right Haar weights and prove their invariance. In comparison with the construction of the comultiplication, this is not that
   complicated in concrete examples.

Let us list examples constructed this way:

1. Quantum $E(2)$: [Wor91a], [Wor91b], [VDWo], [Baa95], [Baa92].
2. Quantum $ax + b$ and Quantum $az + b$: [WZ], [Wor01], [VD01]. These are
   the first examples of a quantum group where the scaling constant is not
   equal to 1.
3. Quantum $\widetilde{SU}(1,1)$: [KK]

4. Quantum $GL(2, \mathbb{C})$: [PuW]

We will discuss the example of Quantum $\widetilde{SU}(1,1)$ in the next subsection.

Method 2

Device a general theoretical construction procedure to generate quantum groups using certain fairly general mathematical structures (like locally compact groups, or quantum groups) as ingredients.

1. The crossed product construction of a group with a quantum group, and generalizations thereof where the group is replaced by a quantum group 'acting' on another quantum group.
2. Bicrossed product constructions.

For these kind of constructions we refer to [BS], [BV], [VV]. The advantage of this method lies in the fact that one generates a multitude of examples by varying the ingredients but sometimes these methods preserve too much of the properties of the original ingredients one starts from. Up till now these construction procedures have not generated an example of a quantum group with a non-trivial scaling constant, whereas the first method has produced such an example. However, the second procedure above has generated examples to disprove some important conjectures (see [BSV]).

Let us also mention that the quantum Lorentz group is constructed in [PoW] as a double crossed product of the compact quantum group $SU_q(2)$ and its discrete Pontryagin dual.

6.1 Quantum $\widetilde{SU}(1,1)$

The quantum group that we present in this subsection is an example constructed following method 1 described above. Recall that $SU(1,1)$ is the Lie group

$$SU(1,1) = \{ \, X \in SL(2, \mathbb{C}) \mid X^*UX = U \, \} \,, \qquad \text{where } U = \begin{pmatrix} 1 & 0 \\ 0 & -1 \end{pmatrix} \,.$$

The equality $X^*UX = U$ is equivalent to saying that X is invariant under the canonical Lorentzian inner product on $\mathbb{C} \oplus \mathbb{C}$. Woronowicz has shown in [Wor91b] that $SU(1,1)$ can not be deformed into a locally compact quantum group (the problems lie in the coassociativity of the comultiplication on the operator algebra level). In order to resolve the problems surrounding quantum $SU(1,1)$, Korogodskii proposed in [Kor] to construct the quantum version of an extension $\widetilde{SU}(1,1)$ of $SU(1,1)$. To be more precise,

$$\widetilde{SU}(1,1) = \{ \, X \in SL(2, \mathbb{C}) \mid X^*UX = U \text{ or } X^*UX = -U \, \} \,.$$

Woronowicz studied the construction of a locally compact quantum group version of quantum $\widetilde{SU}(1,1)$ (without the Haar weight) in [Wor00] but a gap remained in the proof of the coassociativity of the comultiplication. Using the theory of q-hypergeometric functions quantum $\widetilde{SU}(1,1)$ was introduced as a full blown locally compact quantum group in [KK].

The Hopf *-algebra underlying quantum $\widetilde{SU}(1,1)$

In [Kor], Korogodskii implicitly suggested the use of the following Hopf *-algebra. The Hopf *-algebra itself was explicitly introduced by Woronowicz in [Wor00].

Throughout this discussion, we fix a number $0 < q < 1$. In this subsection we will introduce a quantum group that is a deformation, depending on the deformation parameter q, of $\widetilde{SU}(1,1)$. We will refer to this still to be defined quantum group as $\widetilde{SU}_q(1,1)$ or quantum $\widetilde{SU}(1,1)$.

Define $\mathcal{A}$ to be the unital universal *-algebra generated by elements α_0, γ_0 and e_0 and relations

$$\alpha_0^\dagger \alpha_0 - \gamma_0^\dagger \gamma_0 = e_0 \qquad \alpha_0 \alpha_0^\dagger - q^2\, \gamma_0^\dagger \gamma_0 = e_0$$
$$\gamma_0^\dagger \gamma_0 = \gamma_0\, \gamma_0^\dagger$$
$$\alpha_0\, \gamma_0 = q\, \gamma_0\, \alpha_0 \qquad e_0^\dagger = e_0$$
$$\alpha_0\, \gamma_0^\dagger = q\, \gamma_0^\dagger \alpha_0 \qquad e_0^2 = 1$$
$$\alpha_0\, e_0 = e_0\, \alpha_0$$
$$\gamma_0\, e_0 = e_0\, \gamma_0\ ,$$

where $\dagger$ denotes the *-operation on $\mathcal{A}$. By universality of $\mathcal{A}$, there exists a unique unital *-homomorphism $\Delta_0 : \mathcal{A} \to \mathcal{A} \odot \mathcal{A}$ such that

$$\Delta_0(\alpha_0) = \alpha_0 \otimes \alpha_0 + q\,(e_0\, \gamma_0^\dagger) \otimes \gamma_0$$
$$\Delta_0(\gamma_0) = \gamma_0 \otimes \alpha_0 + (e_0 \alpha_0^\dagger) \otimes \gamma_0 \qquad (6.1)$$
$$\Delta_0(e_0) = e_0 \otimes e_0\ .$$

The pair $(\mathcal{A}, \Delta_0)$ turns out to be a Hopf *-algebra with counit ε_0 and antipode S_0 determined by

$$S_0(\alpha_0) = e_0\, \alpha_0^\dagger \qquad\qquad \varepsilon_0(\alpha_0) = 1$$
$$S_0(\alpha_0^\dagger) = e_0\, \alpha_0 \qquad\qquad \varepsilon_0(\gamma_0) = 0$$
$$S_0(\gamma_0) = -q\, \gamma_0 \qquad\qquad \varepsilon_0(e_0) = 1$$
$$S_0(\gamma_0^\dagger) = -\tfrac{1}{q}\, \gamma_0^\dagger$$
$$S_0(e_0) = e_0\ .$$

One obtains the Hopf *-algebra of quantum $SU(1,1)$ by taking $e_0 = 1$ in the above considerations, but this is only a side remark.

If one takes $q = 1$ in the above description, one gets the Hopf *-algebra of polynomial functions on $\widetilde{SU}(1,1)$ as explained below. A simple calculation reveals that

$$\widetilde{SU}(1,1) = \left\{ \begin{pmatrix} a & c \\ \epsilon \bar{c} & \epsilon \bar{a} \end{pmatrix} \mid a,b \in \mathbb{C}, \epsilon \in \{-1,1\} \text{ s.t. } |a|^2 - |c|^2 = \epsilon \right\}.$$

The elements α_0, γ_0 and e_0 can then be realized as the complex valued functions on $\widetilde{SU}(1,1)$ given by

$$\alpha_0 \begin{pmatrix} a & c \\ \epsilon \bar{c} & \epsilon \bar{a} \end{pmatrix} = a, \qquad \gamma_0 \begin{pmatrix} a & c \\ \epsilon \bar{c} & \epsilon \bar{a} \end{pmatrix} = c, \qquad e_0 \begin{pmatrix} a & c \\ \epsilon \bar{c} & \epsilon \bar{a} \end{pmatrix} = \epsilon$$

and $\mathcal{A}$ is the unital *-algebra of complex valued functions on $\widetilde{SU}(1,1)$ generated by α_0, γ_0 and e_0.

Let us now go back to the case $0 < q < 1$. As mentioned in the beginning of this section we want to represent this Hopf *-algebra $\mathcal{A}$ by possibly unbounded operators in some Hilbert space in order to produce a locally compact quantum group in the sense of Definition 5.1. Korogodskii classified the well-behaved irreducible representations of $\mathcal{A}$ in [Kor, Prop. 2.4]. Roughly speaking, our representation of $\mathcal{A}$ is obtained by gluing together these irreducible representations. The representation we use here is a slight variation of the one introduced by Woronowicz in [Wor00]. For this purpose we define

$$I_q = \{ -q^k \mid k \in \mathbb{N} \} \cup \{ q^k \mid k \in \mathbb{Z} \}.$$

Let $\mathbb{T}$ denote the group of complex numbers of modulus 1. We will consider the counting measure on I_q and the normalized Haar measure on $\mathbb{T}$. Our *-representation of $\mathcal{A}$ will act in the Hilbert space $\mathcal{H}$ defined by

$$\mathcal{H} = L^2(\mathbb{T}) \otimes L^2(I_q).$$

In these discussions we will denote for any set J the space of complex functions on J by $\mathcal{F}(J)$ whereas the space of complex functions on J with finite support will be denoted by $\mathcal{K}(J)$.

If $p \in -q^{\mathbb{Z}} \cup q^{\mathbb{Z}}$, we define $\delta_p \in \mathcal{F}(I_q)$ such that $\delta_p(x) = \delta_{x,p}$ for all $x \in I_q$ (note that $\delta_p = 0$ if $p \notin I_q$). The family $(\delta_p \mid p \in I_q)$ is the natural orthonormal basis of $L^2(I_q)$. We let ζ denote the identity function on $\mathbb{T}$. Recall the natural orthonormal basis $(\zeta^m \mid m \in \mathbb{Z})$ for $L^2(\mathbb{T})$.

Instead of looking at the algebra $\mathcal{A}$ as the abstract algebra generated by generators and relations we will use an explicit realization of this algebra as linear operators on the dense subspace E of $\mathcal{H}$ defined by $E = \langle \zeta^m \otimes \delta_x \mid m \in \mathbb{Z}, x \in I_q \rangle \subseteq \mathcal{H}$. Of course, E inherits the inner product from $\mathcal{H}$. Let $\mathcal{L}^+(E)$ denote the *-algebra of adjointable operators on E (see [Sch, Prop. 2.1.8]), i.e.

$$\mathcal{L}^+(E) = \{ T \in \text{End}(E) \mid \exists T^\dagger \in \text{End}(E), \forall v, w \in E : \langle Tv, w \rangle = \langle v, T^\dagger w \rangle \},$$

so $\dagger$ denotes the *-operation in $\mathcal{L}^+(E)$. Here, $\text{End}(E)$ is the space of linear operators on E.

If $T \in \mathcal{L}^+(E)$, $T^\dagger \subseteq T^*$ where T^* is the usual adjoint of T as an operator in the Hilbert space $\mathcal{H}$. Since T^* has dense domain, it follows that T is a closable operator in $\mathcal{H}$.

Define linear operators α_0, γ_0, e_0 in $\mathcal{L}^+(E)$ such that

$$\alpha_0(\zeta^m \otimes \delta_p) = \sqrt{\operatorname{sgn}(p) + p^{-2}} \; \zeta^m \otimes \delta_{qp}$$
$$\gamma_0(\zeta^m \otimes \delta_p) = p^{-1} \; \zeta^{m+1} \otimes \delta_p$$
$$e_0(\zeta^m \otimes \delta_p) = \operatorname{sgn}(p) \; \zeta^m \otimes \delta_p$$

for all $p \in I_q$, $m \in \mathbb{Z}$.

Then $\mathcal{A}$ is the *-subalgebra of $\mathcal{L}^+(E)$ generated by α_0, γ_0 and e_0. Since $\mathcal{L}^+(E) \odot \mathcal{L}^+(E)$ is canonically embedded in $\mathcal{L}^+(E \odot E)$, we obtain $\mathcal{A} \odot \mathcal{A}$ as a *-subalgebra of $\mathcal{L}^+(E \odot E)$. As such, $\Delta_0(\alpha_0)$, $\Delta_0(\gamma_0)$ and $\Delta_0(e_0)$ defined in Eqs. (6.1) belong to $\mathcal{L}^+(E \odot E)$.

The von Neumann algebra underlying quantum $\widetilde{SU}(1,1)$

In this subsection we introduce the von Neumann algebra acting on $\mathcal{H}$ that underlies the von Neumann algebraic version of the quantum group $\widetilde{SU}_q(1,1)$. In order to get into the framework of operator algebras, we need to introduce the topological versions of the algebraic objects α_0, γ_0 and e_0 as possibly unbounded operators in the Hilbert space $\mathcal{H}$. So let α denote the closure of α_0, γ the closure of γ_0 and e the closure of e_0, all as linear operators in $\mathcal{H}$. So e is a bounded linear operator on $\mathcal{H}$, whereas α and γ are unbounded, closed, densely defined linear operators in $\mathcal{H}$. Note that α^* is the closure of $\alpha_0^\dagger$ and that γ^* is the closure of $\gamma_0^\dagger$. Note also that γ is normal.

Define a reflection operator T on $\mathcal{F}(\mathbb{T} \times I_q)$ such that for $f \in \mathcal{F}(\mathbb{T} \times I_q)$, $\lambda \in \mathbb{T}$ and $x \in I_q$, we have that $(Tf)(\lambda, x) = f(\lambda, -x)$ if $-x \in I_q$ and $(Tf)(\lambda, x) = 0$ if $-x \notin I_q$. If $t \in I_q$ and $g \in \mathcal{F}(\mathbb{T})$, then $T_p(g \otimes \delta_t) = g \otimes \delta_{-t}$, thus, $T(g \otimes \delta_t) = 0$ if $-t \notin I_q$.

Define the self-adjoint partial isometry $u \in B(\mathcal{H})$ as the one that is induced by T.

Let us recall the following natural terminology. If $T_1, \ldots, T_n$ are closed, densely defined linear operators in $\mathcal{H}$, the von Neumann algebra N on $\mathcal{H}$ generated by $T_1, \ldots, T_n$ is the one such that

$$N' = \{\, x \in B(\mathcal{H}) \mid xT_i \subseteq T_i x \text{ and } xT_i^* \subseteq T_i^* x \text{ for } i = 1, \ldots, n \,\}.$$

Almost by definition, N is the smallest von Neumann algebra acting on $\mathcal{H}$ so that $T_1, \ldots, T_n$ are affiliated with M in the von Neumann algebraic sense.

It is now very tempting to define the von Neumann algebra underlying quantum $\widetilde{SU}_q(1,1)$ as the von Neumann algebra on $\mathcal{H}$ generated by α, γ and e. However, for reasons that will become clear later (see the comments after Proposition 6.6), the underlying von Neumann algebra will be the one generated by α, γ, e and u (the necessity of the element u was first observed by Woronowicz in [Wor00]).

Proposition 6.1. *We define M to be the von Neumann algebra on $\mathcal{H}$ generated by α, γ, e and u. Then $M = L^\infty(\mathbb{T}) \otimes B(L^2(I_q))$.*

The following picture of M turns out to be the most useful one. For every $p, t \in I_q$ and $m \in \mathbb{Z}$ we define $\Phi(m, p, t) \in B(\mathcal{H})$ so that for $x \in I_q$ and $r \in \mathbb{Z}$,

$$\Phi(m, p, t)\,(\zeta^r \otimes \delta_x) = \delta_{x,t}\,\zeta^{m+r} \otimes \delta_p \ .$$

Define $M^\circ = \langle\, \Phi(m, p, t) \mid m \in \mathbb{Z}, p, t \in I_q \,\rangle$. Using the above equation, it is obvious that $(\,\Phi(m, p, t) \mid m \in \mathbb{Z}, p, t \in I_q\,)$ is a linear basis of M°.

The multiplication and *-operation are easily expressed in terms of these basis elements:

$$\Phi(m_1, p_1, t_1)\,\Phi(m_2, p_2, t_2) = \delta_{p_2, t_1}\ \Phi(m_1 + m_2, p_1, t_2)$$
$$\Phi(m, p, t)^* = \Phi(-m, t, p)$$

for all $m, m_1, m_2 \in \mathbb{Z}$, $p, p_1, p_2, t, t_1, t_2 \in I_q$. So we see that M° is a σ-strongly* dense sub*-algebra of M.

A special function

The construction of quantum $\widetilde{SU}(1, 1)$ and the study of its Pontryagin dual hinges on the theory of q-hypergeometric functions. Let us therefore fix the necessary notation and terminology involved.

Fix a number $0 < u < 1$. Let $a \in \mathbb{C}$. If $k \in \mathbb{N}_0 \cup \{\infty\}$, the q-shifted factorial $(a; u)_k \in \mathbb{C}$ is defined as $(a; u)_k = \prod_{i=0}^{k-1}(1 - u^i a)$, so $(a; u)_0 = 1$.

If $a, b, z \in \mathbb{C}$, we define

$$\Psi\left(\begin{matrix} a \\ b \end{matrix}; u, z\right) = \sum_{n=0}^{\infty} \frac{(a; u)_n\,(b\,u^n; u)_\infty}{(u\,; u)_n} (-1)^n\,u^{\frac{1}{2}n(n-1)}\,z^n \ . \qquad (6.2)$$

This function is analytic in a, b and c. If you are familiar with q-hypergeometric functions note that if $b \notin u^{-\mathbb{N}_0}$, then $\Psi\left(\begin{matrix} a \\ b \end{matrix}; u, z\right) = (b; u)_\infty\ {}_1\varphi_1\left(\begin{matrix} a \\ b \end{matrix}; u, \ z\right)$.

See [GR] for an extensive treatment on q-hypergeometric functions.

The comultiplication on quantum $\widetilde{SU}(1, 1)$

In this subsection we introduce the comultiplication of $\widetilde{SU}_q(1, 1)$. In the first part we start with a motivation for the formulas appearing in Definition 6.2. Although the discussion is not really needed in the build up of $\widetilde{SU}_q(1, 1)$, it is important and clarifying to know how we arrived at the formulas in Definition 6.2.

Our purpose is to define a comultiplication $\Delta : M \to M \otimes M$. Assume for the moment that this has already been done. It is natural to require Δ to be closely related to the comultiplication Δ_0 on $\mathcal{A}$ as defined in Eqs. (6.1). The least that we expect is $\Delta_0(T_0) \subseteq \Delta(T)$ and $\Delta_0(T_0^\dagger) \subseteq \Delta(T)^*$ for $T = \alpha, \gamma, e$. In the rest of this discussion we will focus on the inclusion $\Delta_0(\gamma_0^\dagger \gamma_0) \subseteq \Delta(\gamma^* \gamma)$, where $\Delta_0(\gamma_0^\dagger \gamma_0) \in \mathcal{L}^+(E \odot E)$.

Because $\gamma^* \gamma$ is self-adjoint, the element $\Delta(\gamma^* \gamma)$ would also be self-adjoint. So the hunt is on for self-adjoint extensions of the explicit operator $\Delta_0(\gamma_0^\dagger \gamma_0)$. Unlike in the case of quantum $E(2)$ (see [Wor91b]), the operator $\Delta_0(\gamma_0^\dagger \gamma_0)$ is not essentially self-adjoint. But it was already known in [Kor] that $\Delta_0(\gamma_0^\dagger \gamma_0)$ has self-adjoint extensions (this follows easily because the operator in (6.3) commutes with complex conjugation, implying that the deficiency spaces are isomorphic).

Although $\Delta_0(\gamma_0^\dagger \gamma_0)$ has a self-adjoint extension, it is not unique. We have to make a choice for this self-adjoint extension, but we cannot extract the information necessary to make this choice from α, γ and e alone. This is why we do not work with the von Neumann algebra M° that is generated by α, γ and e alone but with M which has the above extra extension information contained in the element u. These kind of considerations were already present in [WZ] and were also introduced in [Wor00] for quantum $\widetilde{SU}(1,1)$. In [KK], this principle is only lurking in the background but it is treated in a fundamental and rigorous way in [Wor00]. In order to deal with this, Woronowicz develops a nice theory of balanced extensions of operators that is comparable to the theory of self-adjoint extensions of symmetric operators.

Now we get into slightly more detail in our discussion about the extension of $\Delta_0(\gamma_0^\dagger \gamma_0)$. But first we introduce the following auxiliary function

$$\kappa : \mathbb{R} \to \mathbb{R} : x \mapsto \kappa(x) = \operatorname{sgn}(x)\, x^2 \ .$$

Define a linear map $L : \mathcal{F}(\mathbb{T} \times I_q \times \mathbb{T} \times I_q) \to \mathcal{F}(\mathbb{T} \times I_q \times \mathbb{T} \times I_q)$ such that

$$(Lf)(\lambda, x, \mu, y) =$$
$$[\, x^{-2}(\operatorname{sgn}(y) + y^{-2}) + (\operatorname{sgn}(x) + q^2\, x^{-2})\, y^{-2}\,]\, f(\lambda, x, \mu, y)$$
$$+\ \operatorname{sgn}(x)\, q^{-1} \bar{\lambda}\mu\, x^{-1} y^{-1}\, \sqrt{(\operatorname{sgn}(x) + x^{-2})(\operatorname{sgn}(y) + y^{-2})}\, f(\lambda, qx, \mu, qy)$$
$$+\ \operatorname{sgn}(x)\, q\, \lambda\bar{\mu}\, x^{-1} y^{-1}\, \sqrt{(\operatorname{sgn}(x) + q^2 x^{-2})(\operatorname{sgn}(y) + q^2 y^{-2})}\, f(\lambda, q^{-1}x, \mu, q^{-1}y)$$

for all $\lambda, \mu \in \mathbb{T}$ and $x, y \in I_q$. A straightforward calculation shows that $\Delta_0(\gamma_0^\dagger \gamma_0)\, f = L(f)$ for all $f \in E \odot E$. From this, it is a standard exercise to check that $f \in D(\Delta_0(\gamma_0^\dagger \gamma_0)^*)$ and $\Delta_0(\gamma_0^\dagger \gamma_0)^* f = L(f)$ if $f \in \mathcal{L}^2(\mathbb{T} \times I_q \times \mathbb{T} \times I_q)$ and $L(f) \in \mathcal{L}^2(\mathbb{T} \times I_q \times \mathbb{T} \times I_q)$ (without any difficulty, one can even show that $D(\Delta_0(\gamma_0^\dagger \gamma_0)^*)$ consists precisely of such elements f).

If $\theta \in -q^{\mathbb{Z}} \cup q^{\mathbb{Z}}$, we define $\ell'_\theta = \{ (\lambda, x, \mu, y) \in \mathbb{T} \times I_q \times \mathbb{T} \times I_q \mid y = \theta x \}$ and consider $L^2(\ell'_\theta)$ naturally embedded in $L^2(\mathbb{T} \times I_q \times \mathbb{T} \times I_q)$. It follows easily from the above discussion that $\Delta_0(\gamma_0^\dagger \gamma_0)^*$ leaves $L^2(\ell'_\theta)$ invariant. Thus, if T is a self-adjoint extension of $\Delta_0(\gamma_0^\dagger \gamma_0)$, the obvious inclusion $T \subseteq \Delta_0(\gamma_0^\dagger \gamma_0)^*$ implies that T also leaves $L^2(\ell'_\theta)$ invariant.

Therefore we construct a self-adjoint extension T of $\Delta_0(\gamma_0^\dagger \gamma_0)$ by choosing a self-adjoint extension T_θ of the restriction of $\Delta_0(\gamma_0^\dagger \gamma_0)$ to $L^2(\ell_\theta')$ for every $\theta \in -q^{\mathbb{Z}} \cup q^{\mathbb{Z}}$ and setting $T = \oplus_{\theta \in -q^{\mathbb{Z}} \cup q^{\mathbb{Z}}} T_\theta$.

Fix $\theta \in -q^{\mathbb{Z}} \cup q^{\mathbb{Z}}$. Define $J_\theta = \{ z \in I_{q^2} \mid \kappa(\theta)\, z \in I_{q^2} \}$ which is a q^2-interval around 0 (bounded or unbounded towards ∞). On J_θ we define a measure ν_θ such that $\nu_\theta(\{x\}) = |x|$ for all $x \in J_\theta$.

Define the linear operator $L_\theta : \mathcal{F}(J_\theta) \to \mathcal{F}(J_\theta)$ such that

$$
\begin{aligned}
\theta^2 x^2\, (L_\theta f)(x) = {}& -\sqrt{(1+x)(1+\kappa(\theta)\, x)}\, f(q^2 x) \\
& - q^2 \sqrt{(1+q^{-2}x)(1+q^{-2}\kappa(\theta)\, x)}\, f(q^{-2}x) \\
& + [(1+\kappa(\theta)\, x) + q^2(1+q^{-2}x)]\, f(x)
\end{aligned}
\tag{6.3}
$$

for all $f \in \mathcal{F}(J_\theta)$ and $x \in J_\theta$.

Then, an easy verification reveals that $\Delta_0(\gamma_0^\dagger \gamma_0){\upharpoonright}_{\mathcal{K}(\ell_\theta')}$ is unitarily equivalent to $1 \odot L_\theta{\upharpoonright}_{\mathcal{K}(J_\theta)}$. So our problem is reduced to finding self-adjoint extensions of $L_\theta{\upharpoonright}_{\mathcal{K}(J_\theta)}$. This operator $L_\theta{\upharpoonright}_{\mathcal{K}(J_\theta)}$ is a second order q-difference operator for which eigenfunctions in terms of q-hypergeometric functions are known.

We can use a reasoning similar to the one in [ES03, Sec. 2] to get hold of the self-adjoint extensions of $L_\theta{\upharpoonright}_{\mathcal{K}(J_\theta)}$: Let $\beta \in \mathbb{T}$. Then we define a linear operator $L_\theta^\beta : D(L_\theta^\beta) \subseteq L^2(J_\theta, \nu_\theta) \to L^2(J_\theta, \nu_\theta)$ such that $D(L_\theta^\beta)$ consists of all $f \in L^2(J_\theta, \nu_\theta)$ for which

$$
L_\theta(f) \in L^2(J_\theta, \nu_\theta), \quad f(0+) = \beta\, f(0-) \quad \text{and} \quad (D_q f)(0+) = \beta\, (D_q f)(0-)
$$

and L_θ^β is the restriction of L_θ to $D(L_\theta^\beta)$. Here, D_q denotes the Jackson derivative, that is, $(D_q f)(x) = (f(qx) - f(x))/(q-1)x$ for $x \in J_\theta$. Also, $f(0+) = \beta\, f(0-)$ is an abbreviated form of saying that the limits $\lim_{x \uparrow 0} f(x)$ and $\lim_{x \downarrow 0} f(x)$ exist and $\lim_{x \downarrow 0} f(x) = \beta \lim_{x \uparrow 0} f(x)$.

Then L_θ^β is a self-adjoint extension of $L_\theta{\upharpoonright}_{\mathcal{K}(J_\theta)}$.

It is tempting to use the extension L_θ^1 to construct our final self-adjoint extension for $\Delta_0(\gamma_0^\dagger \gamma_0)$ (although there is no apparent reason for this choice). However, in order to obtain a coassociative comultiplication, it turns out that we have to use the extension $L_\theta^{\mathrm{sgn}(\theta)}$ to construct our final self-adjoint extension. This is reflected in the fact that the expression $s(x, y)$ appears in the formula for a_p in Definition 6.2.

This all would be only a minor achievement if we could not go any further. But the results and techniques used in the theory of q-hypergeometric functions will even allow us to find an explicit orthonormal basis consisting of eigenvectors of $L_\theta^{\mathrm{sgn}(\theta)}$. These eigenvectors are, up to a unitary transformation, obtained by restricting the functions a_p in Definition 6.2 to ℓ_θ, which is introduced after this definition. The special case $\theta = 1$ was already known to Korogodskii (see [Kor, Prop. A.1]).

In order to compress the formulas even further, we introduce three extra auxiliary functions.

(1) $\chi : -q^{\mathbb{Z}} \cup q^{\mathbb{Z}} \to \mathbb{Z}$ such that $\chi(x) = \log_q(|x|)$ for all $x \in -q^{\mathbb{Z}} \cup q^{\mathbb{Z}}$,

(2) $\nu : -q^{\mathbb{Z}} \cup q^{\mathbb{Z}} \to \mathbb{R}^+$ such that $\nu(t) = q^{\frac{1}{2}(\chi(t)-1)(\chi(t)-2)}$ for all $t \in -q^{\mathbb{Z}} \cup q^{\mathbb{Z}}$.

(3) the function $s : \mathbb{R}_0 \times \mathbb{R}_0 \to \{-1, 1\}$ defined as

$$s(x,y) = \begin{cases} -1 & \text{if } x > 0 \text{ and } y < 0 \\ 1 & \text{if } x < 0 \text{ or } y > 0 \end{cases}$$

We will also use the normalization constant $c_q = (\sqrt{2}\, q\, (q^2, -q^2; q^2)_\infty)^{-1}$. Recall the special function introduced in Eq. (6.2).

Definition 6.2. *If $p \in I_q$, we define a function $a_p : I_q \times I_q \to \mathbb{R}$ such that for all $x, y \in I_q$, the value $a_p(x,y)$ is given by*

$$c_q\, s(x,y)\, (-1)^{\chi(p)}\, (-sgn(y))^{\chi(x)}\, |y|\, \nu(py/x)$$

$$\times\; \sqrt{\frac{(-\kappa(p), -\kappa(y); q^2)_\infty}{(-\kappa(x); q^2)_\infty}}\; \Psi\left(\begin{matrix} -q^2/\kappa(y) \\ q^2\, \kappa(x/y) \end{matrix}; q^2,\; q^2\kappa(x/p)\right)$$

if $sgn(xy) = sgn(p)$ and $a_p(x,y) = 0$ if $sgn(xy) \neq sgn(p)$.

The extra vital information that we need is contained in the following proposition (see [ASC] and [CKK]). For $\theta \in -q^{\mathbb{Z}} \cup q^{\mathbb{Z}}$ we define $\ell_\theta = \{ (x,y) \in I_q \times I_q \mid y = \theta x \}$.

Proposition 6.3. *Consider $\theta \in -q^{\mathbb{Z}} \cup q^{\mathbb{Z}}$. Then the family $(a_p \restriction_{\ell_\theta} \mid p \in I_q$ such that $sgn(p) = sgn(\theta))$ is an orthonormal basis for $\ell^2(\ell_\theta)$.*

This proposition is used to define the comultiplication on M. It is also essential to the proof of the left invariance of the Haar weight.

Let us also mention the nice symmetry in $a_p(x,y)$ with respect to interchanging x, y and p:

Proposition 6.4. *If $x, y, p \in I_q$, then*

$$a_p(x,y) = (-1)^{\chi(yp)}\, sgn(x)^{\chi(x)}\, |y/p|\, a_y(x,p)$$
$$a_p(x,y) = sgn(p)^{\chi(p)}\, sgn(x)^{\chi(x)}\, sgn(y)^{\chi(y)}\, a_p(y,x)$$
$$a_p(x,y) = (-1)^{\chi(xp)}\, sgn(y)^{\chi(y)}\, |x/p|\, a_x(p,y)\; .$$

Now we produce the eigenvectors of our self-adjoint extension of $\Delta_0(\gamma_0^\dagger\gamma_0)$ (see the remarks after Proposition 6.7). We will use these eigenvectors to define a unitary operator that will induce the comultiplication. The dependence of $F_{r,s,m,p}$ on r, s and p is chosen in such a way that Proposition 6.7 is true.

Definition 6.5. *Consider $r, s \in \mathbb{Z}$, $m \in \mathbb{Z}$ and $p \in I_q$. We define the element $F_{r,s,m,p} \in \mathcal{H} \otimes \mathcal{H}$ such that*

$$F_{r,s,m,p}(\lambda, x, \mu, y) = \begin{cases} a_p(x,y) \, \lambda^{r+\chi(y/p)} \, \mu^{s-\chi(x/p)} & \text{if } y = sgn(p) \, q^m \, x \\ 0 & \text{otherwise} \end{cases}$$

for all $x, y \in I_q$ and $\lambda, \mu \in \mathbb{T}$.

Now we are ready to introduce the comultiplication of quantum $\widetilde{SU}_q(1,1)$.

Proposition 6.6. *Define the unitary transformation $V : \mathcal{H} \otimes \mathcal{H} \to L^2(\mathbb{T}) \otimes L^2(\mathbb{T}) \otimes \mathcal{H}$ such that $V(F_{r,s,m,p}) = \zeta^r \otimes \zeta^s \otimes \zeta^m \otimes \delta_p$ for all $r, s \in \mathbb{Z}$, $m \in \mathbb{Z}$ and $p \in I_q$. Then there exists a unique injective normal *-homomorphism $\Delta : M \to M \otimes M$ such that $\Delta(a) = V^*(1_{L^2(\mathbb{T})} \otimes 1_{L^2(\mathbb{T})} \otimes a)V$ for all $a \in M$.*

The requirement that $\Delta(M) \subseteq M \otimes M$ is the primary reason for introducing the extra generator u. We cannot work with the von Neumann algebra M° that is generated by α, γ and e alone, because $\Delta(M^\circ) \not\subseteq M^\circ \otimes M^\circ$.

This definition of Δ and the operators α and γ imply easily that the space $\langle F_{r,s,m,p} \mid r, s \in \mathbb{Z}, m \in \mathbb{Z}, p \in I_q \rangle$ is a core for $\Delta(\alpha)$, $\Delta(\gamma)$ and

$$\Delta(\alpha) \, F_{r,s,m,p} = \sqrt{sgn(p) + p^{-2}} \; F_{r,s,m,pq}$$
$$\Delta(\gamma) \, F_{r,s,m,p} = p^{-1} \, F_{r,s,m+1,p} \, . \tag{6.4}$$

for $r, s \in \mathbb{Z}$, $m \in \mathbb{Z}$ and $p \in I_q$.

Recall the linear operators $\Delta_0(\alpha_0)$, $\Delta_0(\gamma_0)$ acting on $E \odot E$ (Eqs. (6.1)). Also recall the distinction between $*$ and $\dagger$. The next proposition shows that Δ and Δ_0 are related in a natural way.

Proposition 6.7. *The following inclusions hold: $\Delta_0(\alpha_0) \subseteq \Delta(\alpha)$, $\Delta_0(\alpha_0)^\dagger \subseteq \Delta(\alpha)^*$, $\Delta_0(\gamma_0) \subseteq \Delta(\gamma)$ and $\Delta_0(\gamma_0)^\dagger \subseteq \Delta(\gamma)^*$. Moreover $\Delta(e) = e \otimes e$.*

This proposition implies also that $\Delta(\gamma^*\gamma)$ is an extension of $\Delta_0(\gamma_0^\dagger \gamma_0)$. We also know that $\langle F_{r,s,m,p} \mid r, s \in \mathbb{Z}, m \in \mathbb{Z}, p \in I_q \rangle$ is a core for $\Delta(\gamma^*\gamma)$ and $\Delta(\gamma^*\gamma) \, F_{r,s,m,p} = p^{-2} \, F_{r,s,m,p}$ for $r, s, m \in \mathbb{Z}$, $p \in I_q$. Using this information one can indeed show that $\Delta(\gamma^*\gamma)\restriction_{L^2(\ell_\theta')}$ is unitarily equivalent to $1 \otimes L_\theta^{sgn(\theta)}$ for all $\theta \in -q^{\mathbb{Z}} \cup q^{\mathbb{Z}}$, but we will not make any use of this fact in these notes.

Quantum $\widetilde{SU}(1,1)$ as a locally compact quantum group.

Now we can state the main result of [KK]. Verifying the coassociativity of Δ as in Definition 5.1 turns out to be the most difficult property to check. Producing the Haar weight is not that difficult (and goes back to [ES01]) but proving its invariance requires some work.

Theorem 6.8. *The pair* (M, Δ) *is a unimodular locally compact quantum group.*

We define $\widetilde{SU}_q(1,1) = (M, \Delta)$ and refer to $\widetilde{SU}_q(1,1)$ as quantum $\widetilde{SU}(1,1)$.

Let us give an explicit formula for the Haar weight. Since $M = L^\infty(\mathbb{T}) \otimes B(L^2(I_q))$ we can consider the trace Tr on M given by $\text{Tr} = \text{Tr}_{L^\infty(\mathbb{T})} \otimes \text{Tr}_{B(L^2(I_q))}$, where $\text{Tr}_{L^\infty(\mathbb{T})}$ and $\text{Tr}_{B(L^2(I_q))}$ are the canonical traces on $L^\infty(\mathbb{T})$ and $B(L^2(I_q))$ which we choose to be normalized in such a way that $\text{Tr}_{L^\infty(\mathbb{T})}(1) = 1$ and $\text{Tr}_{B(L^2(I_q))}(P) = 1$ for every rank one projection P in $B(L^2(I_q))$. Next we introduce a GNS-construction for the trace Tr. Define

$$\mathcal{H}_\varphi = \mathcal{H} \otimes L^2(I_q) = L^2(\mathbb{T}) \otimes L^2(I_q) \otimes L^2(I_q) .$$

If $m \in \mathbb{Z}$ and $p, t \in -q^{\mathbb{Z}} \cup q^{\mathbb{Z}}$, we set $f_{m,p,t} = \zeta^m \otimes \delta_p \otimes \delta_t \in \mathcal{H}_\varphi$ if $p, t \in I_q$ and $f_{m,p,t} = 0$ otherwise. Now define

(1) a linear map $\Lambda_{\text{Tr}} : \mathcal{N}_{\text{Tr}} \to \mathcal{H}_\varphi$ such that $\Lambda_{\text{Tr}}(a) = \sum_{p \in I_q} (a \otimes 1_{L^2(I_q)}) f_{0,p,p}$ for $a \in \mathcal{N}_{\text{Tr}}$.
(2) a unital *-homomorphism $\pi_\varphi : M \to B(\mathcal{H}_\varphi)$ such that $\pi_\varphi(a) = a \otimes 1_{L^2(I_q)}$ for all $a \in M$.

Then $(\mathcal{H}_\varphi, \pi_\varphi, \Lambda_{\text{Tr}})$ is a GNS-construction for Tr.

Now we are ready to define the weight that will turn out to be left- and right invariant with respect to Δ. Use the remarks before Proposition 4.27 to define a linear map $\Lambda_\varphi = (\Lambda_{\text{Tr}})_{\gamma^*\gamma} : D(\Lambda_\varphi) \subseteq M \to \mathcal{H}_\varphi$.

Definition 6.9. *We define the faithful normal semi-finite weight* φ *on* M *as* $\varphi = \text{Tr}_{\gamma^*\gamma}$. *By definition,* $(\mathcal{H}_\varphi, \pi_\varphi, \Lambda_\varphi)$ *is a GNS-construction for* φ.

So, on a formal level, $\varphi(x) = \text{Tr}(x\,\gamma^*\gamma)$ and $\Lambda_\varphi(x) = \Lambda_{\text{Tr}}(x\,|\gamma|)$. So we already know that the modular automorphism group σ^φ of φ is such that $\sigma_s^\varphi(x) = |\gamma|^{2is}\, x\, |\gamma|^{-2is}$ for all $x \in M$ and $s \in \mathbb{R}$.

As for any locally compact quantum group we can consider the polar decomposition $S = R\tau_{-\frac{i}{2}}$ of the antipode S of (M, Δ). Thus, R is an anti-*-automorphism of M and τ is a strongly continuous one parameter group of *-automorphisms on M so that R and τ commute. In this example, the following formulas hold:

$$S(\Phi(m, p, t)) = \text{sgn}(p)^{\chi(p)} \text{sgn}(t)^{\chi(t)} (-1)^m\, q^m\, \Phi(m, t, p)$$
$$R(\Phi(m, p, t)) = \text{sgn}(p)^{\chi(p)} \text{sgn}(t)^{\chi(t)} (-1)^m\, \Phi(m, t, p)$$
$$\tau_s(\Phi(m, p, t)) = q^{2mis}\, \Phi(m, p, t)$$
$$\sigma_s^\varphi(\Phi(m, p, t)) = |p^{-1}t|^{2is}\, \Phi(m, p, t)$$

for all $m \in \mathbb{Z}$, $p, t \in I_q$ and $s \in \mathbb{R}$.

To any locally compact quantum group one can associate a multiplicative unitary through the left invariance of the left Haar weight. In this example (and this happens also in other examples) we go the other way around. First we use the orthogonality relations involving the functions a_p (see Proposition 6.3) to produce a partial isometry.

Proposition 6.10. *There exists a unique surjective partial isometry W on $\mathcal{H}_\varphi \otimes \mathcal{H}_\varphi$ such that*

$$W^*(f_{m_1,p_1,t_1} \otimes f_{m_2,p_2,t_2})$$
$$= \sum_{\substack{y,z \in I_q \\ sgn(p_2 t_2)(yz/p_1)q^{m_2} \in I_q}} |t_2/y|\, a_{t_2}(p_1,y)\, a_{p_2}(z, sgn(p_2 t_2)(yz/p_1)q^{m_2})$$
$$\times\ f_{m_1+m_2-\chi(p_1 p_2/t_2 z),z,t_1} \otimes f_{\chi(p_1 p_2/t_2 z),sgn(p_2 t_2)(yz/p_1)q^{m_2},y}$$

for all $m_1, m_2 \in \mathbb{Z}$ and $p_1, p_2, t_1, t_2 \in I_q$.

In a next step one connects this partial isometry with the weight φ by showing that $(\omega\pi_\varphi \otimes \iota)\Delta(a) \in \mathcal{N}_\varphi$ and $\Lambda_\varphi((\omega\pi_\varphi \otimes \iota)\Delta(a)) = (\omega \otimes \iota)(W^*)\Lambda_\varphi(a)$ for all $\omega \in B(\mathcal{H}_\varphi)_*$ and $a \in \mathcal{N}_\varphi$. In turn, this is used to prove the left invariance of φ so that (M,Δ) is indeed a locally compact quantum group and W is the multiplicative unitary naturally associated to (M,Δ):

$$W^*(\Lambda_\varphi(x) \otimes \Lambda_\varphi(y)) = (\Lambda_\varphi \otimes \Lambda_\varphi)(\Delta(y)(x \otimes 1))$$

for all $x,y \in \mathcal{N}_\varphi$. In fact, this formula was used in [KK] to obtain the defining formula for W in Proposition 6.10.

From the general theory of locally compact quantum groups we know that all of the information concerning (M,Δ) is contained in W in the following way:

(1) $\pi_\varphi(M)$ is the σ-strong* closure of $\{\,(\iota\otimes\omega)(W^*) \mid \omega \in B(\mathcal{H}_\varphi)_*\,\}$, in $B(\mathcal{H}_\varphi)$.
(2) $(\pi_\varphi \otimes \pi_\varphi)\Delta(x) = W^*(1 \otimes \pi_\varphi(x))W$ for all $x \in M$.

As a matter of fact, if $m \in \mathbb{Z}$ and $p,t \in I_q$, a concrete element $\omega \in B(\mathcal{H}_\varphi)_*$ can be produced so that $\Phi(m,p,t) = (\iota \otimes \omega)(W^*)$.

Recall that one associates a C*-algebraic quantum group (A_r, Δ_r) to (M,Δ) by requiring that $\pi_\varphi(A_r)$ is the norm closure of the algebra $\{\,(\iota \otimes \omega)(W^*) \mid \omega \in B(\mathcal{H}_\varphi)_*\,\}$ and simply restricting the comultiplication Δ from M to A_r.

In order to describe the C*-algebra A_r in this specific case, we will use the following notation. For $f \in C(\mathbb{T} \times I_q)$ and $x \in I_q$ we define $f_x \in C(\mathbb{T})$ so that $f_x(\lambda) = f(\lambda, x)$ for all $\lambda \in \mathbb{T}$.

For $f \in C_b(\mathbb{T} \times I_q)$, the operator $M_f \in B(\mathcal{H})$ is by definition the left multiplication operator by f on $L^2(\mathbb{T} \otimes I_q)$.

Consider $p \in -q^{\mathbb{Z}} \cup q^{\mathbb{Z}}$. We define a translation operator T_p on $\mathcal{F}(\mathbb{T} \times I_q)$ such that for $f \in \mathcal{F}(\mathbb{T} \times I_q)$, $\lambda \in \mathbb{T}$ and $x \in I_q$, we have that $(T_p f)(\lambda, x) = f(\lambda, px)$ if $px \in I_q$ and $(T_p f)(\lambda, x) = f(\lambda, px) = 0$ if $px \notin I_q$. If $p,t \in I_q$ and $g \in \mathcal{F}(\mathbb{T})$, then $T_p(g \otimes \delta_t) = g \otimes \delta_{p^{-1}t}$, thus, $T_p(g \otimes \delta_t) = 0$ if $p^{-1}t \notin I_q$. We let ρ_p denote the partial isometry in $B(\mathcal{H})$ induced by T_p.

Proposition 6.11. *Denote by $\mathcal{C}$ the C*-algebra of all functions $f \in C(\mathbb{T} \times I_q)$ such that (1) f_x converges uniformly to 0 as $x \to 0$ and (2) f_x converges uniformly to a constant function as $x \to \infty$. Then, A_r is the norm closed linear span, in $B(\mathcal{H})$, of the set $\{ \rho_p M_f \mid f \in \mathcal{C}, p \in -q^{\mathbb{Z}} \cup q^{\mathbb{Z}} \}$.*

If $p, t \in I_q$ and $m \in \mathbb{Z}$, then $\Phi(m, p, t) = \rho_{p^{-1}t} M_{\zeta^m \otimes \delta_t} \in A_r$. Thus, each operator $\Phi(m, p, t)$ belongs to A_r but the C*-algebra A_r is not generated by these operators!

6.2 The bicrossed product of groups

The quantum groups that we present in this subsection are examples generated by construction method 2 and we follow [VV], [BSV] and [BS].

Definition 6.12. *Consider a σ-compact locally compact group G and two closed subgroups G_1, G_2 of G so that $G_1 \cap G_2 = \{e\}$ and such that $G_1 G_2$ has complement of (Haar) measure 0. Then G_1, G_2 is called a matched pair of locally compact groups in G.*

For the next part of this section we fix a matched pair of locally compact groups G_1, G_2 inside a σ-compact locally compact group G.

Note that $G_1 G_2$ is measurable as the countable union of compact sets. Since $G_2 G_1 = (G_1 G_2)^{-1}$, the set $G_2 G_1$ is also measurable and has complement of measure 0.

Define the injective continuous map $\theta : G_1 \times G_2 \to G_1 G_2 : (g_1, g_2) \to g_1 g_2$. Since for all compact subsets $K_1 \subseteq G_1$ and $K_2 \subseteq G_2$ the restriction of $\theta \restriction_{K_1 \times K_2}$ is a homeomorphism from $K_1 \times K_2 \to K_1 K_2$, it follows that θ is a bi-measurable isomorphism from $G_1 \times G_2$ to $G_1 G_2$.

If we define the injective continuous map $\rho : G_1 \times G_2 \to G_1 G_2 : (g_1, g_2) \to g_2 g_1$, we also get a bi-measurable isomorphism from $G_1 \times G_2$ to $G_2 G_1$.

This matched pair of groups defines a partial action α of G_1 on G_2 and a partial action β of G_2 on G_1 as follows. Let $\mathcal{O}$ be the measurable set in $G_1 \times G_2$ defined by

$$\mathcal{O} = \{ (g_1, g_2) \in G_1 \times G_2 \mid g_1 g_2 \in G_2 G_1 \} = \theta^{-1}(G_2 G_1) \,.$$

and

$$\mathcal{O}' = \{ (g_1, g_2) \in G_1 \times G_2 \mid g_2 g_1 \in G_1 G_2 \} = \rho^{-1}(G_1 G_2) \,.$$

If $(g_1, g_2) \in \mathcal{O}$ we define $\alpha_{g_1}(g_2) \in G_2$ and $\beta_{g_2}(g_1) \in G_1$ in such a way that $g_1 g_2 = \alpha_{g_1}(g_2) \beta_{g_2}(g_1)$. So we get a bi-measurable isomorphism $\rho^{-1}\theta : \mathcal{O} \to \mathcal{O}' : (g_1, g_2) \to (\beta_{g_2}(g_1), \alpha_{g_1}(g_2))$

These are partial actions in the following way.

Lemma 6.13. *1. Consider $(g_1, g_2) \in \mathcal{O}$. Then the following holds.*

- Let $h_1 \in G_1$. Then $(h_1 g_1, g_2) \in \mathcal{O} \Leftrightarrow (h_1, \alpha_{g_1}(g_2)) \in \mathcal{O}$ and in this case

$$\alpha_{h_1 g_1}(g_2) = \alpha_{h_1}(\alpha_{g_1}(g_2)) \quad and \quad \beta_{g_2}(h_1 g_1) = \beta_{\alpha_{g_1}(g_2)}(h_1) \beta_{g_2}(g_1) \ .$$

- Let $h_2 \in G_2$. Then $(g_1, h_2 g_2) \in \mathcal{O} \Leftrightarrow (\beta_{g_2}(g_1), h_2) \in \mathcal{O}$ and in this case

$$\beta_{h_2 g_2}(g_1) = \beta_{h_2}(\beta_{h_1}(g_1)) \quad and \quad \alpha_{g_1}(h_2 g_2) = \alpha_{\beta_{g_2}(g_1)}(h_2) \alpha_{g_1}(g_2) \ .$$

2. Consider $g_1 \in G_1$ and $g_2 \in G_2$. Then (g_1, e) and (e, g_2) belong to $\mathcal{O}$ and

$$\alpha_{g_1}(e) = e \ , \quad \alpha_e(g_2) = g_2 \ , \quad \beta_{g_2}(e) \quad and \quad \beta_e(g_1) = g_1 \ .$$

Check the above lemma. Since these actions are not everywhere defined it is important to know that the sets on which they are defined are big enough. This will follow from the next result. The modular functions of G, G_1 and G_2 will be denoted by δ, δ_1 and δ_2 respectively.

Lemma 6.14. *The left Haar measures on G, G_1 and G_2 can be normalized in such a way that for all positive Borel functions $f : G \to \mathbb{R}^+$,*

$$\int_G f(g)\, dg = \int_{G_2} \int_{G_1} f(g_1 g_2)\, \delta(g_2)\, dg_1\, dg_2$$

$$= \int_{G_2} \int_{G_1} f(g_2 g_1)\, \delta_1(g_1^{-1})\, \delta_2(g_2^{-1})\, dg_1\, dg_2 \ .$$

Exercise 6.15. Prove this lemma. Use the formula in the left hand side of this equation to define an integral on $K(G_1 \times G_2)$ and show that it is left invariant. This makes it possible to get the first integral to be equal to the last one. Deduce the remaining equality from this equality.

The above result guarantees that $\mathcal{O}$ and $\mathcal{O}'$ have complement of measure 0 and that $\rho^{-1}\theta : \mathcal{O} \to \mathcal{O}'$ is a measure isomorphism, i.e. if A is a measurable subset of $\mathcal{O}$, then A has measure 0 if and only if $(\rho^{-1}\theta)(A)$ has measure 0. As a consequence, we get an isomorphism of von Neumann algebras

$$\tau : L^\infty(G_1) \otimes L^\infty(G_2) \to L^\infty(G_1) \otimes L^\infty(G_2) : F \mapsto \tau(F) = F \circ (\rho^{-1}\theta) \ .$$

Thus, $\tau(F)(g_1, g_2) = F(\beta_{g_2}(g_1), \alpha_{g_1}(g_2))$ for $F \in L^\infty(G_1 \times G_2)$ and $(g_1, g_2) \in \mathcal{O}$.

Define injective normal *-homomorphisms $\alpha' : L^\infty(G_2) \to L^\infty(G_1) \otimes L^\infty(G_2)$ and $\beta' : L^\infty(G_1) \to L^\infty(G_1) \otimes L^\infty(G_2)$ by

$$\alpha'(f) = \tau(1 \otimes f) \quad and \quad \beta'(g) = \tau(g \otimes 1)$$

for all $f \in L^\infty(G_2)$ and $g \in L^\infty(G_1)$. Lemma 6.13 implies that α' and β' are coactions with respect to $L^\infty(G_1)$ and $L^\infty(G_2)$, respectively:

$$(\iota \otimes \alpha')\alpha' = (\Delta_{G_1} \otimes \iota)\alpha' \qquad \text{and} \qquad (\beta' \otimes \iota)\beta' = (\iota \otimes \Delta^\circ_{G_2})\beta' \ ,$$

where $\Delta^\circ_{G_2} = \chi \Delta_{G_2}$ is the opposite comultiplication. In this respect we define the von Neumann algebras M and $\hat{M}$ on $L^2(G_1 \times G_2)$ as the following crossed products

$$M = (\alpha'(L^\infty(G_2)) \cup (\mathcal{L}(G_1) \otimes 1))'' \quad \text{and} \quad \hat{M} = (\beta'(L^\infty(G_2)) \cup (1 \otimes \mathcal{L}(G_2)))'' \ .$$

Let W_{G_1}, W_{G_2} be the canonical multiplicative unitaries associated to G_1 and G_2 respectively. Define unitaries $\hat{W}$ and W on $L^\infty(G_1) \otimes L^\infty(G_2) \otimes L^\infty(G_1) \otimes L^\infty(G_2)$ such that $W = \Sigma \hat{W}^* \Sigma$ and

$$\hat{W} = (\beta' \otimes \iota \otimes \iota)(W_{G_1} \otimes 1)(\iota \otimes \iota \otimes \alpha')(1 \otimes \Sigma W^*_{G_2} \Sigma) \ .$$

These unitaries can be used to define normal *-homomorphisms $\Delta : M \to M \otimes M$ and $\hat{\Delta} : \hat{M} \to \hat{M} \otimes \hat{M}$ by

$$\Delta(x) = W^*(1 \otimes x)W \qquad \text{and} \qquad \hat{\Delta}(y) = \hat{W}(y \otimes 1)\hat{W}^*$$

for all $x \in M$ and $y \in \hat{M}$.

Theorem 6.16. *Both (M, Δ) and $(\hat{M}, \hat{\Delta})$ are locally compact quantum groups that are each others Pontryagin dual. We call (M, Δ) the bicrossed product of G_1 and G_2.*

The constructions in [VV] are even more general. They start of with locally compact quantum groups and allow *cocycles* to serve as ingredients but we will not go further into this.

Example 6.17. In this example we show how this procedure can be used to construct a deformation of the $ax + b$-group. Define the group G as $G = \{ (a, b) \mid a \in \mathbb{R} \setminus \{0\}, b \in \mathbb{R} \}$ and define the product on G by $(a, b)(c, d) = (ac, d + cb)$ for all $(a, b), (c, d) \in G$. We embed $\mathbb{R} \setminus \{0\}$ into G in 2 different ways. We set

$$G_1 = \{ (s, s - 1) \mid s \in \mathbb{R} \setminus \{0\} \} \qquad \text{and} \qquad G_2 = \{ (s, 0) \mid s \in \mathbb{R} \setminus \{0\} \} \ .$$

Then, G_1, G_2 is a matched pair inside G. Calculate $\mathcal{O}$ and the partial action α and β.

The quantum group (M, Δ) constructed from this data is not compact, not discrete, and non-unimodular. The scaling group is non-trivial and the left and right Haar weights are not traces. This quantum group is self-dual, i.e. $(M, \Delta) \cong (\hat{M}, \hat{\Delta})$ and the scaling constant is 1.

For any locally compact quantum group one can construct the multiplicative unitary. In [BS] one starts from a multiplicative unitary and looks at the possibility of associating C*-algebras and comultiplications to such a multiplicative unitary.

So let $\mathcal{H}$ be a Hilbert space and $W \in B(\mathcal{H} \otimes \mathcal{H})$ a unitary element that satisfies the Pentagonal equation

$$W_{12}\, W_{13}\, W_{23} = W_{23}\, W_{12}\ .$$

One defines A as the norm closure of the vector space $\{\,(\iota \otimes \omega)(W) \mid \omega \in B(\mathcal{H})_*\,\}$ and a *-homomorphism $\Delta : A \to B(\mathcal{H} \otimes \mathcal{H})$ by $\Delta(x) = W^*(1 \otimes x)W$ for all $x \in A$ (and similarly for left slices). Note that A is a subalgebra of $B(\mathcal{H})$. In retrospect it is a natural problem to look for conditions on W that imply that

1. A is a C*-algebra,
2. $\Delta(A) \subseteq M(A \otimes A)$ and $\Delta : A \to M(A \otimes A)$ is a non-degenerate *-homomorphism,
3. $\Delta(A)(1 \otimes A)$ and $\Delta(A)(A \otimes 1)$ are dense in $A \otimes A$.

In [BS] the authors introduced the set $\mathcal{C}(W)$ as the norm closure of the subspace $\{\,(\omega \otimes \iota)(W^*\Sigma) \mid \omega \in B(\mathcal{H})_*\,\}$ and proved that conditions (1), (2) and (3) above are satisfied if $\mathcal{C}(W) = B_0(\mathcal{H})$. One calls a multiplicative unitary regular if $\mathcal{C}(W) = B_0(\mathcal{H})$.

The paper [Wor96] introduces another condition on a multiplicative unitary, called manageability, that implies among other things that conditions (1), (2) and (3) above are satisfied. A multiplicative unitary associated to a locally compact quantum group is always manageable but not always regular (Baaj showed that this is not the case for quantum $E(2)$).

For quite a while one suspected that if W is manageable, then W is semiregular in the sense that $B_0(\mathcal{H}) \subseteq \mathcal{C}(W)$. But a more sophisticated version of example 6.17 produced a locally compact quantum group for which the multiplicative unitary is not semi-regular (see [BSV]).

7 Appendix : several concepts

Closed linear mappings

Although we would prefer to always work with continuous linear maps that are everywhere defined, reality forces us to work with unbounded linear maps in many situations. Most of the time, these non-continuous linear operators are still controllable to a certain degree because they are *closed*.

Consider topological vector spaces X and Y and $T : D(T) \subseteq X \to Y$ a linear map (so $D(T)$ is by definition a subspace of X). We say that T is *closed* if the graph $G(T) := \{\,(v, T(v)) \mid v \in D(T)\,\}$ of T is closed with respect to the product topology on $X \times Y$.

Sometimes it is easier to work with the following characterization (which follows immediately from the definition): T is closed $\Leftrightarrow$ for all $v \in X$, $w \in Y$ and every net $(v_i)_{i \in I}$ in $D(T)$ the following holds:

$$(v_i)_{i \in I} \to v \quad \text{and} \quad \big(T(v_i)\big)_{i \in I} \to w \quad \Rightarrow \quad v \in D(T) \text{ and } T(v) = w \, .$$

Let $T : D(T) \subseteq X \to Y$ be a closed linear map. In a lot of cases, we do not know the precise domain of T but only the action of T on a sufficiently big subspace of $D(T)$. Let us explain this more carefully. We call V a *core* for T if V is a subspace of $D(T)$ so that $\{ (v, T(v)) \mid v \in V \}$ is dense in $G(T)$. Thus, a subspace $V \subseteq D(T)$ is a core for $T \Leftrightarrow$ for all $v \in D(T)$, there exists a net $(v_i)_{i \in I}$ in V such that $(v_i)_{i \in I} \to v$ and $\big(T(v_i)\big)_{i \in I} \to T(v)$.

Let $S : D(S) \subseteq X \to Y$ and $T : D(T) \subseteq X \to Y$ be two closed linear maps such that there exists a subspace $V \subseteq D(S) \cap D(T)$ so that V is a core for S and for T. Check that if $S(v) = T(v)$ for all $v \in V$, then $S = T$.

Related to the notion of a 'core' we have the following. Let $T : D(T) \subseteq X \to Y$ be a linear map. We call T *closable* if there exists a (necessarily closed, linear and unique) map $\bar{T} : D(\bar{T}) \subseteq X \to Y$ so that $G(\bar{T}) = \overline{G(T)}$, where the last set is the closure of $G(T)$ in $X \times Y$. In this case, we call $\bar{T}$ the *closure* of T. It is clear that $\bar{T}$ extends T and that $D(T)$ is a core for $\bar{T}$.

One easily checks that T is closable if and only if for every $u \in X$, the following holds:

If there exist $v, w \in Y$ so that $(u, v), (u, w) \in \overline{G(T)}$, then $v = w \, .$

This leads to the following characterization: T is closable $\Leftrightarrow$ for every net $(v_i)_{i \in I}$ in $D(T)$ and for every $w \in Y$, the following holds

$$(v_i)_{i \in I} \to 0 \text{ and } \big(T(v_i)\big)_{i \in I} \to w \quad \Rightarrow \quad w = 0 \, .$$

Note that if $T : D(T) \subseteq X \to Y$ is a closed linear operator with core V, the linear operator $T{\restriction}_V$ is closable and has T as its closure.

Let us also recall the following result from functional analysis.

Theorem 7.1 (Closed Graph Theorem). *Consider Banach spaces E and F and $T : E \to F$ a linear map. Then T is bounded if and only if T is closed.*

Injective positive operators and their powers

In the theory of locally compact quantum groups a central role is played by injective, positive, self-adjoint operators. Let us collect the basic results and terminology. Fix a Hilbert space H and recall the following terminology.

Definition 7.2. *Let T be a densely defined, linear operator $T : D(T) \subseteq H \to H$. We define the adjoint operator $T^* : D(T^*) \subseteq H \to H$ as follows. The domain $D(T^*)$ consists of vectors $v \in H$ for which there exists a vector $w \in H$ so that $\langle Tu, v \rangle = \langle u, w \rangle$ for all $u \in D(T)$. If $v \in D(T^*)$, the vector w above is unique and we define $T^*(v) = w$.*

Recall that an operator is densely defined $\Leftrightarrow D(T)$ is dense in H.

Check that T^* is a closed linear operator. The operator T^* does not have to be densely defined. One can show that T^* is densely defined if and only if T is closable. If T is closable, then T^{**} is the closure of T.

Definition 7.3. *Let T be a linear operator $T : D(T) \subseteq H \to H$. Then we call T self-adjoint if T is densely defined and $T^* = T$. The last equality is equivalent to the following two conditions:*

1. *$\langle Tu, v \rangle = \langle u, Tv \rangle$ for all $u, v \in D(T)$.*
2. *Let $v, w \in H$. If $\langle Tu, v \rangle = \langle u, w \rangle$ for all $u \in D(T)$, then $v \in D(T)$ and $T(v) = w$.*

We call T positive if T is self-adjoint and $\langle Tv, v \rangle \geq 0$ for all $v \in D(T)$.

A self-adjoint operator is closed. If T is positive, the spectrum of T is contained in $\mathbb{R}^+$. You probably know that such a (self-adjoint) positive operator T has a spectral decomposition $T = \int_{\mathbb{R}^+} \lambda \, dE(\lambda)$ (see for instance [Con] for a good exposition about the spectral decomposition of normal operators).

The spectral measure E is a σ-additive mapping from the σ-algebra $\mathcal{B}(\mathbb{R}^+)$ of all Borel subsets of $\mathbb{R}^+$ into the set of orthogonal projections on H such that $E(\mathbb{R}^+) = 1$, $E(\emptyset) = 0$ and $E(\Delta_1 \cap \Delta_2) = E(\Delta_1) E(\Delta_2)$. If $\xi, \eta \in H$, one gets an ordinary complex measure $E_{\xi,\eta}$ on $\mathcal{B}(\mathbb{R}^+)$ by defining $E_{\xi,\eta}(\Delta) = \langle E(\Delta)\xi, \eta \rangle$ for all $\Delta \in \mathcal{B}(\mathbb{R}^+)$; it is positive if $\xi = \eta$. If $f : \mathbb{R}^+ \to \mathbb{C}$ is a measurable function, the densely defined closed linear operator $\int f \, dE$ in H is defined such that

$$D\left(\int f \, dE \right) = \{ \xi \in H \mid \int |f|^2 \, dE_{\xi,\xi} < \infty \}$$

and if $\xi \in D(\int f \, dE)$, then f is integrable with respect to $E_{\xi,\eta}$ and

$$\left\langle \left(\int f \, dE \right) \xi, \eta \right\rangle = \int f \, dE_{\xi,\eta}$$

for all $\eta \in H$. If $\xi \in D(\int f \, dE)$, then $\| (\int f \, dE)\xi \|^2 = \int |f|^2 \, dE_{\xi,\xi}$.

The spectral measure E is uniquely determined by the fact that $\int \lambda \, dE(\lambda) = T$.

If $f : \mathbb{R}^+ \to \mathbb{C}$ is a bounded measurable function, then $\int f \, dE$ belongs to $B(H)$ and the function $\mathcal{L}^\infty(\mathbb{R}^+) \to B(H) : f \to \int f \, dE$ is a *-homomorphism. For unbounded measurable complex valued functions this map still has certain multiplicativity and additivity properties, but these have to be carefully stated because the operators $\int f \, dE$ are not everywhere defined.

Since T is injective, $E(\{0\}) = 0$ so we can integrate measurable complex valued functions only defined on $\mathbb{R}^+ \setminus \{0\}$. If $z \in \mathbb{C}$, we define the measurable function $g : \mathbb{R}^+ \setminus \{0\} \to \mathbb{C}$ by $g(\lambda) = \lambda^z$ for all $\lambda \in \mathbb{R}^+ \setminus \{0\}$. The complex power T^z is the closed, densely defined linear operator in H defined as $T^z = \int g \, dE$.

If $t \in \mathbb{R}$, then T^{it} is a unitary operator in H and T^t is an injective positive operator in H is. We have different familiar rules for these powers. For instance $T^{s+t} = T^s T^t$ for all $s, t \in \mathbb{R}^+$ and $T^{-s} = (T^s)^{-1}$ for all $s \in \mathbb{R}$ (where $(T^s)^{-1}$ is the inverse of T^s).

Integrating vector-valued functions

We take a pragmatic approach to integrating vector-valued function in these lecture notes and only integrate over closed intervals (that might be infinite). We will only integrate continuous vector valued functions taking their values in a Banach a space or a von Neumann algebra.

Definition 7.4. *Consider a locally convex topological vector space X and let us denote the space of continuous linear functionals on X by X^*. Let I be a closed interval in $\mathbb{R}$ and $f : I \to X$ a function. We call f integrable if there exists an element $x \in X$ so that for $\omega \in X^*$,*

1. $\omega \circ f : I \to \mathbb{C}$ is integrable and

2. $\omega(x) = \int_I \omega(f(t)) \, dt$

If f is integrable, the element x above is unique and we define $\int_I f = \int_I f(t) \, dt = x$.

The local convexity of X implies that X^* separates points of X. This integral is immediately seen to be linear.

Let us look at the two simple integrability results that we need in these lecture notes.

Proposition 7.5. *Consider a Banach space E and a norm continuous function $f : \mathbb{R} \to E$ so that $\|f\|$ integrable. Then f is integrable.*

Sketch of proof.

(1) Let $n \in \mathbb{N}$. Prove that f is integrable on $[-n, n]$. Define a sequence of partitions $(P_k)_{k=1}^{\infty}$ of $[-n, n]$ as follows: $P_1 = \{-n, n\}$ and for $k \in \mathbb{N}$ construct P_{k+1} from P_k by adding the midpoint of two neighboring points of P_k. Associate to each partition P_k an obvious element x_k that approximates $\int_{-n}^{n} f$. Prove that $(x_k)_{k=1}^{\infty}$ is a Cauchy-sequence with respect to the norm on E. The limit will be of course the integral of f over $[-n, n]$.

(2) Show that $\left(\int_{-n}^{n} f(t) \, dt \right)_{n=1}^{\infty}$ is a Cauchy sequence.

Notice that Hahn-Banach implies that $\| \int_{\mathbb{R}} f \| \leq \int_{\mathbb{R}} \|f\|$. The same is true for the integral in the next proposition.

In the framework of von Neumann algebras a similar existence result follows easily from the natural isometric isomorphism $(M_*)^* \cong M$.

Proposition 7.6. *Consider a von Neumann algebra M and a σ-weakly continuous function $f : \mathbb{R} \to M$ so that $\|f\|$ is integrable. Then f is integrable.*

The 'leg-numbering' notation

Consider a Hilbert space H and an operator $X \in B(H \otimes H)$. Then we define the operators X_{12}, X_{13} and X_{23} in $B(H \otimes H \otimes H)$ as follows:

$$X_{12} = X \otimes 1, \qquad X_{23} = 1 \otimes X \qquad X_{13} = (\Sigma \otimes 1)(1 \otimes X)(\Sigma \otimes 1),$$

where $\Sigma : H \otimes H \to H \otimes H$ is the unitary flip map. One can also look at this in the following way. Firstly, the mappings $B(H \otimes H) \to B(H \otimes H \otimes H)$: $X \mapsto X_{12}$, $B(H \otimes H) \to B(H \otimes H \otimes H)$: $X \mapsto X_{23}$ and $B(H \otimes H) \to B(H \otimes H \otimes H)$: $X \mapsto X_{13}$ are σ-strongly* continuous and for simple tensors

$$(x \otimes y)_{12} = x \otimes y \otimes 1 \qquad (x \otimes y)_{23} = 1 \otimes x \otimes y \qquad (x \otimes y)_{13} = x \otimes 1 \otimes y,$$

where $x, y \in B(H)$. In general, if $i, j \in \{1, 2, 3\}$ and $i \neq j$ and k is the remaining element of $\{1, 2, 3\}$, then $(x \otimes y)_{ij}$ is the element in $B(H \otimes H \otimes H)$ obtained by putting x on the i-th spot, putting y on the j-th spot and putting 1 on the k-th spot.

For instance, if $X \in B(H \otimes H)$ then $X_{32} \in B(H \otimes H \otimes H)$ is defined as $X_{32} = 1 \otimes \Sigma X \Sigma$ and one sees that indeed $(x \otimes y)_{32} = 1 \otimes y \otimes x$ for all $x, y \in B(H)$.

One also uses variations of this notation if there are 3 or more subscripts.

References

[Abe]   E. Abe *Hopf algebras*. Cambridge Tracts in Mathematics 74, Cambridge University Press, Cambridge, 1980.

[ASC]   W.A. Al-Salam and L. Carlitz *Some orthogonal q-polynomials*. Math. Nach., 30:47–61, 1965.

[Baa80] S. Baaj *Multiplicateurs non bornés*. Thèse 3ème cycle, Université Paris 6, 1980.

[Baa92] S. Baaj *Représentation régulière du groupe quantique $E_\mu(2)$*. C.R. Acad. Sci., Paris, Sér. I, 314:1021–1026, 1992.

[Baa95] S. Baaj *Représentation régulière du groupe quantique des déplacements de Woronowicz*. Astérisque, 232:11–48, 1995.

[BJ]   S. Baaj and P. Julg *Théorie bivariante de Kasparov et opérateurs non bornés dans les C*-modules hilbertiens*. C.R Acad. Sci. Paris, 296:875–878, 1983.

[BS]   S. Baaj and G. Skandalis *Unitaires multiplicatifs et dualité pour les produits croisés de C*-algèbres*. Ann. scient. Éc. Norm. Sup., 4e série, 26:425–488, 1993.

[BV]   S. Baaj and S. Vaes *Double crossed products of locally compact quantum groups*. Preprint Jussieu, #math.OA/0211145 (electronic), 2001.

[BSV]   S. Baaj, S. Skandalis and S. Vaes *Non-semi-regular quantum groups coming from number theory*. Comm. Math. Phys., 235(1):139–167, 2003.

[Ban96] T. Banica *Theorie des representations du groupé quantique compact libre $O(n)$*. C. R. Acad. Sci., Paris, Série. I, 322(2):241–244, 1996.

[Ban97] T. Banica *Le groupe quantique compact libre U(n)*. Commun. Math. Phys., 190(1):143–172, 1997.

[CKK] N. Ciccoli, E. Koelink and T. Koornwinder *q-Laguerre polynomials and big q-Bessel functions and their orthogonality relations*. Methods and Applications of Analysis., 6(1):109–127, 1999.

[Coh] D.L. Cohn *Measure Theory*. Birkhäuser, 1980.

[Con] J.B. Conway *A Course in Functional Analysis*. Springer-Verlag, 1991.

[ER] E. Effros and Z.-J. Ruan *Discrete quantum groups I: The Haar measure*. Int. J. Math., 5(5):681–723, 1994.

[Eno77] M. Enock *Produit croisé d'une algèbre de von Neumann par une algèbre de Kac*. J. Funct. Anal., 26(1):16–47, 1977.

[Eno98] M. Enock *Inclusions irréductibles de facteurs et unitaires multiplicatifs II*. J. Funct. Anal., 154(1):67–109, 1998.

[Eno99] M. Enock *Sous-facteurs intermédiaires et groupes quantiques mesurés*. J. Operator Theory, 42(2):305–330, 1999.

[EN] M. Enock and R. Nest *Irreducible inclusions of factors, multiplicative unitaries and Kac algebras*. J. Funct. Anal., 137(2):466–543, 1996.

[ES73] M. Enock and J.-M. Schwartz *Une dualité dans les algèbres de von Neumann*. C. R. Acad. Sc. Paris, 277:683–685, 1973.

[ES75] M. Enock and J.-M. Schwartz *Une dualité dans les algèbres de von Neumann*. Supp. Bull. Soc. Math. France, Mémoire 44, 103(4):1–144, 1975.

[ES80] M. Enock and J.-M. Schwartz *Produit croisé d'une algèbre de von Neumann par une algèbre de Kac II*. Publ. RIMS, 16(1):189-232, 1980.

[ES92] M. Enock and J.-M. Schwartz *Kac algebras and duality of locally compact groups*. Springer-Verlag, 1992.

[EV] M. Enock and J.-M. Vallin *C*-algèbres de Kac et algèbres de Kac*. Proc. London Math. Soc. (3), 66(3):619–650, 1993.

[GR] G. Gasper and M. Rahman *Basic hypergeometric series*. Cambridge University Press, 1990.

[Haa] U. Haagerup *The standard form of von Neumann algebras*. Math. Scand., 37(2):271–283, 1976.

[Kac61] G.I. Kac *A generalization of the principle of duality for groups*. Soviet Math. Dokl., 138:275–278, 1961.

[Kac63] G.I. Kac *Ring groups and the principle of duality, I, II*. Trans. Moscow Math. Soc., 12:259–301,1963 and 13:84–113, 1965.

[KaV73] G.I. Kac and L.I. Vainerman *Nonunimodular ring-groups and Hopf-von Neumann algebras*. Soviet Math. Dokl., 14:1144–1148, 1973.

[KaV74] G.I. Kac and L.I. Vainerman *Nonunimodular ring-groups and Hopf-von Neumann algebras*. Math. USSR, Sbornik, 23:185–214, 1974.

[KR1] R.V. Kadison and J.R. Ringrose *Fundamentals of the theory of operator algebras I*. Graduate Studies in Mathematics, 15. Am. Math. Soc., 1997.

[KR2] R.V. Kadison and J.R. Ringrose *Fundamentals of the theory of operator algebras II*. Graduate Studies in Mathematics, 15. Am. Math. Soc., 1997.

[Kir] E. Kirchberg Lecture on the conference 'Invariants in operator algebras, Copenhagen'. 1992.

[Kor] L.I. Korogodskii *Quantum Group $SU(1,1) \rtimes \mathbb{Z}_2$ and super tensor products*. Commun. Math. Phys., 163(3):433–460, 1994.

[KK] E. Koelink and J. Kustermans *A locally compact quantum group analogue of the normalizer of $SU(1,1)$ in $SL(2,\mathbb{C})$*. Communications Math. Phys., 233(2):231-296, 2003.

[ES01]   E. Koelink and J.V. Stokman, with an appendix by M. Rahman *Fourier transforms on the quantum SU(1, 1) group.* Publ. Res. Inst. Math. Sci., 37(4):621–715, 2001.

[ES03]   E. Koelink and J.V. Stokman *The big q-Jacobi function transform.* Constr. Approx., 19(2):191–235, 2003.

[Kus01]   J. Kustermans *Locally compact quantum groups in the universal setting.* Int. J. Math., 12(3):289–338, 2001.

[Kus02]   J. Kustermans *The analytic structure of algebraic quantum groups.* Journal Algebra, 259(2):415-450, 2003.

[KuV99]   J. Kustermans and S. Vaes *A simple definition for locally compact quantum groups.* C.R. Acad. Sci., Paris, Sér. I, 328(10):871–876, 1999.

[KuV00a] J. Kustermans and S. Vaes *Locally compact quantum groups.* Ann. scient. Éc. Norm. Sup., 4è série 33:837–934, 2000.

[KuV00b] J. Kustermans and S. Vaes *The operator algebra approach to quantum groups.* Proc. Natl. Acad. Sci. USA, 97(2):547–552, 2000.

[KuV03]   J. Kustermans and S. Vaes *Locally compact quantum groups in the von Neumann algebraic setting.* Math. Scand., 92(1):68–92, 2003.

[KVD]   J. Kustermans and A. Van Daele *C*-algebraic quantum groups arising from algebraic quantum groups.* Int. J. Math., 8(8):1067–1139, 1997.

[Lan]   C. Lance *Hilbert C*-modules. A toolkit for operator algebraists.* London Math. Soc. Lect. Note Series 210, Cambridge University Press, Cambridge, 1995.

[MVD]   A. Maes and A. Van Daele *Notes on compact quantum groups.* Nieuw Arch. Wiskd. IV., 16(1-2):73–112, 1998.

[MN]   T. Masuda and Y. Nakagami *A von Neumann algebra framework for the duality of the quantum groups.* Publ. RIMS, Kyoto University, 30(5):799–850, 1994.

[MNW]   T. Masuda, Y. Nakagami and S. L. Woronowicz *A C*-algebraic framework for quantum groups.* Int. J. Math., 14(9):903–1002, 2003.

[Mur]   G. Murphy *C*-algebras and operator theory.* Academic Press, 1990.

[PT]   G.K. Pedersen and M. Takesaki *The Radon-Nikodym theorem for von Neumann algebras.* Acta Math., 130:53–87, 1973.

[PoW]   P. Podleś and S.L. Woronowicz *Quantum deformation of Lorentz group.* Commun. Math. Phys., 130(2):381–431, 1990.

[PuW]   W. Pusz and S.L. Woronowicz *A quantum GL(2, ℂ) group at roots of unity.* Rep. Math. Phys., 47(3):431–462, 2001.

[Ros]   M. Rosso *Algèbres enveloppantes quantifiées, groupes quantiques compacts de matrices et calcul différentiel non commutatif.* Duke Math. J., 61(1):11–40, 1990.

[Rud]   W. Rudin *Real and complex analysis.* McGraw-Hill, 1987.

[Sch]   K. Schmüdgen *Unbounded operator algebras and representation theory.* Operator Theory 37, Birkhäuser, 1990.

[Tak79]   M. Takesaki *Theory of Operator Algebras I.* Springer-Verlag, 1979.

[Tak02a] M. Takesaki *Theory of Operator Algebras II.* Encyclopaedia of Mathematical Sciences, Springer-Verlag, 2002.

[Tak02b] M. Takesaki *Theory of Operator Algebras III.* Encyclopaedia of Mathematical Sciences, Springer-Verlag, 2002.

[Va00]   S. Vaes *Examples of locally compact quantum groups through the bicrossed product construction.* Proceedings of the XIIIth Int. Conf. Math. Phys. London, 2000, Int. Press, Boston, 341–348, 2000.

[Va01a]   S. Vaes *Locally compact quantum groups.* Ph.D. Thesis K.U.Leuven, 2001.

[Va01b]   S. Vaes *A Radon-Nikodym theorem for von Neumann algebras.* J. Operator Theory, 46(3):477–489, 2001.

[Va01c]   S. Vaes *The unitary implementation of a locally compact quantum group action.* J. Funct. Anal., 180(2):426–480, 2001.

[Va02]   S. Vaes *Strictly outer actions of groups and quantum groups.* Preprint Jussieu, #math.OA/0211272 (electronic), 2002.

[VV]   S. Vaes and L. Vainerman *Extensions of locally compact quantum groups and the bicrossed product construction.* Adv. Math., 175(1):1–101, 2003.

[Val]   J.-M. Vallin *C*-algèbres de Hopf et C*-algèbres de Kac.* Proc. London Math. Soc. (3), 50(1):131–174, 1985.

[VD94]   A. Van Daele *Multiplier Hopf Algebras.* Trans. Amer. Math. Soc., 342(2):917–932, 1994.

[VD95]   A. Van Daele *The Haar measure on a compact quantum group.* Proc. Amer. Math. Soc., 123(10):3125-3128, 1995.

[VD96]   A. Van Daele *Discrete quantum groups.* Journal of Algebra, 180:431–444, 1996.

[VD98]   A. Van Daele *An algebraic framework for group duality.* Adv. Math., 140(2):323–366, 1998.

[VD01]   A. Van Daele *The Haar measure on some locally compact quantum groups.* Preprint KU Leuven, #math.OA/0109004 (electronic), 2001.

[VDWa]   A. Van Daele and S.Z. Wang *Universal quantum groups.* Int. J. Math., 7(2):255–263, 1996.

[VDWo]   A. Van Daele and S.L. Woronowicz *Duality for the quantum E(2) group.* Pac. J. Math., 173(2):375–385, 1996.

[Wan95]   S.Z. Wang *Free Products of compact quantum groups.* Commun. Math. Phys., 167(3):671–692, 1995.

[Wan97]   S.Z. Wang *Krein duality for compact quantum groups.* J. Math. Phys., 38(1):524–534, 1997.

[Wor87a] S.L. Woronowicz *Compact matrix pseudogroups.* Comm. Math. Phys., 111(4):613–665, 1987.

[Wor87b] S.L. Woronowicz *Twisted SU(2) group. An example of a non-commutative differential calculus.* Publ. RIMS, Kyoto University, 23(1):117–181, 1987.

[Wor88]   S.L. Woronowicz   *Tannaka-Krein duality for compact matrix pseudogroups. Twisted SU(N) groups.* Invent. Math., 93(1):35–76, 1988.

[Wor89]   S.L. Woronowicz *Differential calculus on compact matrix pseudogroups (quantum groups).* Commun. Math. Phys., 122(1):125–170, 1989.

[Wor91a] S.L. Woronowicz *Quantum E(2) group and its Pontryagin dual.* Lett. Math. Phys., 23(4):251–263, 1991.

[Wor91b] S.L. Woronowicz *Unbounded elements affiliated with C*-algebras and non-compact quantum groups.* Commun. Math. Phys., 136(2):399–432, 1991.

[Wor95]   S.L. Woronowicz *C*-algebras generated by unbounded elements.* Rev. Math. Phys., 7(3):481–521, 1995.

[Wor96]   S.L. Woronowicz *From multiplicative unitaries to quantum groups.* Int. J. Math., 7(1):127–149, 1996.

[Wor98]   S.L. Woronowicz *Compact quantum groups, in 'Symétries quantiques' (Les Houches, 1995).* North-Holland,Amsterdam, 845–884, 1998.

[Wor00]   S.L. Woronowicz *Extended SU(1,1) quantum group. Hilbert space level.* Preprint KMMF, not completely finished yet., 2000.

[Wor01]   S.L. Woronowicz *Quantum az + b group on complex plane.* Internat. J.
          Math., 12(4):461–503, 2001.
[WZ]   S.L. Woronowicz and S. Zakrzewski *Quantum ax + b group.* Rev. Math.
          Phys., 14(7-8):797–828, 2002.

Quantum Stochastic Analysis – an Introduction

J. Martin Lindsay

School of Mathematical Sciences
University of Nottingham, University Park
Nottingham, NG7 2RD, UK
martin.lindsay@nottingham.ac.uk

Introduction

By *quantum stochastic analysis* is meant the analysis arising from the natural operator filtration of a symmetric Fock space over a Hilbert space of square-integrable vector-valued functions on the positive half-line. Current texts on quantum stochastics are the monograph [Par], the lecture notes [Mey], the St. Flour lectures [Bia], and the Grenoble lectures [Hud]. Excellent background together with a wealth of examples may be found in these, each of which has its own emphasis. The point of view of these notes is closest to [Bia], as far as the basic construction of quantum stochastic integrals goes. Beyond that, particular emphasis is given to *Markovian cocycles*.

Below is an outline of the course. The first section collects some general background material, including reviews of symmetric Fock space and operator-theoretic positivity, and an introduction to operator spaces with particular emphasis on an analogue of $M_{n,m}(V)$ in which $\mathbb{C}^n$ and $\mathbb{C}^m$ are replaced by Hilbert spaces. The operator spaces V appearing here will be 'concrete', that is closed subspaces of $B(H; K)$ for some Hilbert spaces H and K; $M_{n,m}(V)$ is thereby viewed as a closed subspace of $B(H^m; K^n)$. Quantum stochastic processes are introduced in the second section where exponential domains, adaptedness, quantum Brownian motion, martingales and the fundamental process of creation, number/exchange and annihilation are all defined. In section three quantum stochastic integration is founded on abstract Wiener space analysis, in particular the divergence and gradient of Malliavin calculus and the nonadapted integral of Hitsuda and Skorohod. The quantum Itô formula, whose crudest form is $dA_t dA_t^* = dt$ (cf. $(dB_t)^2 = dt$, for Brownian motion), is then derived from the 'Skorohod isometry'. The fourth section is the heart of the course. The meaning of *solution* for a quantum stochastic differential equation is explained there, and how Picard iteration yields the solution for a natural class of coefficients is also described. Operator spaces provide a natural and efficacious context for considering these equations. Section five begins by describing how classes of Markovian cocycles have an infinitesimal description, as solution of a quantum stochastic differential equation, and then poses the question: how can a property of the cocycle (such as positivity, contractivity or being *-homomorphic) be recognised from the infinitesimal description, i.e. from its *stochastic generator*? The sixth section shows how Markovian cocycles may be constructed using quantum stochastic calculus to provide *-homomorphic *stochastic dilations* of completely positive contraction semigroups (also called quantum dynamical semigroups) on a C^*-algebra or von Neumann algebra, and also how solving quantum stochastic differential equations allows the realisation of Markovian cocycles as perturbations of cocycles with simpler stochastic generators. This is a key step in the classification of E-semigroups arising from Fock space Markovian cocycles (cf. Rajarama Bhat's lectures in this volume). Section six is followed by a brief Afterword containing a taste of non-introductory material. Bibliographical references are largely confined to notes at the end of each section.

Notations and conventions

A glossary of notations and conventions used in these notes may be found at the end; here are the main ones. All linear spaces here are complex, unless declared otherwise, and inner products are linear in their *second* argument (unlike in Johan Kustermans' notes in this volume). Justification for the convention used here is heightened when operator spaces are in play, due to the efficacy of the following Dirac-inspired *bra- and -ket* notation. For Hilbert-space vectors $u \in \mathsf{k}$ and $x \in \mathsf{h}$, the prescriptions

$$\lambda \mapsto \lambda u \text{ and } y \mapsto \langle x, y \rangle \tag{0.1}$$

define operators $|u\rangle \in B(\mathbb{C}; \mathsf{k})$ and $\langle x| \in B(\mathsf{h}; \mathbb{C})$ and thus also an operator $|u\rangle\langle x| \in B(\mathsf{h}; \mathsf{k})$. The map $u \mapsto |u\rangle$ is an isometric isomorphism and, due to the Riesz-Fréchet Theorem, $x \mapsto \langle x|$ defines a conjugate-linear isometric isomorphism. For a vector v in a Hilbert space H, the following notation is used

$$\widehat{v} := \begin{pmatrix} 1 \\ v \end{pmatrix} \in \widehat{\mathsf{H}}, \text{ where } \widehat{\mathsf{H}} := \mathbb{C} \oplus \mathsf{H}. \tag{0.2}$$

Apart from algebraic tensor products, here denoted $\otimes$, spatial and ultra-weak tensor products, denoted $\otimes_{\mathrm{sp}}$ and $\overline{\otimes}$ respectively, are used; these are defined on page 194. A *matrix-space tensor product* $\otimes_{\mathrm{M}}$ will also be introduced.

The following generalisation of the standard indicator function notation is used throughout. For a vector-valued function F and subset I of its domain (here always a subinterval of $\mathbb{R}_+$),

$$F_I : t \mapsto \begin{cases} F(t) & \text{if } t \in I, \\ 0 & \text{otherwise,} \end{cases}$$

defines a function with the same domain and codomain, generalising the standard indicator-function notation. This also applies to vectors, by viewing them as constant functions defined on $\mathbb{R}_+$, thus for example if c is a vector then $c_{[0,t[}$ denotes the function equal to c on $[0, t[$ and 0 on $[t, \infty[$.

Finally, there is *no* (noncommutative) *significance* to be attached to the fact that integrals over subsets of $\mathbb{R}$ are often written $\int ds \cdots$, rather than $\int \cdots ds$.

Warning. Coefficients of operator quantum stochastic differential equations are operators on the Hilbert space $\widehat{\mathsf{k}} \otimes \mathsf{h} \otimes \mathcal{F}$ (k being a *noise dimension space*, h a *system space* and $\mathcal{F}$ a Fock space), contrary to the usual convention $\mathsf{h} \otimes \widehat{\mathsf{k}} \otimes \mathcal{F}$.

1 Spaces and Operators

This section collects together some background material. It begins with a discussion of linear identifications for matrices whose entries are vectors or linear maps, and the lifting of linear maps to such matrices which is fundamental for operator space theory. This is followed by a review of positivity in the context of Hilbert-space operators and C^*-algebras, including complete positivity and the Kolmogorov map for (Hilbert-space operator-valued) nonnegative-definite kernels. The basics of operator space theory are outlined next, with emphasis on a particular class of operator spaces, called matrix spaces, which play a key role in the construction and analysis of quantum stochastic processes on C^*-algebras. This is followed by a quick summary of essential facts about unbounded operators, integration for vector-valued functions, and one-parameter semigroups (cf. appendix material in Johan Kustermans' notes in this volume, which has the different emphasis required for his purposes). The notes at the end of the section contain suggestions for further reading.

1.1 Matrices

$M_{n,m}$ denotes the Banach space of $n \times m$ complex matrices with the norm arising from its usual linear identification with $B(\mathbb{C}^m; \mathbb{C}^n)$; in particular $M_n := M_{n,n}$ has its C^*-norm.

Consider an $n \times m$ matrix A each of whose entries is a $p \times q$ matrix with entries in a vector space V, in other words an element of the vector space $M_{n,m}(M_{p,q}(V))$. By ignoring the 'edges' of each $p \times q$ submatrix we may view A simply as an $np \times mq$ matrix with entries in V, giving the linear isomorphism

$$M_{n,m}(M_{p,q}(V)) \cong M_{np,mq}(V). \tag{1.1}$$

Sesquilinear maps as matrices

For vector spaces U, V and X, let $M(V, U; X)$ denote the vector space of sesquilinear maps $q : V \times U \to X$, thus q is linear in its *second* argument and conjugate linear in its first. Such maps may be thought of as abstract matrices in the following sense. Suppose that V and U are linear spans of orthonormal bases, $(f_\gamma)_{\gamma \in \Gamma}$ and $(e_\lambda)_{\lambda \in \Lambda}$, of two Hilbert spaces, then

$$q \mapsto \big[q(f_\gamma, e_\lambda)\big]_{\substack{\gamma \in \Gamma \\ \lambda \in \Lambda}}$$

defines a linear isomorphism $M(V, U; X) \to M_{\Gamma \times \Lambda}(X)$—in particular, an isomorphism $M(V, U; X) \cong M_{n,m}(X)$ when V and U have finite dimensions n and m respectively.

In these notes much use will be made of a natural subspace of $M(k, h; V)$, when k and h are Hilbert spaces and V is an operator space (see Subsection 1.3).

Matrices of linear maps

Linear algebra provides natural isomorphisms

$$M_{n,m}(L(U;V)) \to L(U^m;V^n)$$

$$[T_{ij}] \mapsto T \text{ where } (T\mathbf{u})_i = \sum_{j=1}^{m} T_{ij}u_j \quad (i = 1,\dots,n), \qquad (1.2)$$

for vector spaces U and V. These restrict to linear isomorphisms

$$M_{n,m}(B(\mathsf{h};\mathsf{k})) \to B(\mathsf{h}^m;\mathsf{k}^n), \qquad (1.3)$$

when h and k are Hilbert spaces. In this way, for any closed subspace V of $B(\mathsf{h};\mathsf{k})$, $M_{n,m}(\mathsf{V})$ may always be identified with a closed subspace of $B(\mathsf{h}^m;\mathsf{k}^n)$ — in particular $M_{n,m}(\mathsf{V})$ is thereby *endowed with a Banach space norm*.

This elementary observation is an important generalisation of the fact that if a C^*-algebra $\mathcal{A}$ acts on a Hilbert space h then $M_n(\mathcal{A})$ may be viewed as a C^*-algebra acting on h^n. Abstractly $M_n(\mathcal{A})$ is the C^*-algebra $\mathcal{A} \otimes M_n$, with identification $[a_{ij}] \mapsto \sum_{i,j} a_{ij} \otimes e_{ij}$ where (e_{ij}) is the standard basis of M_n — there being only one C^*-norm satisfying $\|a \otimes T\| = \|a\| \|T\|$ for $a \in \mathcal{A}$ and $T \in M_n$.

The following linear identification is also useful

$$M_{n,m}(L(U;V)) \to L(U;M_{n,m}(V))$$
$$[T_{ij}] \mapsto T \text{ where } Tu = [T_{ij}u], \qquad (1.4)$$

for vector spaces U and V.

Matrix liftings

For $\phi \in L(U;V)$, the linear map

$$M_{n,m}(U) \to M_{n,m}(V), \quad [x_{ij}] \mapsto [\phi(x_{ij})], \qquad (1.5)$$

is denoted $\phi^{(n,m)}$, or simply $\phi^{(n)}$ when $m = n$.

Let V be a closed subspace of $B(\mathsf{h};\mathsf{k})$, W a closed subspace of $B(\mathsf{h}';\mathsf{k}')$ and ϕ a bounded operator $\mathsf{V} \to \mathsf{W}$. Then the operator

$$\phi^{(n,m)} : M_{n,m}(\mathsf{V}) \to M_{n,m}(\mathsf{W})$$

is bounded.

Exercise. *Prove the estimate*

$$\|\phi^{(n,m)}\| \leq \sqrt{nm}\,\|\phi\|. \qquad (1.6)$$

Question. What happens to $\|\phi^{(n)}\|$ as $n \to \infty$? We return to this vital question later.

1.2 Positivity

Recall that, for an operator $T \in B(\mathsf{H})$ where H is a Hilbert space, T is *positive* (or, more correctly, *nonnegative*) if $\langle \xi, T\xi \rangle \geq 0$ for all $\xi \in \mathsf{H}$. In this case we write $T \geq 0$. The following result gives a useful characterisation of positivity in $B(\mathsf{H} \oplus \mathsf{K})$. In particular it points to the close connection between positivity and contractivity (take A and D to be identity operators).

Proposition 1.1. *Let* $T \in B(\mathsf{H} \oplus \mathsf{K})$ *for Hilbert spaces* H *and* K. *Then the following are equivalent:*

(i) $T \geq 0$;
(ii) T *has block matrix form*

$$\begin{bmatrix} A & A^{1/2}VD^{1/2} \\ D^{1/2}V^*A^{1/2} & D \end{bmatrix} \tag{1.7}$$

where $A \in B(\mathsf{H})$, $D \in B(\mathsf{K})$ *and* $V \in B(\mathsf{K}; \mathsf{H})$ *satisfy* $A, D \geq 0$ *and* $\|V\| \leq 1$.

Remarks. The operator $V \in B(\mathsf{K}; \mathsf{H})$ may be chosen so that $\operatorname{Ker} D \subset \operatorname{Ker} V$ and $\operatorname{Ran} V \subset \overline{\operatorname{Ran} A}$, in which case it is unique.

If $\mathsf{K} = \mathsf{H}$ and $T \in M_2(\mathcal{A})$, for a C^*-algebra $\mathcal{A}$ acting nondegenerately on H, then this unique operator V lies in $\overline{\mathcal{A}}^{\text{uw}}$, the ultraweak closure of $\mathcal{A}$.

For an element a of a C^*-algebra $\mathcal{A}$, *positivity* means any of the following equivalent properties:

(i) $a = x^*x$ for some $x \in \mathcal{A}$;
(ii) $a = \sum_{i=1}^{n} x_i^* x_i$ for some $n \in \mathbb{N}$ and $x_1, \ldots, x_n \in \mathcal{A}$;
(iii) $a^*a = aa^*$ and the spectrum of a is contained in $\mathbb{R}_+$;
(iv) $\pi(a) \geq 0$ for some faithful representation (π, K) of $\mathcal{A}$.

Clearly $\pi(a) \geq 0$ for *all* representations (π, K) of $\mathcal{A}$ when a is positive. Again this is written $a \geq 0$, and the cone of such elements is denoted $\mathcal{A}_+$.

Complete positivity

For a linear map $\phi : \mathcal{A} \to \mathcal{C}$ between C^*-algebras *positivity* means positivity preservation:

$$\phi(\mathcal{A}_+) \subset \mathcal{C}_+.$$

Positive maps are bounded and, when $\mathcal{A}$ is unital, satisfy $\|\phi\| = \|\phi(1)\|$.

Definition. A linear map $\phi : \mathcal{A} \to \mathcal{C}$ between C^*-algebras is *n-positive* if $\phi^{(n)} : M_n(\mathcal{A}) \to M_n(\mathcal{C})$ is positive, and is *completely positive* (abbreviated CP) if it is n-positive for each $n \geq 1$.

Exercise. *Verify that *-homomorphisms between C^*-algebras are completely positive and if $T \in B(\mathsf{K}; \mathsf{H})$ then $A \mapsto T^*AT$ defines a CP map $B(\mathsf{H}) \to B(\mathsf{K})$.*

Example 1.11 below shows the sense in which these two examples are exhaustive.

Proposition 1.2. *If either source or target is abelian then complete positivity follows from positivity.*

In particular a *state* (= positive linear functional) on a C^*-algebra is automatically completely positive.

Example 1.3. Let $\mathcal{A} = M_2$ and let $\phi : \mathcal{A} \to \mathcal{A}$ be the transpose map $\begin{bmatrix} a & b \\ c & d \end{bmatrix} \mapsto \begin{bmatrix} a & c \\ b & d \end{bmatrix}$. If p is the projection $\begin{bmatrix} 1 & 0 \\ 0 & 0 \end{bmatrix}$ and v is the partial isometry $\begin{bmatrix} 0 & 0 \\ 1 & 0 \end{bmatrix}$, and if $x = \begin{pmatrix} 1 \\ 0 \end{pmatrix}$ and $y = \begin{pmatrix} 0 \\ -1 \end{pmatrix}$, then

$$v^*v = p, \ vv^* = I - p, \ vp = v, \ v^*y = -x \text{ and } py = 0.$$

It follows that, in $M_2(\mathcal{A})$,

$$\begin{bmatrix} 1 & v \\ v^* & p \end{bmatrix} \geq 0 \text{ but } \left\langle \begin{pmatrix} x \\ y \end{pmatrix}, \begin{bmatrix} 1 & v^* \\ v & p \end{bmatrix} \begin{pmatrix} x \\ y \end{pmatrix} \right\rangle = -\|y\|^2 < 0.$$

Thus the transpose map on M_2 is an example of a positive linear map which fails to be 2-positive.

The next result is known as the *operator-Schwarz inequality*.

Proposition 1.4 (Kadison). *2-positive maps $\phi : \mathcal{A} \to \mathcal{C}$ satisfy*

$$\|\phi\|\phi(a^*a) \geq \phi(a)^*\phi(a). \tag{1.8}$$

In the unital case this may be seen by faithfully representing $\mathcal{A}$ and $\mathcal{C}$ on Hilbert spaces h and k, and applying Proposition 1.1 in turn to $\begin{bmatrix} a^*a & a^* \\ a & 1 \end{bmatrix}$ and to its image under $\phi^{(2)}$; an approximate identity may be used to prove it in the nonunital case.

The following result is proved using the operator-Schwarz inequality.

Proposition 1.5. *Let $\phi : \mathcal{A} \to \mathcal{C}$ be a completely positive map into a unital C^*-algebra. Then*

$$(a, \lambda) \mapsto \phi(a) + \lambda\|\phi\|1_\mathcal{C} \tag{1.9}$$

defines a completely positive map extending ϕ to the unitisation of $\mathcal{A}$, with the same norm.

Kolmogorov map

For a set S and Hilbert space H, a $B(\mathsf{H})$-*valued nonnegative-definite kernel on* S is a map $k : S \times S \to B(\mathsf{H})$ satisfying the following condition: for all $n \in \mathbb{N}$ and $\mathbf{s} \in S^n$,

$$[k(s_i, s_j)] \geq 0 \text{ in } \mathrm{M}_n(B(\mathsf{H})) = B(\mathsf{H}^n).$$

Example 1.6. For any Hilbert space K and map $\eta : S \to B(\mathsf{H}; \mathsf{K})$,

$$k(s, t) = \eta(s)^* \eta(t) \tag{1.10}$$

defines such a kernel because

$$[\eta(s_i)^* \eta(s_j)] = T^* T \text{ where } T = [\eta(s_1) \cdots \eta(s_n)] \in B(\mathsf{H}^n; \mathsf{K}).$$

This example is general, as the following result shows. Loosely speaking, the result says that, by means of a nonnegative-definite kernel, any set may be 'linearised' to a Hilbert space.

Theorem 1.7. *If k is a $B(\mathsf{H})$-valued nonnegative-definite kernel on S then there is a Hilbert space K and mapping $\eta : S \to B(\mathsf{H}; \mathsf{K})$ such that* (1.10) *holds and*

$$\overline{\mathrm{Lin}}\, \eta(S)\mathsf{H} = \mathsf{K}. \tag{1.11}$$

If $\eta' : S \to B(\mathsf{H}; \mathsf{K}')$ is another map satisfying (1.10) *then there is a unique isometry $V : \mathsf{K} \to \mathsf{K}'$ such that*

$$V\eta(s) = \eta'(s) \quad \text{for all } s \in S. \tag{1.12}$$

Proof. Let K_{00} be the subspace of the vector space $\mathrm{Map}(S; \mathsf{H})$ consisting of the functions $f : S \to \mathsf{H}$ of finite support: $\#\{s \in S : f(s) \neq 0\} < \infty$. Thus

$$\mathsf{K}_{00} = \mathrm{Lin}\,\{u\delta_s : u \in \mathsf{H}, s \in S\}, \text{ where } u\delta_s(t) = \begin{cases} u & \text{if } t = s \\ 0 & \text{otherwise} \end{cases}.$$

Nonnegative definiteness of the kernel k implies that

$$q(f, g) := \sum_{s,t \in S} \langle f(s), k(s, t)g(t) \rangle$$

defines a nonnegative sesquilinear form on K_{00}. The Schwarz inequality implies that $U := \{f \in \mathsf{K}_{00} | q(f, f) = 0\}$ is a subspace of K_{00}, and that

$$\langle [f], [g] \rangle := q(f, g)$$

defines an inner product on the linear quotient space $\mathsf{K}_0 := \mathsf{K}_{00}/U$. Next let $\iota : \mathsf{K}_0 \to \mathsf{K}$ be a completion of K_0. Then

$$\eta(s)u := \iota[u\delta_s]$$

defines linear maps $\eta(s) : \mathsf{H} \to \mathsf{K}$ which satisfy

$$\langle \eta(s)u, \eta(s')u' \rangle = q(u\delta_s, u'\delta_{s'}) = \langle u, k(s, s')u' \rangle,$$

and so are bounded, and moreover (1.10) holds. Now

$$\operatorname{Lin} \eta(S)\mathsf{H} = \operatorname{Lin} \{ \iota[u\delta_s] : u \in \mathsf{H}, s \in S \} = \iota(\mathsf{K}_0),$$

which is dense in K, so (1.11) holds too.

If $\eta' : S \to B(\mathsf{H}; \mathsf{K}')$ is another map satisfying (1.10) then,

$$\langle \eta'(s)u, \eta'(t)v \rangle = \langle u, k(s, t)v \rangle = \langle \eta(s)u, \eta(t)v \rangle$$

for all $s, t \in S$ and $u, u' \in \mathsf{H}$. Since $\eta(S)\mathsf{H}$ is total in K it follows that there is a unique isometry $V : \mathsf{K} \to \mathsf{K}'$ such that

$$V\eta(s)u = \eta'(s)u \quad \text{for all } s \in S, u \in \mathsf{H}.$$

Therefore there is a unique isometry $V : \mathsf{K} \to \mathsf{K}'$ such that (1.12) holds.   □

Definition. A map $\eta : S \to B(\mathsf{H}; \mathsf{K})$ such that

$$\eta(s)^*\eta(t) = k(s, t)$$

is called a *Kolmogorov map* for the $B(\mathsf{H})$-valued nonnegative-definite kernel k. It is called a *minimal Kolmogorov map* if also

$$\eta(S)\mathsf{H} \text{ is total in } \mathsf{K}.$$

Thus, for any nonnegative-definite kernel k, minimal Kolmogorov maps exist and enjoy the universal property summarised in the commutative diagram below:

They are unique in the same sense that completions and tensor products are unique, namely up to isomorphism *of maps*.

Remark. If $\mathsf{H} = \mathbb{C}$ then $B(\mathsf{H}; \mathsf{K})$ is identified with K, so a Kolmogorov map takes the form

$$\eta : S \to \mathsf{K} \text{ and } k(s, t) = \langle \eta(s), \eta(t) \rangle.$$

Exercise. *Let* $\eta : S \to B(\mathsf{H}; \mathsf{K})$ *be a minimal Kolmogorov map for a nonnegative-definite kernel* k, *and let* k' *be a nonnegative-definite kernel dominated by* k—*in the sense that there is a constant* C *for which* $[k'(s_i, s_j)] \leq C[k(s_i, s_j)]$ *for all* $\mathbf{s} \in S^n$ *and* $n \in \mathbb{N}$. *Show that there is a bounded operator* T *on* K *for which* $T\eta(\cdot)$ *is a Kolmogorov map for* k'.

Example 1.8 (Hilbert space tensor product). For Hilbert spaces $h_1, \ldots, h_n$,

$$(\mathbf{u}, \mathbf{v}) \mapsto \langle u_1, v_1 \rangle \cdots \langle u_n, v_n \rangle$$

defines a nonnegative-definite kernel on $h_1 \times \ldots \times h_n$, because the *Schur product* of nonnegative-definite matrices is nonnegative-definite:

$$\text{if } c_{ij} = a_{ij}b_{ij} \text{ and } [a_{ij}], [b_{ij}] \geq 0 \text{ then } [c_{ij}] \geq 0.$$

The minimal Kolmogorov map for this kernel *is* the Hilbert space tensor product:

$$K = h_1 \otimes \cdots \otimes h_n, \quad \eta(u_1, \ldots u_n) = u_1 \otimes \cdots \otimes u_n.$$

Example 1.9 (Infinite tensor products). For a sequence of Hilbert spaces (h_n), and a sequence of unit vectors $(e_n \in h_n)_{n \geq 1}$, let

$$S = \left\{ \xi \in \prod_{n \geq 1} h_n \; \middle| \; \exists_{N \geq 1} \; \xi_n = e_n \text{ for } n \geq N \right\}.$$

Then $k : S \times S \to \mathbb{C}$, $(\xi, \eta) \mapsto \prod_{n \geq 1} \langle \xi_n, \eta_n \rangle$, defines a nonnegative-definite kernel on S. Its minimal Kolmogorov map is the infinite tensor product of (h_n) with *stabilising sequence* (e_n). Notation:

$$\bigotimes^{(e_n)} h_n \text{ for } K \text{ and } \otimes \xi_n \text{ for } \eta(\xi).$$

Example 1.10 (GNS construction). Let ω be a state on a C^*-algebra $\mathcal{A}$. Let $\tilde{\mathcal{A}}$ denote $\mathcal{A}$ when the algebra is unital, and its unitisation otherwise, and write $\tilde{\omega}$ for the extension of ω to $\tilde{\mathcal{A}}$ defined in (1.9). Then there is a unital representation $(\tilde{\pi}, K)$ of $\tilde{\mathcal{A}}$ and a unit vector $\xi \in K$ such that

$$\tilde{\omega}(x) = \langle \xi, \tilde{\pi}(x)\xi \rangle, \text{ and } \pi(\mathcal{A})\xi \text{ is dense in } K,$$

where $\pi = \tilde{\pi}|_{\mathcal{A}}$. These are obtained by letting $\eta : \tilde{\mathcal{A}} \to K$ be the minimal Kolmogorov map for the kernel

$$\tilde{\mathcal{A}} \times \tilde{\mathcal{A}} \to \mathbb{C}, \quad (x, y) \mapsto \tilde{\omega}(x^*y),$$

setting $\xi = \eta(1)$, and exploiting minimality to verify that for each $x \in \tilde{\mathcal{A}}$ the map $\eta(y) \mapsto \eta(xy)$ extends to a bounded operator $\tilde{\pi}(x)$ on K which defines a unital representation $\tilde{\pi}$ of $\tilde{\mathcal{A}}$. The fact that $\tilde{\mathcal{A}}$ is the linear span of its unitary elements may be used here.

Example 1.11 (Stinespring Theorem). Let $\phi : \mathcal{A} \to B(H)$ be a completely positive map defined on a C^*-algebra $\mathcal{A}$. Define $\tilde{\mathcal{A}}$ as above and let $\tilde{\phi} : \tilde{\mathcal{A}} \to B(H)$ denote the extension (1.9). Then there is a representation $(\tilde{\pi}, K)$ of $\tilde{\mathcal{A}}$ and an operator $T \in B(H; K)$ such that

$$\tilde{\phi}(x) = T^*\tilde{\pi}(x)T \text{ and } \pi(\mathcal{A})TH \text{ is total in K,} \qquad (1.13)$$

where $\pi = \tilde{\pi}|_{\mathcal{A}}$. These are obtained by letting $\eta : \tilde{\mathcal{A}} \to B(\mathsf{H}; \mathsf{K})$ be the minimal Kolmogorov map for the kernel $\tilde{\mathcal{A}} \times \tilde{\mathcal{A}} \to B(\mathsf{H})$, $(x, y) \mapsto \tilde{\phi}(x^*y)$, setting $T = \eta(1)$, and verifying that $\tilde{\pi}(x) : \eta(y)\xi \mapsto \eta(xy)\xi$ defines a representation $(\tilde{\pi}, \mathsf{K})$ of $\tilde{\mathcal{A}}$, in fact a unital representation.

Since states on a C^*-algebra are automatically completely positive (by Proposition 1.2), Example 1.10 is a special case of Example 1.11.

Remark. If the representation in (1.13) is chosen to be unital then $\|T\| = \|\phi\|^{1/2}$ and furthermore T is isometric if and only if $\tilde{\phi}$ is unital. In this case we may take $\mathsf{K} = \mathsf{H} \oplus \mathsf{h}$ for a Hilbert space h and $T = \begin{bmatrix} I \\ 0 \end{bmatrix}$ so that $\tilde{\pi}$ has the block matrix form

$$a \mapsto \begin{bmatrix} \phi(a) & * \\ * & * \end{bmatrix}.$$

Exercise. Show that if ϕ is also contractive then it has a Stinespring decomposition (1.13) with K and T as in the above remark. Typically $\tilde{\pi}$ will no longer be unital.

Example 1.12 (Sz. Nagy unitary dilation). Let $C \in B(\mathsf{H})$ be a contraction. Then there is a Hilbert space K, a unitary $U \in B(\mathsf{K})$ and an isometry $J \in B(\mathsf{H}; \mathsf{K})$ such that

$$C^n = J^*U^nJ \text{ for } n \in \mathbb{Z}_+, \text{ and } \{U^nJ\xi : \xi \in \mathsf{H}, n \in \mathbb{Z}\} \text{ is total in } \mathsf{K}$$

This is proved again by first verifying that

$$(j, k) \mapsto C^{k-j}, \quad j, k \in \mathbb{Z} \text{ with } k \geq j,$$

extends to a $B(\mathsf{H})$-valued nonnegative-definite kernel on $\mathbb{Z}$.

Remarks. There is a continuous-parameter version of this in which C and U are replaced by contractive and unitary c_0-*semigroups* respectively (see page 199 for the definition). The isometric version of these, in which $\mathbb{Z}$ is replaced by $\mathbb{Z}_+$ (respectively $\mathbb{R}$ by $\mathbb{R}_+$) and U is isometric rather than unitary, is discussed in the lectures of Rajarama Bhat in this volume.

1.3 Operator spaces

Let V be a closed subspace of $B(\mathsf{h}; \mathsf{k})$, for Hilbert spaces h and k. Note two key properties of the induced norms on matrices over V, arising from the linear identification (1.3):

(OSi) For $S \in M_{n_1, m_1}(\mathsf{V})$ and $T \in M_{n_2, m_2}(\mathsf{V})$, so that $S \oplus T \in M_{n_3, m_3}(\mathsf{V})$ where $n_3 = n_1 + n_2$ and $m_3 = m_1 + m_2$,

$$\|S \oplus T\| = \max\{\|S\|, \|T\|\}.$$

(OSii) For $S \in M_{n,m}(V), \Gamma \in M_{l,n}$ and $\Lambda \in M_{m,p}$, so that $\Gamma S \Lambda \in M_{l,p}(V)$,

$$\|\Gamma S \Lambda\| \le \|\Gamma\| \|S\| \|\Lambda\|.$$

Here $S \oplus T$ denotes the diagonal block matrix $\begin{bmatrix} S & \\ & T \end{bmatrix}$. Note that the multiplication by scalar matrices in (OSii) makes good sense.

Definition. A complex vector space V with complete norms on each $M_n(V)$ satisfying the compatibility conditions (OSi) and (OSii) (for $n_1 = m_1, n_2 = m_2$ and $n = m$) is called an *operator space*.

Remark. If V is an operator space then, viewing $M_{n,m}(V)$ as a subspace of $M_{n+m}(V)$ by occupying the top right-hand corner and filling the remaining entries with zeros, norms are induced on each $M_{n,m}(V)$ too — these necessarily also satisfy (OSi) and (OSii) with differing n's and m's.

Example 1.13. Given an operator space W, each $M_{p,q}(W)$ becomes an operator space itself by using the linear isomorphism (1.1). In other words, the norms on $M_n(M_{p,q}(W)) = M_{np,nq}(W)$, for $n = 1, 2, \ldots$ (arising from W being an operator space) satisfy conditions (OSi) and (OSii).

Definition. A linear map $\phi : V \to W$ between operator spaces is *completely bounded* if $\sup_n \|\phi^{(n)}\| < \infty$; it is called *completely contractive* if $\|\phi^{(n)}\| \le 1$ for each n, and a *complete isometry* if each $\phi^{(n)}$ is isometric.

The space of completely bounded maps $V \to W$ is denoted $CB(V; W)$ and has the complete norm

$$\|\phi\|_{\mathrm{cb}} := \sup_n \|\phi^{(n)}\|.$$

Proposition 1.14. *Let $\phi : V \to W$ be a bounded linear map between operator spaces. In each of the following cases ϕ is automatically completely bounded:*

(i) $\dim V < \infty$;
(ii) $\dim W < \infty$;
(iii) V *is an abelian C^*-algebra.*

This may be compared with Proposition 1.2.

Example 1.15. Any completely positive map ϕ between C^*-algebras is completely bounded and satisfies $\|\phi\|_{\mathrm{cb}} = \|\phi\|$.

Example 1.16. Closed subspaces of C^*-algebras are operator spaces. If ϕ is a *-homomorphism between C^*-algebras then ϕ is a complete contraction, and a complete isometry when injective. A left multiplication operator $L_A : B(H; K) \to B(H; K')$, $T \mapsto AT$, where $A \in B(K; K')$, is completely bounded with $\|L_A\|_{\mathrm{cb}} = \|A\|$, and similarly for right multiplication operators.

Theorems 1.20 and 1.22 below show that Example 1.16 is exhaustive in a sense. However the abstract point of view is immediately vindicated by the next example, which should be contrasted with the fact that, for (nontrivial) Hilbert spaces h and k, $B(h; k)$ is *not itself* a Hilbert space.

Example 1.17. For operator spaces V and W, $CB(V; W)$ is endowed with operator space structure as follows. The map (1.4) restricts to a linear isomorphism

$$M_{n,m}\big(CB(V; W)\big) \to CB(V; M_{n,m}(W)),$$

and the norms induced on matrices over $CB(V; W)$ by the resulting linear identifications satisfy (OSi) and (OSii).

In particular, taking $W = \mathbb{C}$, the Banach space dual of an operator space gains operator space structure through the linear identification $M_n(V^*) = CB(V; M_n)$. This exploits part (ii) of Proposition 1.14.

Example 1.18. Let h be a Hilbert space and recall the bra- -ket notation (0.1). The *column space* $|h\rangle := B(\mathbb{C}; h)$ and the *row space* $\langle h| := B(h; \mathbb{C})$ are important examples of operator spaces. They are mutually dual operator spaces, however in general the natural isometric antiisomorphism $|u\rangle \mapsto \langle u|$ is not a complete isometry. It isn't even completely bounded. Pisier has found an operator space structure on h which *is* completely isometric to its CB-dual, and shown it to be the unique structure enjoying this self-duality.

Example 1.19 (Cf. Example 1.3). Let $\mathcal{K}$ be the C^*-algebra of compact operators on l^2 and let $\phi : \mathcal{K} \to \mathcal{K}$ be the transpose map $[z_{ij}] \mapsto [z_{ji}]$. Then ϕ is isometric but is not completely bounded.

There is a Gelfand-Naimark-type theorem for operator spaces, to the effect that every operator space has a concrete realisation.

Theorem 1.20 (Ruan). *Let V be an operator space. Then there is a Hilbert space H and a complete isometry $\phi : V \to B(H)$.*

Thus every abstract operator space has a concrete realisation. Operator-space theory has been dubbed *quantised functional analysis* by one of its architects, Effros. The following Hahn-Banach-type theorem exemplifies why.

Theorem 1.21 (Arveson). *Let V_0 be a subspace of an operator space V and let $\phi_0 : V_0 \to B(H)$ be a completely bounded map. Then there is a completely bounded map $\phi : V \to B(H)$ satisfying $\|\phi\|_{cb} = \|\phi_0\|_{cb}$ and extending ϕ_0:*

There is also a Stinespring-like decomposition for CB maps.

Theorem 1.22 (Wittstock-Paulsen-Haagerup). *Let $\mathcal{A}$ be a C^*-algebra and let $\phi : \mathcal{A} \to B(\mathsf{H})$ be a completely bounded map. Then there is a representation $\pi : \mathcal{A} \to B(\mathsf{K})$ and operators $R, S \in B(\mathsf{H}; \mathsf{K})$ such that*

$$\phi(a) = R^*\pi(a)S; \quad \|\phi\|_{\mathrm{cb}} = \|R\| \, \|S\|.$$

If $\mathcal{A}$ is a von Neumann algebra and ϕ is ultraweakly continuous then π may be chosen to be normal.

In view of the structure of normal representations ([Tak]), in the von Neumann algebra case there is a Hilbert space k and operators $R, S \in B(\mathsf{H}; \mathsf{h} \otimes \mathsf{k})$ for which

$$\phi(a) = R^*(a \otimes I_{\mathsf{k}})S,$$

where h is the Hilbert space on which $\mathcal{A}$ acts.

Tensor products

For concrete operator spaces V_1 and V_2 their *spatial tensor product* $\mathsf{V}_1 \otimes_{\mathrm{sp}} \mathsf{V}_2$ (respectively, *ultraweak tensor product* $\mathsf{V}_1 \overline{\otimes} \mathsf{V}_2$) is simply the norm closure (respectively, ultraweak closure) of their algebraic tensor product $\mathsf{V}_1 \underline{\otimes} \mathsf{V}_2$. An important feature of CB maps is that they may be 'tensored'. Thus if $\phi_i : \mathsf{V}_i \to \mathsf{W}_i$ $(i = 1, 2)$ are CB maps, then the map $\phi_1 \underline{\otimes} \phi_2 : \mathsf{V}_1 \underline{\otimes} \mathsf{V}_2 \to \mathsf{W}_1 \underline{\otimes} \mathsf{W}_2$ extends uniquely to a CB map, denoted $\phi_1 \otimes \phi_2$, from $\mathsf{V}_1 \otimes_{\mathrm{sp}} \mathsf{V}_2$ to $\mathsf{W}_1 \otimes_{\mathrm{sp}} \mathsf{W}_2$. If the spaces are ultraweakly closed and the maps are ultraweakly continuous then there is further extension to an ultraweakly continuous map $\phi_1 \overline{\otimes} \phi_2 : \mathsf{V}_1 \overline{\otimes} \mathsf{V}_2 \to \mathsf{W}_1 \overline{\otimes} \mathsf{W}_2$.

Matrix spaces

The idea is to consider $\mathsf{M}_{n,m}(\mathsf{V})$, for an operator space V, and to liberate it from its coordinates (i.e. to replace $\mathbb{C}^n$ and $\mathbb{C}^m$ by abstract Hilbert spaces h and k) and also liberate it from its finite dimensions (i.e. to allow h and k to be infinite dimensional). This may be done abstractly, but its concrete form will suffice for present purposes. Earlier this was done at the level of linear algebra; now we require the construction to yield operator spaces.

Notation/Convention. For a Hilbert space vector $e \in \mathsf{h}$, the operator

$$\mathsf{H} \to \mathsf{h} \otimes \mathsf{H}, \quad u \mapsto e \otimes u$$

will be denoted E_e, and its adjoint by E^e, with context dictating the Hilbert space H. Thus $E_e \in B(\mathsf{H}; \mathsf{h} \otimes \mathsf{H})$, $\|E_e\| = \|e\|$ and $E_e E_f = E_{e \otimes f}$; also $E^e E_f = \langle e, f \rangle I$ and $E_e E^f$ is an ampliation of $|e\rangle\langle f|$.

When orthonormal bases (e_λ) and (f_α) are understood, for k and h respectively, and $T \in B(h \otimes H; k \otimes K)$ we write

$$E^{(\alpha)}TE_{(\lambda)} \text{ for } E^f TE_e \text{ where } f = f_\alpha \text{ and } e = e_\lambda. \tag{1.14}$$

Note the strong operator convergence

$$\sum_{\lambda \in \Lambda} E_{(\lambda)}E^{(\lambda)} = \sum_{\lambda \in \Lambda} |e_\lambda\rangle\langle e_\lambda| \otimes I_H = I_{h \otimes H}. \tag{1.15}$$

Definition. Let V be an operator space in $B(H; K)$. The h-k *matrix space over* V is given by

$$M(k, h; V)_b := \left\{ T \in B(h \otimes H; k \otimes K) \middle| E^d TE_e \in V \text{ for all } d \in k, e \in h \right\}.$$

This notation and terminology specialises as follows.

$$\begin{aligned}
M(h; V)_b &:= M(h, h; V)_b &&\text{(square matrix space);} \\
C(k; V)_b &:= M(k, \mathbb{C}; V)_b &&\text{(column matrix space);} \\
R(h; V)_b &:= M(\mathbb{C}, h; V)_b &&\text{(row matrix space).}
\end{aligned}$$

Remark. By continuity and (conjugate-) linearity of the maps $e \mapsto E_e$ (respectively, $d \mapsto E^d$) it suffices to check $E^d TE_e \in V$ for vectors d and e from some total subsets of their respective Hilbert spaces.

Properties.

(i) $M(k, h; V)_b$ is an operator space.
(ii) If $h = \mathbb{C}^m$ and $k = \mathbb{C}^n$ then, with respect to standard bases,

$$T \mapsto [E^{(i)}TE_{(j)}]$$

defines a completely isometric isomorphism $M(k, h; V)_b \to M_{n,m}(V)$.
(iii) If $W = M(k_1, h_1; V)_b$ then $M(k_2, h_2; W)_b = M(k, h; V)_b$ where $h = h_2 \otimes h_1$ and $k = k_2 \otimes k_1$.
(iv) The following inclusions hold:

$$B(h; k) \otimes_{sp} V \subset M(k, h; V)_b \subset B(h; k)\overline{\otimes}V.$$

The former follows from the fact that $T \otimes x \in M(k, h; V)_b$ for $T \in B(h; k)$ and $x \in V$; the latter may be verified by applying (1.15) and heeding the above remark. The first inclusion is an equality if *either* h and k are both finite dimensional *or* V is finite dimensional; the second is an equality if and only if V is ultraweakly closed.
(v) For a C^*-algebra $\mathcal{A}$, $M(h; \mathcal{A})_b$ is typically *not* a C^*-algebra.

Example 1.23. Let $\mathcal{A} = c_0$, the commutative C^*-algebra of complex sequences converging to 0 — represented on the Hilbert space l^2 by diagonal matrices, and let $\mathsf{h} = l^2$. Consider the operator $T \in B(\mathsf{h} \otimes l^2) = B(\bigoplus_{n \geq 1} l^2)$ given by the matrix

$$\begin{bmatrix} e_1 \ 0 \ 0 \ \cdots \\ e_2 \ 0 \ 0 \ \cdots \\ e_3 \ 0 \ 0 \ \cdots \\ \vdots \ \vdots \ \vdots \ \ddots \end{bmatrix}$$

in which $e_k = \mathrm{diag}[0, \ldots, 0, 1, 0, \ldots]$ with 1 in the kth place and zeros elsewhere, so that T^*T has matrix

$$\begin{bmatrix} e \ 0 \ \cdots \\ 0 \ 0 \ \cdots \\ \vdots \ \vdots \ \ddots \end{bmatrix}$$

where $e = \mathrm{diag}[1, 1, 1, \ldots]$. Then $e_k \in \mathcal{A}$ for each k and so (by the remark following the definition) $T \in M(\mathsf{h}; \mathcal{A})_{\mathrm{b}}$. However $e \notin \mathcal{A}$ so $T^*T \notin M(\mathsf{h}; \mathcal{A})_{\mathrm{b}}$. Note that this also implies that T *cannot* belong to the C^*-algebra $B(\mathsf{h}) \otimes_{\mathrm{sp}} \mathcal{A}$.

This example illustrates properties (v) and (iv) above.

Proposition 1.24. *Let $\phi \in CB(\mathsf{V}; \mathsf{W})$, for operator spaces V and W. Then, for any Hilbert spaces h and k, there is a unique completely bounded map*

$$\phi^{(\mathsf{k}, \mathsf{h})} : M(\mathsf{k}, \mathsf{h}; \mathsf{V})_{\mathrm{b}} \to M(\mathsf{k}, \mathsf{h}; \mathsf{W})_{\mathrm{b}}$$

satisfying

$$E^d \phi^{(\mathsf{k}, \mathsf{h})}(T) E_e = \phi(E^d T E_e), \quad \text{for all } d \in \mathsf{k}, e \in \mathsf{h}.$$

Remarks. (i) If $\mathsf{h} = \mathbb{C}^m$ and $\mathsf{k} = \mathbb{C}^n$ then $\phi^{(\mathsf{k}, \mathsf{h})}$ is the matrix lifting $\phi^{(n,m)}$ defined in (1.5).

(ii) $\phi^{(\mathsf{k}, \mathsf{h})}$ extends the map $\mathrm{id} \otimes \phi : B(\mathsf{h}; \mathsf{k}) \otimes_{\mathrm{sp}} \mathsf{V} \to B(\mathsf{h}; \mathsf{k}) \otimes_{\mathrm{sp}} \mathsf{W}$.

(iii) If V and W are ultraweakly closed and ϕ is ultraweakly continuous then $\phi^{(\mathsf{k}, \mathsf{h})}$ coincides with the map

$$\mathrm{id} \,\overline{\otimes}\, \phi : B(\mathsf{h}; \mathsf{k}) \overline{\otimes} \mathsf{V} \to B(\mathsf{h}; \mathsf{k}) \overline{\otimes} \mathsf{W}.$$

(iv) $\|\phi^{(\mathsf{k}, \mathsf{h})}\|_{\mathrm{cb}} = \|\phi\|_{\mathrm{cb}}$ (cf. (1.6)).

Matrix-space tensor products; left and right

The convention adopted above is

$$M(\mathsf{k}, \mathsf{h}; \mathsf{V})_{\mathrm{b}} \subset B(\mathsf{h}; \mathsf{k}) \overline{\otimes} B(\mathsf{H}; \mathsf{K})$$

with $B(\mathsf{h}; \mathsf{k})$ on the left. The right convention is also needed. The *matrix-space tensor product* notations

$$\text{V} \otimes_M B(\text{h}; \text{k}) \text{ and } \phi \otimes_M \text{id}_{B(\text{h};\text{k})} \qquad (1.16)$$

will be used, for the *right matrix spaces* in $B(\text{H}; \text{K})\overline{\otimes}B(\text{h}; \text{k}) = B(\text{H} \otimes \text{h}; \text{K} \otimes \text{k})$ and *right liftings*. We shall also adopt this tensor notation for the left matrix spaces. The two are compatible, so that no ambiguity arises, in expressions such as

$$B(\text{h}_1; \text{k}_2) \otimes_M \text{V} \otimes_M B(\text{h}_2; \text{k}_2).$$

Matrix spaces arise in several ways in quantum stochastics. Firstly it is natural to view QS processes as being (right) matrix space valued. In this case the Hilbert space is the Fock space carrying the quantum noise. Secondly the generators of QS flows are naturally maps into a (left) matrix space over the Hilbert space which specifies the noise. The E^e and E_e notations will also be used for mapping between H and $\text{H} \otimes \text{h}$; again context will avoid confusion.

1.4 Operators, integrals and semigroups

Part of the appendix to Johan Kustermans' notes in this volume is devoted to a summary of facts about unbounded operators needed for understanding his lectures. Here is a further summary focusing on the Hilbert-space case, and some of the specific properties needed for quantum stochastics.

When a Hilbert-space operator $T : \text{H} \to \text{K}$ is unbounded its domain $\mathcal{D} = \text{Dom} \, T$ is typically a *proper subspace* of H. Operators of interest have two properties: they are *densely defined*, i.e. $\mathcal{D}$ is dense in H, and they are *closable*, which means T has an extension to a domain on which it is *closed*. Closed, for an operator $T : \text{H} \to \text{K}$ with domain $\mathcal{D}$, means that its graph is a closed subspace of $\text{H} \oplus \text{K}$. In this case the *graph norm* $\|\xi\|_T := \left(\|\xi\|^2 + \|T\xi\|^2\right)^{1/2}$ makes $\mathcal{D}$ into a Hilbert space, h say, and $\xi \mapsto T\xi$ then defines a *bounded* operator $\text{h} \to \text{K}$. An operator T is closable precisely if the closure of its graph is the graph of an operator; that operator is then written $\overline{T}$ and is called the *closure* of T. A densely defined closable operator T is bounded if and only if $\text{Dom} \, \overline{T} = \text{H}$, thus closed and everywhere defined implies bounded.

If $\mathcal{D}$ is a subspace of the domain of a closed operator T such that the graph of $T|_{\mathcal{D}}$ is dense in the graph of T (i.e. $\overline{T|_{\mathcal{D}}} = T$) then $\mathcal{D}$ is called a *core* for T. A desirable property for an unbounded operator is that it have a 'nice' core.

Every densely defined operator $T : \text{H} \to \text{K}$ has an *adjoint operator* T^* : $\text{K} \to \text{H}$ defined as follows. The domain of T^* is

$$\left\{\eta \in \text{K} : \text{the (densely defined linear) functional } \xi \mapsto \langle \eta, T\xi \rangle \text{ is bounded}\right\},$$

with $T^*\eta$ being the unique vector satisfying $\langle T^*\eta, \xi \rangle = \langle \eta, T\xi \rangle$, for all $\xi \in \text{Dom} \, T$, given by the Riesz-Fréchet Theorem. Adjoint operators are closed; moreover T^* is densely defined if and only if T is closable, and in this case $\overline{T} = T^{**}$. A densely defined operator T satisfying $T^* = T$ is called *self-adjoint*.

For operators $T : \text{H} \to \text{H}'$ and $S : \text{H}' \to \text{H}''$ the operator $ST : \text{H} \to \text{H}''$ has domain $\{\xi \in \text{Dom} \, T : T\xi \in \text{Dom} \, S\}$. If S is closed and T is bounded then ST

is necessarily closed, but it need not be densely defined even if S is. A useful notation for unbounded operators is $S \subset T$, meaning $\mathrm{Dom}\, S \subset \mathrm{Dom}\, T$ and $S = T|_{\mathrm{Dom}\, S}$.

Exercise. *Show that if operators S, T and their product ST are all densely defined then $(ST)^* \supset T^* S^*$, with equality if S is bounded.*

A densely defined operator T is *symmetric* if it satisfies $T^* \supset T$ and *essentially self-adjoint* if furthermore $T^* = \overline{T}$, equivalently (since the adjoint of an operator coincides with the adjoint of its closure) $\overline{T}$ is self-adjoint.

Positivity extends to unbounded operators, thus $T : \mathsf{H} \to \mathsf{H}$ with domain $\mathcal{D}$ is *positive* if $\langle \xi, T\xi \rangle \geq 0$ for all $\xi \in \mathcal{D}$. Every densely defined positive operator has a distinguished self-adjoint extension called its Friedrichs extension. For a closed densely defined operator T, the positive operator $T^* T$ is self-adjoint. Both of these facts draw on the theory of quadratic forms on a Hilbert space.

If T and T' are closable densely defined operators, with domains $\mathcal{D}$ and $\mathcal{D}'$ respectively, then the operator $T \otimes T'$ with domain $\mathcal{D} \otimes \mathcal{D}'$ is also (densely defined and) closable; its closure is denoted $\overline{T} \otimes \overline{T'}$, or simply $T \otimes T'$ when T and T' are already closed.

Operator-valued functions

Let H and K be Hilbert spaces. A function F from a measure space into the linear space of operators $\mathsf{H} \to \mathsf{K}$ with given domain $\mathcal{D}$ is said to be *weak operator measurable* (respectively, *strong operator measurable*) if for each vector $\xi \in \mathcal{D}$, the K-valued function $F(\cdot)\xi$ is weakly (respectively, strongly) measurable. We therefore recall the different notions of measurability for Hilbert space-valued functions next.

Let f be a K-valued function defined on a measure space. Then f is *weakly measurable* if, for each vector $\eta \in \mathsf{K}$, the scalar-valued function $\langle \eta, f(\cdot) \rangle$ is measurable; it is *strongly measurable* if f is the almost everywhere (a.e.) limit of a sequence of functions (f_n) where each f_n is of the form $\sum_{j=1}^{N} v_j 1_{E_j}$ for some $N \in \mathbb{N}$, $v_1, \ldots, v_N \in \mathsf{K}$ and measurable sets $E_1, \ldots, E_N$. We also say that f is measurable *in the usual sense* if, for each Borel subset U of K (with respect to the norm topology), the set $f^{-1}(U)$ is measurable. Thus measurability in the weak and usual senses do not refer to the measure, but strong measurability does. Any a.e. limit of a sequence of weakly (respectively, strongly) measurable functions is weakly (respectively, strongly) measurable. Strongly measurable functions are clearly weakly measurable; they are also *a.e. separably valued* meaning that there is a separable subspace K_0 such that $f^{-1}(\mathsf{K} \setminus \mathsf{K}_0)$ is a null set.

Theorem 1.25. *Let f be a Hilbert space-valued function on a measure space, then the following implications hold.*

(a) *If f is strongly measurable then it is measurable in the usual sense.*

(b) *If f is measurable in the usual sense then it is weakly measurable.*
(c) *If f is weakly measurable and a.e. separably valued then f is strongly measurable.*

This result is true also for Banach space-valued functions. Part (c) is known as Pettis' Theorem, and it implies that for separable spaces there is no distinction between weak and strong measurability. It also implies that continuous Hilbert space-valued functions defined on a separable topological space are strongly measurable. Note that if f is measurable in the usual sense then so is the scalar-valued function $\|f(\cdot)\|$.

An integral for weakly measurable functions is discussed in Johan Kustermans' notes in this volume (in the wider context of topological vector space-valued functions). We shall need the integral appropriate to strongly measurable functions which is known as the *Bochner integral*. A function f is Bochner integrable if it is an a.e. limit of a sequence of functions (f_n) as above (with each f_n being zero outside a set of finite measure), which furthermore satisfies

$$\int \|f(s) - f_n(s)\| \mu(ds) \to 0 \text{ as } n \to \infty.$$

Then the sequence of vectors $\left(\int f_n \, d\mu \right)$, with obvious definition, converges. Its limit, which does not depend on the sequence (f_n) chosen, is called the Bochner integral of f, and is written as for ordinary integrals.

Theorem 1.26 (Bochner). *Let f be a strongly measurable Hilbert space-valued function defined on a measure space. Then f is Bochner integrable if and only if the function $\|f(\cdot)\|$ is integrable.*

This is also true for Banach space-valued functions. Bochner-integrable functions satisfy

$$\left\| \int f \, d\mu \right\| \leq \int \|f(s)\| \mu(ds), \text{ and } T\left(\int f d\mu \right) = \int (Tf)(s) \mu(ds),$$

for bounded operators T. The second of these has an extension to closed operators T where f should be strongly measurable as a $\mathcal{D}$-valued map where $\mathcal{D} = \text{Dom}\, T$ carries its graph norm.

Finally there are the Bochner-Lebesgue spaces consisting of (measure equivalence classes of) strongly measurable functions f for which $\|f(\cdot)\|^p$ is integrable ($p \geq 1$).

c_0-semigroups

A family of bounded operators $T = (T_t)_{t\geq 0}$ on a Hilbert space H satisfying

$$T_0 = I, \ T_{s+t} = T_s T_t \text{ and } t \mapsto T_t \xi \text{ is continuous for all } \xi \in \mathsf{H}$$

is called a c_0-*semigroup* on H. For such a semigroup,

$$G\xi = \lim_{t\to 0} t^{-1}(T_t\xi - \xi), \quad \text{Dom } G = \{\xi \in \mathsf{H} : \lim_{t\to 0} t^{-1}(T_t\xi - \xi) \text{ exists}\},$$

defines a closed and densely defined operator, called the *generator* of T, from which the semigroup may be reconstructed as follows. There is $\omega \in \mathbb{R}$ and $M \geq 1$ such that

$$\|T_t\| \leq Me^{\omega t} \text{ for all } t \geq 0.$$

For each $\lambda > \omega$, the closed operator $(\lambda - G)$ is bijective $\text{Dom } G \to \mathsf{H}$ and therefore, by the Closed Graph Theorem (see Johan Kustermans' notes in this volume), has a bounded inverse, moreover

$$\left(1 - \tfrac{t}{n}G\right)^{-n}\xi \to T_t\xi \text{ as } n \to \infty \text{ for all } \xi \in \mathsf{H}.$$

The strong continuity condition is actually equivalent to continuity in the weak operator topology:

$$t \mapsto \langle \eta, T_t\xi \rangle \text{ is continuous for all } \xi, \eta \in \mathsf{H}.$$

In particular, $(T_t^*)_{t\geq 0}$ is also a c_0-semigroup; its generator is G^*.

If T is a contractive c_0-semigroup, so that ω and M may be taken to be 0 and 1 respectively, then its generator G is *dissipative*:

$$\text{Re}\,\langle \xi, G\xi \rangle \leq 0 \text{ for all } \xi \in \text{Dom } G.$$

Any densely defined dissipative operator L is closable, moreover its closure is dissipative; $\overline{L}$ is then the generator of a (contractive) c_0-semigroup if and only if $\text{Ran}\,(\lambda - L)$ is dense in H for some (in which case, all) $\lambda > 0$ — in this situation L is called a *pregenerator* of the semigroup. A densely defined dissipative operator on a Hilbert space is a pregenerator of a contractive c_0-semigroup if (and only if) its adjoint is dissipative too.

Given a dense subspace of the domain of the generator of a contraction semigroup, there is a useful sufficient condition for it to be a core for the generator.

Theorem 1.27. *Let T be a contractive c_0-semigroup on a Hilbert space H with generator G. Suppose that $\mathcal{D}$ is a dense subspace of H contained in $\text{Dom } G$ such that $T_t\mathcal{D} \subset \mathcal{D}$ for all $t \geq 0$. Then $\mathcal{D}$ is a core for G, in other words $G|_{\mathcal{D}}$ is a pregenerator of the semigroup T.*

A contractive c_0-semigroup T is unitary-valued if and only if its generator G is skew-adjoint (so that $G = iH$ where $H = H^*$), and conversely every such operator generates a unitary c_0-semigroup. In this case T_t equals e^{itH}, defined through the functional calculus and Spectral Theorem for self-adjoint operators (cf. the appendix to Johan Kustermans' lectures in this volume), and moreover T extends to a strongly continuous one-parameter group $(T_t)_{t\in\mathbb{R}}$, by $T_{-t} := (T_t)^* = (T_t)^{-1} = e^{-itH}$ for $t > 0$. This circle of ideas includes Stone's Theorem, a cornerstone of mathematical quantum theory. The following result is sometimes useful.

Theorem 1.28 (von Neumann). *let U be a one-parameter group of unitary operators ($U_0 = I$, $U_{s+t} = U_s U_t$ for $s, t \in \mathbb{R}$) on a separable Hilbert space. Then U is strongly continuous if it is weak operator measurable.*

We end this subsection where perhaps we should have begun. For any bounded operator L on a Hilbert space H, $e^{tL} := \sum_{n \geq 0} (n!)^{-1} t^n L^n$ (convergence in norm) defines a c_0-semigroup on H which is norm continuous in the parameter t. Conversely, norm continuity for a c_0-semigroup T implies that its generator G is bounded and that $T_t = e^{tG}$.

1.5 Fock, Cook, Wiener and Guichardet

Let H be a Hilbert space. The symmetric n-fold tensor product of H is the closed subspace of $H^{\otimes n}$ generated by $\{u^{\otimes n} : u \in H\}$, and will be denoted $H^{\vee n}$. The convention here is that $H^{\otimes n} = \mathbb{C}$ when $n = 0$. The orthogonal projection onto $H^{\vee n}$ has the following action on product vectors

$$P^{(n)} : u_1 \otimes \cdots \otimes u_n \mapsto \frac{1}{n!} \sum_{\pi \in \mathfrak{S}_n} u_{\pi(1)} \otimes \cdots \otimes u_{\pi(n)} \qquad (1.17)$$

where $\mathfrak{S}_n$ is the symmetric group of the set $\{1, \ldots, n\}$.

Full Fock space over H is the Hilbert space

$$\Phi(H) := \bigoplus_{n \geq 0} H^{\otimes n} = \mathbb{C} \oplus H \oplus (H \otimes H) \oplus \cdots, \qquad (1.18)$$

and *symmetric Fock space over* H is the Hilbert space

$$\Gamma(H) := \bigoplus_{n \geq 0} H^{\vee n}.$$

There is also an *anti-symmetric Fock space over* H:

$$\Psi(H) := \bigoplus_{n \geq 0} H^{\wedge n}$$

in which $H^{\wedge n}$ is the image of $H^{\otimes n}$ under the orthogonal projection defined by (1.17), modified by including the sign of the permutation in each summand.

Fock space

In this course we shall have only a little use for $\Phi(H)$ and $\Psi(H)$; we therefore speak of *Fock space* taking "symmetric" as understood. Another name for Fock space is *exponential Hilbert space*, with notation e^H — we shall shortly see why.

It is often convenient to identify $H^{\vee n}$ with the corresponding subspace of $\Gamma(H)$:

$$\{0\} \oplus \cdots \oplus \{0\} \oplus \mathsf{H}^{\vee n} \oplus \{0\} \oplus \{0\} \oplus \cdots , \tag{1.19}$$

where the first orthogonal sum is n-fold. Thus

$$\Gamma_{00}(\mathsf{H}) := \mathrm{Lin}\,\{u^{\otimes n} : n \geq 0, u \in \mathsf{H}\}$$

is a useful dense subspace of $\Gamma(\mathsf{H})$. That $\Gamma(\mathsf{H})$ is a subspace of $\Phi(\mathsf{H})$ is sometimes exploited. We shall write P_{sym} for the orthogonal projection

$$\Phi(\mathsf{H}) \to \Gamma(\mathsf{H}). \tag{1.20}$$

The most fundamental Fock space operator of all is defined next.

Definition. The *number operator* on $\Gamma(\mathsf{H})$ is defined by

$$\mathrm{Dom}\,N = \Big\{\xi \in \Gamma(\mathsf{H})\,\Big|\, \sum_{n\geq 0} n^2\|\xi_n\|^2 < \infty \Big\}, \quad N\xi = (n\xi_n)_{n\geq 0}.$$

Thus N is a positive self-adjoint operator on $\Gamma(\mathsf{H})$ which has $\Gamma_{00}(\mathsf{H})$ as an operator core, so that N is the closure of its restriction to $\Gamma_{00}(\mathsf{H})$. In particular, for any function $f : \mathbb{N} \to \mathbb{C}$, $f(N)$ is defined through the functional calculus for self-adjoint operators, by

$$\mathrm{Dom}\,f(N) = \Big\{\xi \in \Gamma(\mathsf{H})\,\Big|\, \sum_{n\geq 0} |f(n)|^2\|\xi_n\|^2 < \infty \Big\}$$

$$f(N)\xi = \big(f(n)\xi_n\big)_{n\geq 0}.$$

Two examples are important to us:

$$\sqrt{N} \text{ and } z^N \quad (z \in \mathbb{C}), \tag{1.21}$$

the latter being bounded if and only if $|z| \leq 1$, in which case it is a contraction.

Definition. For $u \in \mathsf{H}$ the vector

$$\varepsilon(u) := \big((n!)^{-1/2} u^{\otimes n}\big)_{n\geq 0} = \Big(1, u, \frac{1}{\sqrt{2}} u \otimes u, \frac{1}{\sqrt{3!}} u \otimes u \otimes u, \dots \Big)$$

in $\Gamma(\mathsf{H})$ is called the *exponential vector* of u. For any subset S of H define the following subspace of $\Gamma(\mathsf{H})$:

$$\mathcal{E}(S) := \mathrm{Lin}\,\{\varepsilon(u) : u \in S\}. \tag{1.22}$$

Sometimes it is more convenient to use *normalised exponential vectors* for which the terminology *coherent vector* is also used:

$$\varpi(u) := e^{-\frac{1}{2}\|u\|^2} \varepsilon(u). \tag{1.23}$$

Continuity of the exponential map is manifest from the estimates

$$\|\varpi(f) - \varpi(g)\| \leq \|\varepsilon(f) - \varepsilon(g)\| \leq \|f - g\| e^{\frac{1}{2}(\|f\|+\|g\|)^2}, \tag{1.24}$$

(whose proof is an **exercise**); analyticity is exploited next.

Proposition 1.29. *The exponential map* $\varepsilon : H \to \Gamma(H)$ *is a minimal Kolmogorov map for the nonnegative-definite kernel*

$$H \times H \to \mathbb{C}, \quad (u, v) \mapsto e^{\langle u, v \rangle}. \tag{1.25}$$

Proof. From the definition it follows that $\langle \varepsilon(u), \varepsilon(v) \rangle = e^{\langle u, v \rangle}$, and so it remains only to prove minimality, in other words that $\mathcal{E}(H)$ is dense in $\Gamma(H)$. This follows from the following useful observation. For each $u \in H$ the vector-valued map $f : \mathbb{C} \to \Gamma(H)$, $z \mapsto \varepsilon(zu)$, is analytic and

$$f^{(n)}(0) = \sqrt{n!}\, u^{\otimes n}.$$

$\square$

Corollary 1.30. *If S is a dense subset of H then $\mathcal{E}(S)$ is a dense subspace of $\Gamma(H)$.*

Proof. This follows immediately from the continuity of the exponential map.

$\square$

In fact density of $\mathcal{E}(S)$ requires much less of S than it be dense in H. In the next section we shall see an example of this useful for quantum stochastics.

Thus Fock space may also be defined by a universal property, namely if $\eta : H \to K$ is a map into a Hilbert space K satisfying

$$\langle \eta(u), \eta(v) \rangle = e^{\langle u, v \rangle}, \quad u, v \in H,$$

then there is a unique linear isometry $T : \Gamma(H) \to K$ (a Hilbert-space isomorphism if $\operatorname{Ran} \eta$ is total in K) such that $T \circ \varepsilon = \eta$:

Here is a nice illustration of the universal property.

Proposition 1.31. *For Hilbert spaces H_1 and H_2,*

$$\Gamma(H_1 \oplus H_2) = \Gamma(H_1) \otimes \Gamma(H_2). \tag{1.26}$$

Proof (Sketch).

$$\begin{aligned}
\langle \varepsilon(u_1) \otimes \varepsilon(u_2), \varepsilon(v_1) \otimes \varepsilon(v_2) \rangle &= \langle \varepsilon(u_1), \varepsilon(v_1) \rangle \langle \varepsilon(u_2), \varepsilon(v_2) \rangle \\
&= e^{\langle u_1, v_1 \rangle} e^{\langle u_2, v_2 \rangle} \\
&= e^{\langle u_1, v_1 \rangle + \langle u_2, v_2 \rangle} \\
&= e^{\langle (u_1, u_2), (v_1, v_2) \rangle} \\
&= \langle \varepsilon(u_1, u_2), \varepsilon(v_1, v_2) \rangle.
\end{aligned}$$

$\square$

In the notation $\Gamma(\mathsf{H}) = e^{\mathsf{H}}$,

$$e^{\mathsf{H}_1 \oplus \mathsf{H}_2} = e^{\mathsf{H}_1} \otimes e^{\mathsf{H}_2},$$

showing why Fock space has also been called exponential Hilbert space by some authors. We shall content ourselves with refering to (1.26) as the *exponential property* of Fock space. By the same token, for Hilbert spaces $\mathsf{H}_1, \ldots, \mathsf{H}_n$,

$$\Gamma(\mathsf{H}_1 \oplus \cdots \oplus \mathsf{H}_n) = \Gamma(\mathsf{H}_1) \otimes \cdots \otimes \Gamma(\mathsf{H}_n).$$

There is an extension to infinite orthogonal sums too:

$$\Gamma\left(\bigoplus_{n \geq 1} \mathsf{H}_n\right) = \bigotimes^{(\Omega_n)} \Gamma(\mathsf{H}_n),$$

where the stabilising sequence is given by $\Omega_n := \varepsilon(0)$ in $\Gamma(\mathsf{H}_n)$.

The next result has also proved invaluable in the development of quantum stochastic calculus.

Proposition 1.32. *The set $\{\varepsilon(u) : u \in \mathsf{H}\}$ is linearly independent.*

Proof. Let $\xi = \sum_{i=1}^{n} \lambda_i \varepsilon(u_i)$, where $u_1, \ldots u_n \in \mathsf{H}$ are distinct and $\lambda_1, \ldots \lambda_n \in \mathbb{C}$, and suppose that $\xi = 0$. Choose $v \in \mathsf{H}$ such that $z_1 := \langle v, u_1 \rangle, \ldots, z_n := \langle v, u_n \rangle \in \mathbb{C}$ are distinct. (***Exercise.*** Show that this can be done.)

Since the function

$$f : \mathbb{R} \to \mathbb{C}, \quad t \mapsto \langle \varepsilon(tv), \xi \rangle = \sum_{i=1}^{n} \lambda_i e^{t z_i}$$

is identically zero,

$$\sum_{i=1}^{n} \lambda_i z_i^k = f^{(k)}(0) = 0 \quad \text{for } k \geq 0. \tag{1.27}$$

On the other hand, a straightforward induction confirms Vandermonde's identity

$$\det V(\mathbf{z}) = \prod_{i > j} (z_i - z_j),$$

where

$$V(\mathbf{z}) := \begin{bmatrix} 1 & 1 & 1 & \cdots & 1 \\ z_1 & z_2 & z_3 & \cdots & z_n \\ z_1^2 & \ddots & & & \vdots \\ \vdots & & \ddots & & \vdots \\ z_1^{n-1} & & & & z_n^{n-1} \end{bmatrix}.$$

Since $z_1, \ldots, z_n$ are distinct this implies that $V(\mathbf{z})$ is nonsingular. But (1.27) may be read as $V(\mathbf{z})\boldsymbol{\lambda} = \mathbf{0}$, so $\lambda_1 = \cdots = \lambda_n = 0$. $\quad\square$

Alternative argument. Being eigenfunctions of the differential operator $f \mapsto f'$ with distinct eigenvalues, the set of functions $\{t \mapsto e^{zt} : z \in \mathbb{C}\}$ is linearly independent.

Corollary 1.33. *Let $u_1, \ldots, u_n \in \mathsf{H}$ be distinct and let $x_1, \ldots x_n \in \mathsf{h}$, for another Hilbert space h. If*

$$\sum_{i=1}^{n} x_i \otimes \varepsilon(u_i) = 0 \ in \ \mathsf{h} \otimes \Gamma(\mathsf{H})$$

then $x_1 = \cdots = x_n = 0$.

Thus each element of $\mathsf{h} \underline{\otimes} \mathcal{E}(\mathsf{H})$ is uniquely expressible in the form

$$\sum_{i=1}^{n} x_i \otimes \varepsilon(u_i) \text{ for some } n \geq 0, \mathbf{x} \in (\mathsf{h} \setminus \{0\})^n, \mathbf{u} \in \mathsf{H}^n,$$

where $u_1, \ldots u_n$ are distinct—the empty sum where $n = 0$ yielding 0. This fact is exploited for defining operators in the calculus.

Real Fock space

The Fock space construction applies equally to a *real Hilbert space* h. Thus $u \mapsto \varepsilon(u)$, defined in the same way, is a minimal Kolmogorov map for the (real-valued) nonnegative-definite kernel (1.25), in which $\mathsf{H} = \mathsf{h}$. Moreover complexification commutes with the Fock space construction: if H is the Hilbert space complexification of h then $\Gamma(\mathsf{H})$ is the complexification of (the real Hilbert space) $\Gamma(\mathsf{h})$, in particular $\{\varepsilon(u) : u \in \mathsf{h}\}$ is total in $\Gamma(\mathsf{H})$. This is relevant here since classical probability naturally yields the Hilbert spaces of real-valued square-integrable random variables.

Wiener space

Let $\mathsf{k}^{\mathbb{R}}$ be a finite dimensional real Hilbert space with complexification k. The linear space $\mathcal{C} := C(\mathbb{R}_+; \mathsf{k}^{\mathbb{R}})$ carries a metric defined as follows:

$$d(f, g) = p(g - f) \text{ and } p(h) := \sum_{n \geq 1} 2^{-n} (p_n(h)) / (1 + p_n(h)),$$

where for each $n \in \mathbb{N}$, p_n is the seminorm on $\mathcal{C}$ given by $h \mapsto \sup \{\|h(t)\| : t \in [0, n]\}$. With respect to this metric $\mathcal{C}$ is complete and separable, and thus a Fréchet space, and has the *path space*

$$\mathcal{C}_0 := \{\xi \in \mathcal{C} : \xi(0) = \mathbf{0}\} \tag{1.28}$$

as a closed subset. Therefore, as a topological space $\mathcal{C}_0$, is Polish, meaning separable and metrisable by a metric with respect to which it is complete.

Exercise. Show that its Borel σ-algebra coincides with the σ-algebra generated by the evaluations:

$$\text{Borel}(\mathcal{C}_0) = \sigma\{B_t : t \geq 0\} \ \text{where} \ B_t(\xi) := \xi(t).$$

There is a unique probability measure on this σ-algebra such that, for each finite collection $(t_1, E_1), \ldots, (t_n, E_n)$ in $\mathbb{R}_+ \times \text{Borel}(\mathsf{k}^{\mathbb{R}})$ with $0 \leq t_1 \leq \cdots \leq t_n$ (and the understanding $t_0 := 0$ and $\mathbf{x}^0 := \mathbf{0}$),

$$\mathbb{P}\left(\bigcap_{i=1}^{n}\{B_{t_i} \in E_i\}\right) = \int_{E_1 \times \cdots \times E_n} \prod_{i=1}^{n} p(t_i - t_{i-1}, \mathbf{x}^i - \mathbf{x}^{i-1}) \, d\mathbf{x}^1 \ldots d\mathbf{x}^n$$

where $p(t, \mathbf{x}) = (2\pi t)^{-d/2} \exp(-\|\mathbf{x}\|^2 / 2t)$ for $t > 0$, and $p(0, \mathbf{x}) \, d\mathbf{x}$ stands for the Dirac measure at $\mathbf{0}$. The process $B = (B_t)_{t \geq 0}$ is called the standard Wiener process, or canonical Brownian motion, on $\mathsf{k}^{\mathbb{R}}$. It is of course a Gaussian process, and a Lévy process; much more detailed information with guidance into the literature may be found in David Applebaum's lectures in this volume. Let us agree to call $L^2(\mathcal{C}_0)$ *Wiener space* for $\mathsf{k}^{\mathbb{R}}$; also write $\mathsf{K}^{\mathbb{R}}$ for the real Hilbert space $L^2(\mathbb{R}_+; \mathsf{k}^{\mathbb{R}})$, and K for its complexification $L^2(\mathbb{R}_+; \mathsf{k})$.

Simple properties of B show that the prescription

$$1_{[a,c[} \otimes \mathbf{e} \mapsto E^{\mathbf{e}}(B_c - B_a) \ \text{for} \ \mathbf{e} \in \mathsf{k}^{\mathbb{R}}, 0 \leq a < c,$$

extends uniquely to a linear isometry $\mathsf{K}^{\mathbb{R}} \to L^2(\mathcal{C}_0)$, denoted $f \mapsto b(f)$. A further map $\mathsf{K}^{\mathbb{R}} \to L^2(\mathcal{C}_0)$ is defined by

$$e(f) := \exp\left\{b(f) - \tfrac{1}{2}\|f\|^2\right\}.$$

Exercise. Show that $f \mapsto e(f)$ defines a minimal Kolmogorov map for the nonnegative-definite kernel (1.25), in which H is the real Hilbert space $\mathsf{K}^{\mathbb{R}}$. (The part requiring work is the proof of minimality.)

In view of the remarks above on real Fock space, it follows that there is a unique Hilbert space isomorphism

$$\mathcal{F}_{\mathsf{k}} \to L^2(\mathcal{C}_0), \tag{1.29}$$

mapping $\varepsilon(f)$ to $e(f)$ for $f \in \mathsf{K}^{\mathbb{R}}$. Here we are anticipating the notation (2.1) This is sometimes called the *duality transform*.

Exercise. What is the image of $\varepsilon(f)$ for a general $f \in \mathsf{K}$?

The above construction extends nicely to infinite-dimensional $\mathsf{k}^{\mathbb{R}}$.

Fock-space operators

Returning to Fock space over the (complex) Hilbert space H, let $u, v \in$ H, and $T, L \in B($H$)$. The prescriptions

1a. $a_0(u)\varepsilon(x) = \langle u, x \rangle \varepsilon(x)$;

1b. $a_0^\dagger(u)\varepsilon(x) = f'(0)$ where $f : \mathbb{R} \to \Gamma(H)$ is the function $s \mapsto \varepsilon(x + su)$;

2. $n_0(L)\varepsilon(x) = g'(0)$ where $g : \mathbb{R} \to \Gamma(H)$ is the function $s \mapsto \varepsilon(e^{sL}x)$;

3. $\Gamma_0(T)\varepsilon(x) = \varepsilon(Tx)$;

4. $W_0(u)\varepsilon(x) = \exp\{-\frac{1}{2}\|u\|^2 - \langle u, x \rangle\}\varepsilon(x + u)$;

5. $\Gamma_0(z, u, T, v)\varepsilon(x) = \exp\{z + \langle v, x \rangle\}\varepsilon(Tx + u)$;

define closable operators on $\Gamma($H$)$ with domain $\mathcal{E}($H$)$.

Remarks. In fact the more appropriate condition on L is that it be a c_0-semigroup generator, for example a skew-adjoint operator and thus the generator of a unitary group. Such operators have domain $\mathcal{E}(\mathcal{D})$ where $\mathcal{D} = \mathrm{Dom}\, L$. See the last section of David Applebaum's notes in this volume for a discussion of the Lévy-Khintchine formula from a Fock-space viewpoint.

Creation and annihilation operators. *Basic facts*:

$$a_0^\dagger(u) \subset a_0(u)^*, \quad \overline{a_0^\dagger(u)} = a_0(u)^*, \quad \overline{a_0(u)} = a_0^\dagger(u)^*$$

and

$$u \mapsto a_0^\dagger(u) \text{ is linear.}$$

Definition. For $u \in$ H define

$$a(u) := a_0^\dagger(u)^* \quad \text{and} \quad a^*(u) := a_0(u)^*.$$

These operators are mutually adjoint and have $\Gamma_{00}($H$)$ as a core, where their actions are determined by

$$a(u)v^{\otimes n} = \sqrt{n}\langle u, v \rangle v^{\otimes(n-1)} \text{ and}$$
$$a^*(u)v^{\otimes n} = \sqrt{n+1}P^{(n+1)}(u \otimes v^{\otimes n}),$$

$P^{(n+1)}$ denoting the symmetrising projection defined in (1.17). The *canonical commutation relations* (CCR)

$$a(u)a^*(v) = a^*(v)a(u) + \langle u, v \rangle I$$

are easily verified both on $\Gamma_{00}($H$)$ and weakly on $\mathcal{E}($H$)$ — *weakly* because creation operators do not leave $\mathcal{E}($H$)$ invariant. Some further useful identities follow:

$$a(u) = \sqrt{N+1}d(u) \supset d(u)\sqrt{N}; \quad a^*(u) = \sqrt{N}d(u)^* \supset d(u)^*\sqrt{N+1};$$
$$a^*(u)a(u) = Nd(u)^*d(u) = \|u\|^2 NP_u;$$

where, with the understanding $I_{-1} := 0$ and viewing $\Gamma(\mathsf{H})$ as a subspace of $\Phi(\mathsf{H})$,

$$d(u) = \bigoplus_{n \geq 0} \langle u | \otimes I_{n-1}, \text{ so that } d(u)^* = \bigoplus_{n \geq 0} P^{(n)}\left(|u\rangle \otimes I_{n-1}\right),$$

and P_u is the orthogonal projection onto

$$\overline{\mathrm{Lin}}\{P^{(n)}(u \otimes v^{\otimes(n-1)}) : n \geq 1, v \in \mathsf{H}\}.$$

Differential second quantisation. *Basic facts*:

$$n_0(L^*) \subset n_0(L)^*, \overline{n_0(L^*)} = n_0(L)^*$$

and

$$L \mapsto n_0(L) \text{ is linear.}$$

Definition. For $L \in B(\mathsf{H})$ define

$$d\Gamma(L) := \overline{n_0(L)}.$$

These operators have $\Gamma_{00}(\mathsf{H})$ as a core, where their actions are determined by

$$d\Gamma(L)v^{\otimes n} = \sum_{i=1}^{n} v^{\otimes(i-1)} \otimes Lv \otimes v^{\otimes(n-i)}$$
$$= nP^{(n)}(Lv \otimes v^{\otimes(n-1)}).$$

In particular, $d\Gamma(L)\varepsilon(0) = 0$ and $d\Gamma(I) = N$. The commutation relations

$$d\Gamma(L)a^*(u) = a^*(u)d\Gamma(L) + a^*(Lu) \tag{1.30}$$

are easily verified on $\Gamma_{00}(\mathsf{H})$.

Second quantisation. *Basic facts*:

$$\Gamma_0(T^*) \subset \Gamma_0(T)^*, \overline{\Gamma_0(T^*)} = \Gamma_0(T)^*$$

and

$$T \mapsto \Gamma_0(T) \text{ is multiplicative.}$$

Definition. For $T \in B(\mathsf{H})$ define

$$\Gamma(T) := \overline{\Gamma_0(T)}.$$

These operators have $\Gamma_{00}(\mathsf{H})$ as a core where their actions are determined by

$$\Gamma(T)v^{\otimes n} = (Tv)^{\otimes n}.$$

In particular,

$$\Gamma(T)\varepsilon(0) = \varepsilon(0),\ \Gamma(I) = I,\ \Gamma(0) = |\varepsilon(0)\rangle\langle\varepsilon(0)|,$$

and if T is a contraction then $\Gamma(T)$ is a contraction too. Moreover $\Gamma(T)$ is isometric (respectively, coisometric, a projection) if T is. In general, for $T \neq 0$,

$$\Gamma(T) = z^N \Gamma(\widetilde{T})$$

where $z = \|T\|$ and $\widetilde{T}$ is the contraction $\|T\|^{-1}T$. This said, the notation here is generally reserved for *contractive* operators T. The identity

$$\Gamma(T)a^*(u) = a^*(Tu)\Gamma(T)$$

is easily verified on both $\Gamma_{00}(\mathsf{H})$ and (through adjoints) on $\mathcal{E}(\mathsf{H})$, cf. (1.30).

Fock Weyl operator. *Basic fact*: In view of the identity

$$\langle W_0(u)\varepsilon(v), W_0(u)\varepsilon(w)\rangle = \langle \varepsilon(v), \varepsilon(w)\rangle,$$

$W_0(u)$ extends uniquely to an isometric operator on $\Gamma(\mathsf{H})$.

Definition. For $u \in \mathsf{H}$ define

$$W(u) := \overline{W_0(u)}.$$

In terms of the normalised exponential vectors defined in (1.23),

$$W(u)\varpi(v) = e^{-i\,\mathrm{Im}\,\langle u,v\rangle}\varpi(v+u) \text{ and } \varpi(v) = W(v)\varepsilon(0),$$

from which the *Weyl commutation relations* are easily seen:

$$W(u)W(v) = e^{-i\,\mathrm{Im}\,\langle u,v\rangle}W(u+v),\quad W(0) = I.$$

In particular, $W(u)$ is unitary and $W(u)^* = W(-u)$.

Exponential operator. *Basic facts*:

$$\Gamma_0(\bar{z}, v, T^*, u) \subset \Gamma_0(z, u, T, v)^*,\ \Gamma_0(0, 0, I, 0) \subset I$$

and

$$\Gamma_0(z_1, u_1, T_1, v_1)\Gamma_0(z_2, u_2, T_2, v_2) = \Gamma_0(z, u, T, v)$$

where

$$z = z_1 + z_2 + \langle v_1, u_2\rangle,\quad u = u_1 + T_1 u_2,\quad T = T_1 T_2,\quad \text{and}\quad v = T_2^* v_1 + v_2.$$

The linear span of this family of operators therefore forms a unital *-algebra within $L(\mathcal{E}(\mathsf{H}))$, which includes the previous two classes:

$$\Gamma_0\big(-\tfrac{1}{2}\|u\|^2, u, I, -u\big) = W_0(u) \text{ and } \Gamma_0(0,0,T,0) = \Gamma_0(T).$$

The decomposition

$$\Gamma_0(z, u, T, v) = e^z e^{a^*(u)} \Gamma_0(T) e^{a(v)}$$

is a nice example of *Wick ordering*, in which creation is left-most and annihilation is right-most.

Definition. For $z \in \mathbb{C}$, $u, v \in \mathsf{H}$ and $T \in B(\mathsf{H})$ define

$$\Gamma(z, u, T, v) := \overline{\Gamma_0(z, u, T, v)}.$$

Note the identity

$$\langle \varepsilon(x), \Gamma(z, u, T, v)\varepsilon(y)\rangle = \exp\big\{z + \langle v, y\rangle + \langle x, Ty\rangle + \langle x, u\rangle\big\}, \tag{1.31}$$

which may also be written

$$\langle \varepsilon(x), \Gamma(z, u, T, v)\varepsilon(y)\rangle = \exp\langle \hat{x}, L\hat{y}\rangle,$$

in the notation (0.2), where

$$L = \begin{bmatrix} z & \langle v| \\ |u\rangle & T \end{bmatrix} \in B(\widehat{\mathsf{H}}).$$

It is easily checked that Γ is *isometric* if and only if L takes the form

$$\begin{bmatrix} i\theta - \tfrac{1}{2}\|u\|^2 & \langle -V^*u| \\ |u\rangle & V \end{bmatrix} \tag{1.32}$$

where $\theta \in \mathbb{R}$ and V is isometric.

Exercise. *Show that the operator $\Gamma(z, u, T, v)$ is a contraction if and only if T is a contraction and $(v + T^*u) = (1 - T^*T)^{1/2}x$ for some vector x satisfying $\|x\|^2 \leq -\tfrac{1}{2}\|u\|^2 - \operatorname{Re} z$.*

An invariant domain

For the 'Fock-space part' of our operators we shall use exponential domains almost exclusively. We have mentioned one other useful dense subspace of Fock space, namely $\Gamma_{00}(\mathsf{H})$. Here is another:

$$K_{\mathsf{H}} := \bigcap_{z \in \mathbb{C}} \operatorname{Dom} z^N \tag{1.33}$$

$$= \Big\{\xi \in \Gamma(\mathsf{H}) \Big| \sum_{n \geq 0} a^n \|\xi_n\|^2 < \infty \quad \text{for all } a > 0\Big\}, \tag{1.34}$$

z^N being the operator defined in (1.21). A nice feature of this subspace which contains both $\mathcal{E}(\mathsf{H})$ and $\Gamma_{00}(\mathsf{H})$ is that, as well as lying in the domains of all the closed operators we have met so far, unlike $\mathcal{E}(\mathsf{H})$ or $\Gamma_{00}(\mathsf{H})$ it is also left invariant by them.

Guichardet space

There is an alternative view of Fock space which has been profitable in QS analysis. For a set S and nonnegative integer n, define

$$\Gamma_S := \{\sigma \subset S : \#\sigma < \infty\} \text{ and } \Gamma_S^{(n)} := \{\sigma \subset S : \#\sigma = n\}.$$

Thus $\Gamma_S^{(0)} = \{\emptyset\}$ and Γ_S is the disjoint union $\bigcup_{n \geq 0} \Gamma_S^{(n)}$. Any function $f : S \to \mathbb{C}$ determines a *product function*

$$\pi_f : \Gamma_S \to \mathbb{C}, \quad \pi_f(\sigma) = \prod_{s \in \sigma} f(s). \tag{1.35}$$

These enjoy obvious properties

$$\overline{\pi_f} = \pi_{\overline{f}}, \ \pi_f \pi_g = \pi_{fg} \text{ and } |\pi_f|^p = \pi_{|f|^p} \ (p > 0). \tag{1.36}$$

Moreover, if $<$ is a total order on S and we write $S^n_<$ for $\{\mathbf{s} \in S^n : s_1 < \cdots < s_n\}$, then we have a bijection

$$(s_1, \ldots, s_n) \mapsto \{s_1, \ldots, s_n\}, \quad S^n_< \to \Gamma_S^{(n)}.$$

Now let $S = I$, a subinterval of $\mathbb{R}_+$. Restricting Lebesgue measure on I^n induces a measure on $\Gamma_I^{(n)}$, for each n, and thereby a measure on Γ_I by letting $\emptyset$ be an atom of unit measure. Integration with respect to this measure is denoted $\int \cdots d\sigma$. If f and f' are measurable functions $I \to \mathbb{C}$ which agree almost everywhere then π_f and $\pi_{f'}$ are measurable and agree almost everywhere too. If $f : I \to \mathbb{C}$ is integrable then so is π_f, moreover

$$\int \pi_f(\sigma) \, d\sigma = e^{\int f(s)ds}.$$

Combining these two facts with (1.36) shows that $f \mapsto \pi_f$ defines maps $L^p(I) \to L^p(\Gamma_I)$, for $1 \leq p \leq \infty$, satisfying

$$\int \overline{\pi_f(\sigma)} \pi_g(\sigma) \, d\sigma = e^{\int \overline{f(s)}g(s)ds} \text{ for } f \in L^p(I), g \in L^{p'}(I),$$

when p and p' are conjugate exponents. In particular,

$$\langle \pi_f, \pi_g \rangle = e^{\langle f, g \rangle} \text{ for } f, g \in L^2(I).$$

Exercise. Show that $\{\pi_f : f \in L^2(I)\}$ is total in $\mathcal{G}_I := L^2(\Gamma_I)$.

For analysis in Guichardet space the following identity, known as the *integral-sum formula*, is fundamental:

$$\int_\Gamma d\sigma \sum_{\alpha \subset \sigma} F(\alpha, \overline{\alpha}) = \int d\alpha \int d\beta \, F(\alpha, \beta)$$

for integrable, or nonnegative and measurable, functions F, where $\overline{\alpha}$ denotes $\sigma \setminus \alpha$.

Exercise. *Prove it (first) for product functions.*

We next address the question of how to extend this to vector-valued functions. Let $K_I = L^2(I; k) = L^2(I) \otimes k$, for a Hilbert space k. Recalling the notation (1.18) for full Fock space, define

$$\mathcal{G}_{k,I} := \{F \in L^2(\Gamma_I; \Phi(k)) : F(\sigma) \in k^{\otimes \# \sigma} \text{ for a.a. } \sigma\},$$

and, for $f \in K_I$ and $\sigma \in \Gamma_I$,

$$\pi_f(\sigma) = \begin{cases} f(s_1) \otimes \cdots \otimes f(s_n) & \text{if } \sigma = \{s_1 < \cdots < s_n\} \\ 1 & \text{if } \sigma = \emptyset. \end{cases}$$

As for $\mathcal{F}$ (see (2.1) below) we drop subscripts when $k = \mathbb{C}$, respectively $I = \mathbb{R}_+$.

Exercise. *Show that $\pi_f \in \mathcal{G}_{k,I}$ and that $f \mapsto \pi_f$ defines a minimal Kolmogorov map for the nonnegative-definite kernel (1.25), now for $H = K_I$.*

The map $\pi_f \to \varepsilon_f$ ($f \in K_I$) therefore extends uniquely to a Hilbert space isomorphism

$$\mathcal{G}_{k,I} \cong \mathcal{F}_{k,I} \tag{1.37}$$

where $\mathcal{F}_{k,I}$ denotes the symmetric Fock space $\Gamma(K_I)$.

Combining the above isomorphism with the duality transform (1.29), in case $k = \mathbb{C}$ and $I = \mathbb{R}_+$, one may ask how multiplication of random variables in $L^2(\mathcal{C}_0)$ looks when transformed to $\mathcal{G}$. This has the following elegant answer. Let $F, G \in K$, the domain (1.33) for $H = L^2(\mathbb{R}_+)$. Then the *integral-sum convolution*

$$\sigma \mapsto \sum_{\alpha \subset \sigma} \int_\Gamma d\omega \, F(\alpha \cup \omega) G(\omega \cup \bar{\alpha})$$

defines an element $F * G \in K$ and, if $\hat{\ }$ denotes the duality transform (1.29), but now viewed as a map $\mathcal{G} \to L^2(\mathcal{C}_0)$, then

$$\widehat{F * G} = \hat{F} \, \hat{G}.$$

Exercise. *(Open-ended.) Deconstruct the following statement of duality. Every element of Wiener space is expressible in the form*

$$\int_\Gamma F(\sigma) \, dB_\sigma$$

for a unique $F \in \mathcal{G}$.

In fact this may be done for *Poisson space*, and more generally for normal martingales enjoying the *chaotic representation property*, each giving a product on Guichardet space.

Notes

Positivity is key, both in noncommutative probability and in the theory of operator algebras. There are now many texts on operator algebras: [Mur₁], [Sun] and [Weg] are particularly accessible; [Dix₁,₂], [Ped], [Sak], [StZ] and [Tak] are the classics; and [KR₁,₂] is very much geared to the student, with carefully worked solutions to all of its exercises provided in [KR₃,₄]. Johan Kustermans has provided a nice introduction in the first section of his notes in this volume; for another see [Sau].

The use of Kolmogorov maps in the quantum theory of open systems was forcefully advocated in the 'little red book', [EvL]. A Hilbert C^*-module generalisation of Kolmogorov maps may be found in [Mur₂]; this is useful in the context of C^*-algebraic dilations.

For the theory of operator spaces there are two fine books that have recently appeared: [EfR], [Pis]. Also an early book on the subject has recently appeared in expanded and updated form: [Pau]. Matrix spaces were introduced in [LW₃]; Example 1.23 has been extracted from there, and modified. When $h = l^2$, $M(h;V)_b$ is completely isometric to an operator space of infinite matrices: $\{x \in M_\mathbb{N}(V) : \|x\| := \sup_N \|x^{[N]}\| < \infty\}$ where $x^{[N]} \in M_N(V)$ denotes the top-left $N \times N$ truncation of x (see [EfR]). Whereas these have been defined for abstract operator spaces but concrete coefficients, the matrix spaces used in these notes involve abstract Hilbert spaces from which the coefficients come, but concrete operator spaces. The two have a satisfactory fusion in the form of a fully abstract matrix space ([LSa]).

Useful information on unbounded operators is collected in the appendix of Johan's notes in this volume; for a thorough treatment of the basics (for Hilbert-space operators) the final chapter of [RS₁] is recommended. For c_0-semigroups [Dav] is recommended, [RS₂] has a useful section and [HiP] is the classic text.

Good sources on Fock-space operators and the canonical commutation relations (CCR) and canonical anticommutation relations (CAR) are [EvL], [BR₂], [Pet] and [Fan]; see also [EvK], which incorporates much of [EvL]. Fock space ([Foc]) was put on a sound mathematical footing in [Coo]. There is also a duality transform for anti-symmetric Fock space ([Seg]) whose image is the (tracial) noncommutative L^2-space of the Clifford process—an anticommuting/Fermionic analogue of the Wiener process having its own stochastic calculus ([BSW₁]). There is now a duality transform for full Fock space too, whose image is the noncommutative L^2-space of the Wigner process—an analogue of the Wiener process in 'free probability' (see [VDN] and the lectures by Ole Barndorff-Nielsen and Steen Thorbjørnsen in the second volume of these notes). This too has its own stochastic calculus (see [BiS] and the lecture notes [Spe]).

Guichardet space (under another name!) is expounded in [Gui] and, for vector-valued functions, in [Sch]. Its basic properties, including the integral-sum convolution formula for Wiener space products ([Maa]), are described in the lecture notes [L7]. These notes include applications to hypercontractivity estimates ([L5], [LMe]; see the lectures of David Applebaum in this volume) and cohomological deformation of the Wiener product ([LP1]). For an introduction to normal martingales enjoying the chaotic representation property, see the last section of the lecture notes [Eme].

2 QS Processes

In this section (Hilbert-space) operator processes and (operator-space) mapping processes are defined, and examples are given to illustrate the definitions. Quantum stochastic martingales are also defined. The first part of the section concerns the choice of domain for quantum stochastic processes. For these notes exponential domains are used exclusively. In this context a useful density result is proved.

We need Hilbert spaces h_1, h_2 for the action and a Hilbert space k governing the dimension of the noise which we *fix from now on* and refer to as the 'noise dimension space'.

For any subinterval I of $\mathbb{R}_+$ we write

$$\mathsf{K}_I := L^2(I; \mathsf{k}), \quad \mathcal{F}_{\mathsf{k},I} := \Gamma(\mathsf{K}_I) \quad \text{and} \quad \Omega_{\mathsf{k},I} := \varepsilon(0) \text{ in } \mathcal{F}_{\mathsf{k},I}, \qquad (2.1)$$

dropping the subscript I when $I = \mathbb{R}_+$, and dropping the subscript k when $\mathsf{k} = \mathbb{C}$. The exponential property (1.26), applied to the orthogonal direct-sum decomposition

$$\mathsf{K} = \mathsf{K}_{[0,s[} \oplus \mathsf{K}_{[s,t[} \oplus \mathsf{K}_{[t,\infty[}$$

for $t \geq s \geq 0$, yields the tensor product decomposition

$$\mathsf{h} \otimes \mathcal{F}_{\mathsf{k}} = \mathsf{h} \otimes \mathcal{F}_{\mathsf{k},[0,s[} \otimes \mathcal{F}_{\mathsf{k},[s,t[} \otimes \mathcal{F}_{\mathsf{k},[t,\infty[}, \qquad (2.2)$$

for each Hilbert space h. We shall use such identifications all the time, and shall take it to infinitesimal extremes: "$t = s + ds, ds > 0$". The family $(\mathcal{F}_{\mathsf{k},[0,t[})_{t\geq 0}$ gives the basic example of an Arveson product system of Hilbert spaces (see Rajarama Bhat's lectures in this volume).

2.1 Exponential domains

Exponential domains have proved highly convenient for the definition, construction and analysis of noncommutative stochastic processes defined on a Fock space; we shall largely use such domains here. For this purpose we need our domains to respect decompositions such as (2.2).

Definition. A k-*admissible set* is a subset S of K such that

(a) $\mathcal{E}(S)$ is dense in $\mathcal{F}_{\mathsf{k}}$, and
(b) $f \in S, t \geq 0 \Rightarrow f_{[0,t[} \in S$.

In particular, $0 \in S$ for a k-admissible set S.

Remark. In the past further properties have been incorporated into the definition of admissibility. Sometimes it is useful to assume that S is linear, or at least closed under small scalar multiples. It is often convenient to assume that S consists of locally essentially bounded functions.

Apart from K itself, and the set $S = \{f \in (L^2 \cap L^\infty_{loc})(\mathbb{R}_+; k) \mid \dim f < \infty\}$, where $\dim f$ denotes the minimum dimension of the linear span of the range of f in k for (measure) representatives of f, a very useful class of admissible sets arises as follows. For any subset A of k define $\mathcal{E}_A := \mathcal{E}(\mathbb{S}_A)$ where

$$\mathbb{S}_A := \{f \in \mathbb{S} : f \text{ is } A\text{-valued}\},$$

and $\mathbb{S}$ denotes the set of step functions, i.e. $\text{Lin}\{e_{[0,t[} : e \in k, t \geq 0\}$.

Proposition 2.1. *Let* T *be a total subset of the Hilbert space* k *containing* 0. *Then* $\mathcal{E}_T$ *is dense in* $\mathcal{F}_k$.

Proof. Denote the collection of bounded subintervals of $\mathbb{R}_+$ by $\mathcal{I}$ and let A be the closed convex hull of T. First note that, for $t \in [0,1]$ and $c, d \in k$, if $J_n = \bigcup_k J_{k,n}$ where $J_{k,n} = [k2^{-n}, (k+t)2^{-n}[$ then the sequence $(cl_{J_n} + dl_{J_n^c})$ converges weakly in K to $(tc + (1-t)d) 1_{[0,1]}$. By rescaling, shifting and taking sums and limits, it follows that each element of $\mathbb{S}_A$ may be weakly approximated by a sequence from $\mathbb{S}_T$. It follows from the exponential relation $\langle \varepsilon(f), \varepsilon(g) \rangle = e^{\langle f, g \rangle}$, and the density of $\mathcal{E}$ (K) in $\mathcal{F}_k$, that the exponential map $f \mapsto \varepsilon(f)$ is weakly continuous on bounded sets. Therefore, since the weak and strong closure of convex subsets of a Hilbert space coincide, $\overline{\mathcal{E}_T} \supset \mathcal{E}_A$. Thus

$$\varepsilon(t_1 c_1 1_{I_1} + \cdots + t_n c_n 1_{I_n}) \in \overline{\mathcal{E}_T}$$

for any $n \in \mathbb{N}$ and $(\mathbf{t}, \mathbf{c}, \mathbf{I}) \in [0,1]^n \times T^n \times \mathcal{I}^n$ with $\{I_j\}$ disjoint. Now $\{cl_I : c \in T, I \in \mathcal{I}\}$ is obviously total in K and, since the generalised diagonal $\{\mathbf{s} \in \mathbb{R}^n : s_i = s_j \text{ for some } i \neq j\}$ is Lebesgue null, it is not hard to see that the set

$$\{c_1 1_{I_1} \otimes \cdots \otimes c_n 1_{I_n} : (\mathbf{c}, \mathbf{I}) \in T^n \times \mathcal{I}^n, \{I_j\} \text{ disjoint}\}$$

is total in $K^{\otimes n}$. The result therefore follows from the identity

$$\sqrt{n!} P^{(n)} (c_1 1_{I_1} \otimes \cdots \otimes c_n 1_{I_n}) = \frac{\partial^n}{\partial t_1 \cdots \partial t_n}\bigg|_{t=0} \varepsilon(t_1 c_1 1_{I_1} + \cdots + t_n c_n 1_{I_n}),$$

where $P^{(n)}$ is the symmetrising projection (1.17). $\qquad\square$

Example 2.2. Let $k = \mathbb{C}$, so that $K = L^2(\mathbb{R}_+)$. Then elements of $K^{\vee n}$ may be viewed as symmetric functions of n variables from $\mathbb{R}_+$. Consider the function defined on $\{(x, y, z) \in (\mathbb{R}_+)^3 \mid 0 \leq x \leq y \leq z\}$ by

$$g(x, y, z) = \begin{cases} 1 & \text{if } y \leq (x+z)/2 \\ -1 & \text{otherwise.} \end{cases}$$

Exercise. *Show that symmetric extension of g yields an element of $\mathsf{K}^{\vee 3}$ which is orthogonal to all vectors of the form $\varepsilon(\mathbf{1}_{[a,b[})$.*

This example shows that indicator functions of intervals alone will not do.

Convention. When the admissible set is such that each element has a right-continuous version, such as for $\mathbb{S}_T$, we shall use *that* representative.

2.2 Operator processes

We are ready for the basic definition.

Definition. For a k-admissible set S, dense subspace $\mathcal{D}$ of h_1 and second Hilbert space h_2, an h_1-h_2-*process with domain* $\mathcal{D}\underline{\otimes}\mathcal{E}(S)$ is a family $X = (X_t)_{t \geq 0}$ of operators $\mathsf{h}_1 \otimes \mathcal{F}_\mathsf{k} \to \mathsf{h}_2 \otimes \mathcal{F}_\mathsf{k}$, each having domain $\mathcal{D}\underline{\otimes}\mathcal{E}(S)$, which is (Lebesgue) weak operator measurable in t satisfies the *adaptedness condition*:

$$X_t = X(t)\underline{\otimes}I' \tag{2.3}$$

for an operator $X(t) : \mathsf{h}_1 \otimes \mathcal{F}_{\mathsf{k},[0,t[} \to \mathsf{h}_2 \otimes \mathcal{F}_{\mathsf{k},[0,t[}$ with domain $\mathcal{D}\underline{\otimes}\mathcal{E}(S|_{[0,t[})$, where I' is the restriction of $I_{\mathsf{k},[t,\infty[}$ to $\mathcal{E}(S|_{[t,\infty[})$.

Thus $E^\Omega X_t E_\Omega = X(t)$ where $\Omega = \Omega_{\mathsf{k},[t,\infty[}$, and

$$\langle u\varepsilon(f), X_t v\varepsilon(g)\rangle = \langle u\varepsilon(f_{[0,t[}), X_t v\varepsilon(g_{[0,t[})\rangle \exp\langle f_{[t,\infty[}, g_{[t,\infty[}\rangle \tag{2.4}$$

for all $u \in \mathsf{h}_2$, $v \in \mathcal{D}$, $f \in \mathsf{K}$ and $g \in S$.

Remark. For weak operator measurability it suffices to check that $\langle \xi, X_t\varepsilon\rangle$ is Lebesgue measurable in t for ($\varepsilon \in \mathcal{D}\underline{\otimes}\mathcal{E}(S)$ and) ξ running through any total subset of $\mathsf{h}_2 \otimes \mathcal{F}_\mathsf{k}$, such as $\mathcal{D}'\underline{\otimes}\mathcal{E}(S')$ where $\mathcal{D}'$ is dense and S' is k-admissible.

Two h_1-h_2-processes X and Y with domain $\mathcal{D}\underline{\otimes}\mathcal{E}(S)$ are *identified* if

$$\forall \xi \in \mathsf{h}_2 \otimes \mathcal{F}_\mathsf{k}, \varepsilon \in \mathcal{D}\underline{\otimes}\mathcal{E}(S) \quad \langle \xi, X_t\varepsilon\rangle = \langle \xi, Y_t\varepsilon\rangle \text{ for a.a. } t.$$

Thus a process may properly be viewed as an element of the vector space $L(\mathcal{D}\underline{\otimes}\mathcal{E}(S); L_\mathsf{w}^0(\mathbb{R}_+; \mathsf{h}_2 \otimes \mathcal{F}_\mathsf{k}))$, where L_w^0 denotes measure equivalence classes of weakly measurable functions. The collection of such processes is denoted

$$\mathbb{P}(\mathcal{D}\underline{\otimes}\mathcal{E}(S); \mathsf{h}_2 \otimes \mathcal{F}_\mathsf{k}), \tag{2.5}$$

or simply $\mathbb{P}(\mathcal{D}\underline{\otimes}\mathcal{E}(S))$ when $\mathsf{h}_2 = \mathsf{h}_1$.

Adjoint processes

Suppose that $X \in \mathbb{P}(\mathcal{D}\underline{\otimes}\mathcal{E}(S); \mathsf{h}_2 \otimes \mathcal{F}_\mathsf{k})$ and $X^\dagger \in \mathbb{P}(\mathcal{D}'\underline{\otimes}\mathcal{E}(S'); \mathsf{h}_1 \otimes \mathcal{F}_\mathsf{k})$ for subspaces $\mathcal{D}$ of h_1 and $\mathcal{D}'$ of h_2, and k-admissible sets S and S'. If, for all $\varepsilon' \in \mathcal{D}'\underline{\otimes}\mathcal{E}(S')$, $\varepsilon \in \mathcal{D}\underline{\otimes}\mathcal{E}(S)$,

$$\langle X_t^\dagger \varepsilon', \varepsilon\rangle = \langle \varepsilon', X_t\varepsilon\rangle \text{ for a.a. } t$$

then $(X^\dagger, X)$ is referred to as an adjoint pair of processes.

Process types

Let $X \in \mathbb{P}(\mathcal{D}\underline{\otimes}\mathcal{E}(S); h_2 \otimes \mathcal{F}_k)$ for some dense subspace $\mathcal{D}$ of h_1, k-admissible subset S and Hilbert spaces h_1, h_2 and k. Recall the operator measurability definitions on page 198. The process X is called *measurable* (respectively, *continuous*) if it is a (Lebesgue) strong operator measurable (respectively continuous) function of t. Thus a process X is measurable if and only if $t \mapsto X_t\varepsilon$ is (a.e.) separably valued, for each $\varepsilon \in \mathcal{D}\underline{\otimes}\mathcal{E}(S)$. In particular, measurability is automatic if both h_2 and k have countable dimensions; in any case all continuous processes are measurable.

A process X is *bounded* (respectively, contractive, isometric, etc.) if each X_t enjoys that property. We use the notation $\mathbb{P}_b(h_1 \otimes \mathcal{F}_k; h_2 \otimes \mathcal{F}_k)$, or simply $\mathbb{P}_b(h \otimes \mathcal{F}_k)$ when $h_2 = h_1 = h$, for the space of bounded processes.

Example 2.3 (Time reversal process). Let $(r_t)_{t\geq 0}$ be the family of operators on K given by

$$(r_t f)(s) = \begin{cases} f(t-s) & \text{for } 0 \leq s < t \\ f(s) & \text{for } s \geq t \end{cases}.$$

Then second quantisation:

$$R_t = I_h \otimes \Gamma(r_t), \quad t \geq 0,$$

defines a bounded, continuous h-process, called the *reflection process*, or *time reversal process*.

If V is an operator space in $B(h_1; h_2)$ then $\mathbb{P}(\mathcal{D}\underline{\otimes}\mathcal{E}(S); h_2 \otimes \mathcal{F}_k)_V$ denotes the collection of h_1-h_2-processes X with domain $\mathcal{D}\underline{\otimes}\mathcal{E}(S)$ satisfying

$$E^\xi X_t E_\varepsilon \in V \text{ for all } \xi \in \mathcal{F}_k, \varepsilon \in \mathcal{E}(S), \tag{2.6}$$

in other words $X_t \in M(\mathcal{F}_k, \mathcal{E}(S); V)$ (see Subsection 1.1). We refer to these as *operator processes in* V.

Further remarks on domains

The care we are taking with domains reflects two facts of QS life. Firstly we must deal with processes consisting of unbounded operators — for example the fundamental processes to be reviewed shortly, and QS integrals formed from these. Secondly, even when our processes are bounded the fact that exponential domains reflect the continuous tensor product structure of Fock space so precisely makes it often convenient to consider all processes on the same domain.

That said, other aspects — particularly algebraic questions — demand the handling of more general domains. For example we *cannot* suppose that processes of interest leave an exponential domain invariant, so the operator composition of processes is problematic. However, one can get around this limitation to a surprising extent by considering inner products of exponential vectors acted upon by processes, $\langle X_t\varepsilon, Y_t\varepsilon'\rangle$, as a substitute for composition.

Martingales

The time-s *conditional expectation* of an operator $T : h_1 \otimes \mathcal{F}_k \to h_2 \otimes \mathcal{F}_k$ with domain $\mathcal{D} \underline{\otimes} \mathcal{E}(S)$ is defined by

$$\mathbb{E}_s[T] = T(s) \underline{\otimes} I_{\mathcal{F}_{k,[s,\infty[}} |_{\mathcal{D} \underline{\otimes} \mathcal{E}(S)}, \text{ where}$$
$$T(s) = E^\Omega T E_\Omega, \quad \Omega = \Omega_{k,[s,\infty[} \in \mathcal{F}_{k,[s,\infty[}.$$

Example 2.4. Let $T = |\varepsilon(f)\rangle\langle\varepsilon(g)|$ where $f, g \in \mathsf{K}$. Then

$$\mathbb{E}_s[T] = |\varepsilon(f|_{[0,s[})\rangle\langle\varepsilon(g|_{[0,s[})| \otimes I_{[s,\infty[}. \tag{2.7}$$

A process $X \in \mathbb{P}(\mathcal{D} \underline{\otimes} \mathcal{E}(S); h_2 \otimes \mathcal{F}_k)$ is a *martingale* if, for $u \in h_2, v \in \mathcal{D}, f \in k$ and $g \in S$,

$$\langle u\varepsilon(f_{[0,s[}), (X_t - X_s)v\varepsilon(g_{[0,s[})\rangle = 0 \text{ for } s < t, \tag{2.8}$$

(cf. (2.4)). The collection of h_1-h_2-martingales with given domain forms a subspace of the space of processes.

Complete martingales. In view of the tower property of conditional expectations

$$\mathbb{E}_s \circ \mathbb{E}_t = \mathbb{E}_s \text{ for } s \le t,$$

any operator $T : h_1 \otimes \mathcal{F}_k \to h_2 \otimes \mathcal{F}_k$ with domain $\mathcal{D} \underline{\otimes} \mathcal{E}(S)$ determines a martingale in $\mathbb{P}(\mathcal{D} \underline{\otimes} \mathcal{E}(S); h_2 \otimes \mathcal{F}_k)$ by

$$X_t = \mathbb{E}_t[T].$$

Thus, for example, (2.8) defines a complete martingale (until f and g are replaced by *locally* square-integrable functions).

Annihilation. With $h_1 = \mathbb{C}, h_2 = k$ and $S = K = L^2(\mathbb{R}_+) \otimes k$,

$$X_t \varepsilon(f) = E^{1_{[0,t[}} f \otimes \varepsilon(f)$$

defines a process satisfying

$$(X_t - X_s)\varepsilon(f_{[0,s[}) = E^{1_{[s,t[}} f_{[0,s[} \otimes \varepsilon(f_{[0,s[}) = 0,$$

so (2.8) is satisfied and X is a martingale. As a slight variant of this, if $h_2 = \mathbb{C}$ then

$$a(g_{[0,t[}), \text{ where } g \in L^2_{\text{loc}}(\mathbb{R}_+; k), \tag{2.9}$$

defines a martingale $A^{\langle g|}$. Here L^2_{loc} denotes *locally square-integrable*.

Adjoint martingales. If a martingale X has an adjoint h_2-h_1-process $X^\dagger$ with domain $\mathcal{D}' \underline{\otimes} \mathcal{E}(S')$, then $X^\dagger$ is a martingale too.

Creation processes. As a basic example of an adjoint martingale let $h_1 = h_2 = \mathbb{C}$ and $S = k$ again, then

$$a^*(g_{[0,t[}), \text{ where } g \in L^2_{\text{loc}}(\mathbb{R}_+; \mathsf{k})$$

defines a martingale $A^*_{|g\rangle}$, adjoint to the martingale $A_{\langle g|}$.

Preservation and Number/Exchange processes. With $\mathsf{k} = \mathbb{C}$ so that $\mathsf{K} = L^2(\mathbb{R}_+)$, using Guichardet space (see page 211),

$$(N_t \xi)(\sigma) = \#(\sigma \cap [0,t]) \xi(\sigma), \quad \xi \in \mathsf{h} \underline{\otimes} \mathcal{E}(\mathsf{K}) \tag{2.10}$$

defines an essentially self-adjoint martingale. This generalises as follows. Keeping $\mathsf{h}_1 = \mathsf{h}_2 = \mathbb{C}$ and $S = \mathsf{K}$,

$$d\Gamma(R_{[0,t[}), \text{ where } R \in L^\infty_{\text{loc}}(\mathbb{R}_+; B(\mathsf{k})),$$

defines a martingale N_R. Here L^∞_{loc} denotes essentially locally bounded and strong operator measurable (see page 198).

Quantum stochastic integration provides a very rich source of further examples of martingales. The above ones may then be seen as *quantum Wiener integrals*:

$$A^{\langle g|}_t = \int_0^t \langle g(s)| \, dA_s, \ N_{R,t} = \int_0^t R(s) \, dN_s \text{ and } A^*_{|g\rangle,t} = \int_0^t |g(s)\rangle \, dA^*_s.$$

Quantum Brownian motion

Brownian motion, the prototype continuous-time classical stochastic process, has many symmetries. For example time reversal, spatial reflection, translation and (in multidimensions) rotation too. Quantum Brownian motion is distinguished by having a further one, namely *gauge symmetry*.

Set
$$A^\theta_t = \Gamma^*_\theta A_t \Gamma_\theta \text{ and } Q^\theta_t = \overline{A^\theta_t + (A^\theta_t)^*},$$

where $A_t = a(\mathbf{1}_{[0,t[})$ and $\Gamma_\theta = \Gamma(e^{i\theta} I)$. Writing Q for Q^0 then, under the duality transform (1.29) Q_t corresponds to the multiplication operator M_{B_t} ($t \geq 0$). In this way the commutative algebra of bounded random variables $L^\infty(\mathcal{C}_0)$ is realised as a von Neumann subalgebra of $B(\mathcal{F})$ (case $d = 1$, so $\mathsf{k} = \mathbb{C}$). This begins to justify the following terminology. *Quantum Brownian motion* is the noncommuting family of classical Brownian motions $\{Q^\theta : \theta \in [0, 2\pi[\}$. It is a paradigm example of a quantum Lévy process, although you will meet a somewhat different definition in other lectures in these volumes.

2.3 Mapping processes

Recall the definition of operator processes *in an operator space* (see page 217). For an operator space V in $B(\mathsf{h}_1; \mathsf{h}_2)$ define

$$\mathbb{P}(\mathsf{V}, \mathcal{D} \underline{\otimes} \mathcal{E}(S)) := L(\mathsf{V}; \mathbb{P}(\mathcal{D} \underline{\otimes} \mathcal{E}(S); \mathsf{h}_2 \otimes \mathcal{F}_\mathsf{k})_\mathsf{V}),$$

the space of *processes on* V *with noise dimension space* k. A process k is *measurable* (respectively, *continuous*) if the operator process $t \mapsto k_t(x)$ enjoys that property, for each $x \in$ V.

Let $\mathbb{P}_b(V, k)$ and, more importantly $\mathbb{P}_{cb}(V, k)$, denote respectively the subspaces of *bounded* and *completely bounded processes* k on V, defined as those processes k for which each k_t is bounded, respectively completely bounded $V \to V \otimes_M B(\mathcal{F}_k)$. Also important for us are *completely contractive* processes and, when V is a C^*-algebra, *completely positive* and especially **-homomorphic* processes. The latter will provide dilations of quantum dynamical semigroups; the former provide an excellent framework for building and analysing such dilations.

Example 2.5 (Pure number process). Let k $= \mathbb{C}$ and let $\psi \in B(V)$ for an operator space V. Then a bounded process on V is defined by

$$k_t = \psi^{N_t},$$

where N is the one-dimensional number process (2.10), in the following sense: again using Guichardet space,

$$(k_t(x)f)(\sigma) = \psi^{\#(\sigma \cap [0,t])}(x)(f(\sigma)). \qquad (2.11)$$

For each element $x \in$ V, if $s \leq t$ then $(k_t(x) - k_s(x))$ vanishes on $\mathcal{G}_{[0,s[} \otimes \Omega_{[s,\infty[}$, so $(k_t(x))_{t \geq 0}$ is a martingale. We shall see, in Example 5.3 below, that it is also a 'Markovian cocycle' with a good infinitesimal description, and also that it generalises nicely to multidimensional noise.

In the above example the discrete semigroup $(\psi^n)_{n \geq 0}$ on V is randomised. See page 244 for a discussion of randomising continuous-time groups.

Regularity of processes

Let S and S' be k-admissible sets, $\mathcal{D}$ a dense subspace of a Hilbert space h_1 and V an operator space. Then $X \in \mathbb{P}(\mathcal{D} \underline{\otimes} \mathcal{E}(S); h_2 \otimes \mathcal{F}_k)$ is S'-*weakly regular* (respectively, *strongly regular*) if, for all $g \in S$ and $f \in S'$, $E^{\varepsilon(f)} X_t E_{\varepsilon(g)}$ (respectively, $X_t E_{\varepsilon(g)}$) is bounded, with norms locally bounded in t; similarly, $k \in \mathbb{P}(V, \mathcal{D} \underline{\otimes} \mathcal{E}(S))$ is *weakly regular* (respectively, *strongly regular*) if $E^{\varepsilon(f)} k_t(\cdot) E_{\varepsilon(g)}$ (respectively, $k_t(\cdot) E_{\varepsilon(g)}$) is bounded, with norms locally bounded in t.

These conditions arise naturally in the theory of quantum stochastic differential equations. The processes we are most interested in are *contractive* and thus automatically strongly regular.

Notes

The density result for exponential vectors of indicator functions, Proposition 2.1, was first proved in [PSu] (for dim k $= 1$ and T $= \{0, 1\}$) using a classical martingale convergence argument. A wholly different proof, from a characterisation of minimality for quantum stochastic dilations, was given in [Bha]. The elementary proof given here is an adaptation of the proof in [Ske], exploiting an observation in [Arv]. Example 2.2, showing that indicator functions of intervals won't do, is from [FeL]. A nice interpolation between these facts is given in [AtB], in the form of a characterisation of the orthogonal complement of the set $\{\varepsilon(\mathbf{1}_B)\,|\,B$ is a union of at most n intervals$\}$. For the question of overcoming the limitation of exponential domains not being invariant under the action of most processes, see the notes to the next section.

There is an elegant treatment of symmetries of classical Brownian motion in [Hid]. Quantum Brownian motion is axiomatised in [CoH]; for its origin in the physics literature see [Sen]. The standard one defined here has minimal variance for compliance with Heisenberg uncertainty, and this makes it *degenerate* in various respects (see [HL$_1$], [HL$_3$]). Non-gauge-invariant quantum Brownian motions are important too. These arise from squeezed states in quantum optics (see [HHKKR]). There are now many other processes that may be considered as noncommutative 'Brownian motions', for example the Wigner process in free probability theory (see [BiS]).

For an interesting explanation of why there are only the three quantum noises (in Fock space, with one dimension of noise) of creation, preservation and annihilation see [Coq$_1$].

3 QS Integrals

In this section quantum stochastic integrals are defined and the so-called Fundamental Formulae are established, which include the quantum Itô product formula. These are based on what is here termed the fundamental observation and estimate, involving the Hitsuda-Skorohod integral and its compatibility with exponential domains. The versatility of QS analysis is then illustrated by a demonstration of how Fermi fields may be realised as QS integrals. This is remarkable since QS integrals may naturally be viewed as a generalisation of free Bose fields in which 'smearing' is by operator-valued functions rather than vector-valued test functions. The section ends with a treatment of iterated quantum stochastic integrals, which are applied in the following sections to solve quantum stochastic differential equations and establish algebraic properties of solutions. Iterated QS integrals are finding further fruitful application in current research.

3.1 Abstract gradient and divergence

Let Γ be the symmetric Fock space over a Hilbert space H. Define

$$P_{\mathrm{sym}} : \mathsf{H} \otimes \Gamma \to \Gamma$$

by continuous linear extension of the prescription

$$u \otimes v^{\otimes n} \mapsto (n+1)^{-1} \sum_{i=0}^{n} v^{\otimes i} \otimes u \otimes v^{\otimes(n-i)}.$$

Note that

$$\mathsf{H} \otimes \varGamma = \bigoplus_{n \geq 0} \mathsf{H} \otimes \mathsf{H}^{\vee n} \subset \varPhi(\mathsf{H}),$$

and P_{sym} is the restriction of the orthogonal projection $\varPhi(\mathsf{H}) \to \varGamma(\mathsf{H})$, defined via (1.17).

The *divergence operator*

$$\mathcal{S} := \sqrt{N} P_{\mathrm{sym}}$$

is closed, since it is a closed operator times a bounded operator, and is densely defined, since its domain contains the simple tensors of the form $u \otimes v^{\otimes n}$ $(n \geq 0, u, v \in \mathsf{H})$ which are total in $\mathsf{H} \otimes \varGamma$. The *gradient operator*

$$\nabla := \mathcal{S}^*$$

is therefore also closed and densely defined.

Proposition 3.1. *For a Hilbert space* H,

(a) *The following relations hold:*

$$\mathcal{S} \supset P_{\mathrm{sym}}(I_{\mathsf{H}} \otimes \sqrt{N+1}); \quad \mathrm{Dom}\, \nabla = \mathrm{Dom}\, \sqrt{N}; \quad and \quad \mathcal{S}\nabla = N.$$

(b) *For* $v \in \mathsf{H}$,

$$\nabla \varepsilon(v) = v \otimes \varepsilon(v).$$

(c) *For* $z_1, z_2 \in \mathrm{Dom}\left(I_{\mathsf{H}} \otimes \sqrt{N}\right)$

$$\langle \mathcal{S} z_1, \mathcal{S} z_2 \rangle = \langle z_1, z_2 \rangle + \Big\langle (I_{\mathsf{H}} \otimes \nabla) z_1, (\varPi \otimes I_{\varGamma})(I_{\mathsf{H}} \otimes \nabla) z_2 \Big\rangle, \qquad (3.1)$$

where $\varPi$ *denotes the tensor flip on* $\mathsf{H} \otimes \mathsf{H}$.

Proof (Sketch). Let $u, u_1, \ldots, v_2, w \in \mathsf{H}$.

1. If $z = u \otimes v^{\otimes n}$ then

$$P_{\mathrm{sym}}\left(I_{\mathsf{H}} \otimes \sqrt{N+1}\right) z = \sqrt{n+1} P_{\mathrm{sym}} z = \sqrt{N} P_{\mathrm{sym}} z.$$

2. Note the simple identity

$$\langle \varepsilon(u), N\varepsilon(v) \rangle = \sum_{n \geq 1} \frac{1}{n!} \langle u^{\otimes n}, n v^{\otimes n} \rangle$$

$$= \langle u, v \rangle \langle \varepsilon(u), \varepsilon(v) \rangle = \langle u \otimes \varepsilon(u), v \otimes \varepsilon(v) \rangle.$$

3. If $z = u \otimes \varepsilon(v)$ then

$$Sz = P_{\text{sym}}\left(\sqrt{\frac{n+1}{n!}}\, u \otimes v^{\otimes n}\right)_{n \geq 0},$$

so

$$\langle Sz, \varepsilon(w)\rangle = \sum_{n \geq 0} \sqrt{\frac{n+1}{n!}}\, \frac{1}{\sqrt{(n+1)!}} \langle u \otimes v^{\otimes n}, w^{\otimes(n+1)}\rangle$$

$$= \sum_{n \geq 0} \frac{1}{n!} \langle u, w\rangle \langle v, w\rangle^n$$

$$= \langle z, w \otimes \varepsilon(w)\rangle.$$

4. If $z_i = u_i \otimes \varepsilon(v_i)$ for $i = 1, 2$ then, since

$$\left\langle u_1 \otimes v_1^{\otimes n}, \sum_{i=0}^{n} v_2^{\otimes i} \otimes u_2 \otimes v_2^{\otimes(n-i)}\right\rangle$$

$$= \langle u_1 \otimes v_1^{\otimes n}, u_2 \otimes v_2^{\otimes n}\rangle + n\langle u_1, v_2\rangle \langle v_1, u_2\rangle \langle v_1, v_2\rangle^{n-1},$$

so

$$\langle Sz_1, Sz_2\rangle = \sum_{n \geq 0} \frac{n+1}{n!} \frac{1}{n+1} \left\langle u_1 \otimes v_1^{\otimes n}, \sum_{i=0}^{n} v_2^{\otimes i} \otimes u_2 \otimes v_2^{\otimes(n-i)}\right\rangle$$

$$= \langle z_1, z_2\rangle + \left\langle u_1 \otimes v_1 \otimes \varepsilon(v_1), v_2 \otimes u_2 \otimes \varepsilon(v_2)\right\rangle$$

$$= \langle z_1, z_2\rangle + \left\langle (I_H \otimes \nabla)z_1, (\Pi \otimes I_\Gamma)(I_H \otimes \nabla)z_2\right\rangle$$

since $\nabla \varepsilon(v_i) = v_i \otimes \varepsilon(v_i)$. $\square$

Exercise. Show that C is a core for each of the operators ∇, N and $\sqrt{N}$, and that $H \otimes C$ is a core for S, when C is either $\Gamma_{00}(H)$ or $\mathcal{E}(H)$, and convert the above sketch into a complete proof.

Remark. In less abstract settings (3.1) is known as the *Skorohod isometry* (after *Itô isometry*, which *is* actually an isometric relation!).

Exercise. Let $z \in \text{Dom}\, S$. Show that, for any orthogonal projection Q in $B(H)$,

$$(Q \otimes I_\Gamma)z \in \text{Dom}\, S \text{ and } \|S(Q \otimes I_\Gamma)z\| \leq C(z) + \sqrt{3}\|Sz\|,$$

where $C(z)$ is a constant independent of Q.

3.2 Hitsuda, Skorohod, Malliavin and Bochner

Now take $K := L^2(\mathbb{R}_+; k)$ (as in Section 2) for the Hilbert space H, so that $\Gamma = \mathcal{F}_k$, and add a further Hilbert space h into the fray. Using tensor flips define

$$P_{\text{sym}} : K \otimes h \otimes \mathcal{F}_k \to h \otimes \mathcal{F}_k$$

and

$$\mathcal{S} = (I_h \otimes \sqrt{N})P_{\text{sym}}.$$

Thus $\nabla = \mathcal{S}^*$ now satisfies

$$\nabla(u \otimes \varepsilon(f)) = f \otimes u \otimes \varepsilon(f). \tag{3.2}$$

In this context $\mathcal{S}$ is called the *Hitsuda-Skorohod integral* and ∇ is the gradient operator of Malliavin calculus.

For $z \in \text{Dom}\,\mathcal{S}$, $\xi \in \text{Dom}\,\nabla$ and $t > 0$, the following (a.e.) notation is used:

$$\mathcal{S}_t z \text{ for } \mathcal{S}(z_{[0,t[}) \text{ and } \nabla_t \xi \text{ for } (\nabla \xi)_t,$$

where we view $K \otimes h \otimes \mathcal{F}_k$ as $L^2(\mathbb{R}_+; k \otimes h \otimes \mathcal{F}_k)$, and place the argument of such functions as subscripts. Note that an exercise at the end of the previous subsection ensures that the former is defined, in other words that Skorohod integrability implies local Skorohod integrability. Also define *adapted spaces* $L^p_{\text{ad}}(\mathbb{R}_+; H \otimes \mathcal{F}_k)$, for any Hilbert space H and $p \geq 1$, by

$$\left\{ \varphi \in L^p(\mathbb{R}_+; H \otimes \mathcal{F}_k) \middle| \varphi_t = \varphi(t) \otimes \Omega_{k,[t,\infty[} \text{ where } \varphi(t) \in H \otimes \mathcal{F}_{k,[0,t[} \text{ for a.a. } t \right\},$$

and let P_{ad} be the orthogonal projection onto this subspace of $L^2(\mathbb{R}_+; H \otimes \mathcal{F}_k)$, when $p = 2$.

Let us revisit the Skorohod isometry. Note the identifications

$$K^{\otimes 2} \otimes h \otimes \mathcal{F}_k = L^2((\mathbb{R}_+)^2; k^{\otimes 2}) \otimes h \otimes \mathcal{F}_k = L^2((\mathbb{R}_+)^2; k^{\otimes 2} \otimes h \otimes \mathcal{F}_k).$$

Proposition 3.2. *Let* $z, w \in \text{Dom}\,(I_{K \otimes h} \otimes \sqrt{N})$. *Then*

$$\langle \mathcal{S}z, \mathcal{S}w \rangle = \langle z, w \rangle + \int ds \int dt \, \langle \nabla_t z_s, (\pi \otimes I_{h \otimes \mathcal{F}_k})\nabla_s w_t \rangle \tag{3.3}$$

where π *is the tensor flip on* $k \otimes k$.

Proof. In view of the Skorohod isometry already proved it suffices to check that the tensor flip and integration variables are correctly arranged. For this (by sesquilinearity) we need only check with elementary tensors from $K \otimes h \otimes \mathcal{E}(K)$. If $z = g \otimes u \otimes \varepsilon(f)$ and $w = g' \otimes u' \otimes \varepsilon(f')$ then $(I_K \otimes \nabla)z = g \otimes f \otimes u \otimes \varepsilon(f)$ and similarly for w, so

$$((I_K \otimes \nabla)z)(s,t) = g(s) \otimes f(t) \otimes u \otimes \varepsilon(f) = \nabla_t z_s, \text{ and}$$

$$((\Pi \otimes I_{h \otimes \mathcal{F}_k})(I_K \otimes \nabla)w)(s,t) = f'(s) \otimes g'(t) \otimes u' \otimes \varepsilon(f') = (\pi \otimes I_{h \otimes \mathcal{F}_k})\nabla_s w_t.$$

The result follows. $\qquad\qquad\square$

We are now ready for what now, in retrospect, may be considered the fundamental observation and estimate for quantum stochastic analysis.

Theorem 3.3. *Let* $x \in L^2_{\mathrm{ad}}(\mathbb{R}_+; \mathsf{k} \otimes \mathsf{h} \otimes \mathcal{F}_\mathsf{k})$ *with* $x_t = x(t) \otimes \Omega_{\mathsf{k},[t,\infty[}$ *for a.a.* $t \geq 0$, *let* $f \in \mathsf{K}$, *and let* $z \in L^2(\mathbb{R}_+; \mathsf{k} \otimes \mathsf{h} \otimes \mathcal{F}_\mathsf{k}) = \mathsf{K} \otimes \mathsf{h} \otimes \mathcal{F}_\mathsf{k}$ *be given by*

$$z_t = x(t) \otimes \varepsilon\big(f\big|_{[t,\infty[}\big), \text{ for a.a. } t \geq 0. \tag{3.4}$$

Then $z \in \mathrm{Dom}\,\mathcal{S}$ *and*

$$\|\mathcal{S}z\| \leq C_f \|z\| \tag{3.5}$$

where

$$C_f = \|f\| + \sqrt{1 + \|f\|^2}. \tag{3.6}$$

Proof. Let $x = P_{\mathrm{ad}}x'$ where $x' \in L^2(\mathbb{R}_+; \mathsf{k} \otimes \mathsf{h} \otimes \mathcal{F}_\mathsf{k}) = \mathsf{K} \otimes \mathsf{h} \otimes \mathcal{F}_\mathsf{k}$. First suppose that $x' = G \otimes \varepsilon(h)$ where $G \in \mathsf{K} \otimes \mathsf{h} = L^2(\mathbb{R}_+; \mathsf{k} \otimes \mathsf{h})$ and $h \in \mathsf{K}$. Then z has the form

$$z_t = G(t) \otimes \varepsilon(k^{[t]}) \text{ where } k^{[t]} = h_{[0,t[} + f_{[t,\infty[}.$$

Using the identity $\|\sqrt{N}\varepsilon(v)\| = \|v\|\,\|\varepsilon(v)\|$,

$$\int dt\, \|G(t) \otimes \sqrt{N}\varepsilon(k^{[t]})\|^2 = \int dt\, \|G(t)\|^2 \|k^{[t]}\|^2 \|\varepsilon(k^{[t]})\|^2$$

$$= \int dt\, \|z_t\|^2 \big(\|h_{[0,t[}\|^2 + \|f_{[t,\infty[}\|^2\big)$$

$$\leq \big(\|h\|^2 + \|f\|^2\big)\|z\|^2$$

$$< \infty,$$

so $z \in \mathrm{Dom}\,(I_{\mathsf{K}\otimes\mathsf{h}} \otimes \sqrt{N})$. Next let $x' \in \mathsf{K}_0 \underline{\otimes} \mathsf{h} \otimes \mathcal{E}(\mathsf{K})$ where K_0 is the subspace of K consisting of functions with compact essential support. Then $z \in \mathrm{Dom}\,(I_{\mathsf{K}\otimes\mathsf{h}} \otimes \sqrt{N})$ so $z_{[0,t[} \in \mathrm{Dom}\,(I_{\mathsf{K}\otimes\mathsf{h}} \otimes \sqrt{N}) \subset \mathrm{Dom}\,\mathcal{S}$, and

$$\|\mathcal{S}_t z\| = \|\mathcal{S}z_{[0,t[}\| \leq \big\|(I_{\mathsf{K}\otimes\mathsf{h}} \otimes \sqrt{N+I})z_{[0,t[}\big\| \leq \big\|(I_{\mathsf{K}\otimes\mathsf{h}} \otimes \sqrt{N+I})z\big\|,$$

for all $t \geq 0$. Now, setting $y_t = \sup_{r \leq t} \|\mathcal{S}_r z\|$, Proposition 3.2 and the mutual adjointness of ∇ and $\mathcal{S}$ imply that

$$\|\mathcal{S}_t z\|^2 - \|z_{[0,t[}\|^2$$

$$= \int_0^t dr \int_0^t ds\, \langle \nabla_s z_r, (\pi \otimes I_{\mathsf{h}\otimes\mathcal{F}_\mathsf{k}})\nabla_r z_s \rangle$$

$$= \iint_{0<s<r<t} ds dr\, \langle (\pi \otimes I_{\mathsf{h}\otimes\mathcal{F}_\mathsf{k}})\nabla_s z_r, f(r) \otimes z_s \rangle$$

$$+ \iint_{0<r<s<t} dr ds\, \langle f(s) \otimes z_r, (\pi \otimes I_{\mathsf{h}\otimes\mathcal{F}_\mathsf{k}})\nabla_r z_s \rangle$$

$$= \int_0^t dr \, \langle z_r, f(r) \otimes \mathcal{S}_r z \rangle + \int_0^t ds \, \langle f(s) \otimes \mathcal{S}_s z, \, z_s \rangle$$

$$= \int_0^t dr \, 2\mathrm{Re} \, \langle z_r, f(r) \otimes \mathcal{S}_r z \rangle$$

$$\le 2 \|z_{[0,t[}\| \, \|f_{[0,t[}\| \, y_t.$$

Since $t \mapsto \|z_{[0,t[}\|, \|f_{[0,t[}\|$ are nondecreasing functions, this implies that

$$y_t^2 \le a_t^2 + 2a_t b_t y_t, \ \text{ or } \ (y_t - a_t b_t)^2 \le a_t^2 (1 + b_t^2),$$

for $a_t = \|z_{[0,t[}\|$ and $b_t = \|f_{[0,t[}\|$, and so

$$\|\mathcal{S}_t z\| \le \|z_{[0,t[}\| \left(\|f_{[0,t[}\| + \sqrt{1 + \|f_{[0,t[}\|^2} \right) \le C_f \|z\|,$$

for each t. Therefore, since z has compact support, (3.5) holds when $x = P_{\mathrm{ad}} x'$ for such x'. Finally let $x' \in \mathsf{K} \otimes \mathsf{h} \otimes \mathcal{F}_\mathsf{k}$ be arbitrary. Since $\mathcal{S}$ is a closed operator the result follows by approximating x' from the dense subspace $\mathsf{K}_0 \underline{\otimes} \mathsf{h} \underline{\otimes} \mathcal{E}(\mathsf{K})$. $\qquad \square$

Corollary 3.4. *Let z be as in the previous theorem. Then $z_{[s,t[} \in \mathrm{Dom}\, \mathcal{S}$ for each $s < t$, and the map $t \mapsto \mathcal{S}_t z$ is continuous $\mathbb{R}_+ \to \mathsf{h} \otimes \mathcal{F}_\mathsf{k}$. Moreover*

$$\mathcal{S}_t z = y(t) \otimes \varepsilon(f|_{[t,\infty[})$$

where $y(t) \in \mathsf{h} \otimes \mathcal{F}_{\mathsf{k},[0,t[}$.

Proof. The first part follows from the fact that, for $0 \le s \le t$, $z_{[s,t[}$ also has the form (3.4), and so

$$\|\mathcal{S}_t z - \mathcal{S}_s z\| = \|\mathcal{S}(z_{[s,t[})\| \le C_f \|z_{[s,t[}\| = C_f \left(\int_s^t dr \|z_r\|^2 \right)^{1/2}.$$

The second part follows since

$$\langle u \otimes \varepsilon(h), \mathcal{S}(z_{[0,t[}) \rangle$$

$$= \int_0^t ds \, \langle h(s) \otimes u \otimes \varepsilon(h), x(s) \otimes \varepsilon(f|_{[s,\infty[}) \rangle$$

$$= \int_0^t ds \, \langle h(s) \otimes u \otimes \varepsilon(h|_{[0,t[}), x(s) \otimes \varepsilon(f|_{[s,t[}) \rangle \langle \varepsilon(h_{[t,\infty[}), \varepsilon(f_{[t,\infty[}) \rangle$$

$$= \int_0^t ds \, \langle u \otimes \varepsilon(h|_{[0,t[}), E^{h(s)} x(s) \otimes \varepsilon(f|_{[s,t[}) \rangle \langle \varepsilon(h_{[t,\infty[}), \varepsilon(f_{[t,\infty[}) \rangle.$$

$\qquad \square$

Along with the Skorohod integrals, which are *stochastic*, there are Bochner integrals, which are simply vector-valued integrals in *time*. Thus for $z \in L^1(\mathbb{R}_+; \mathsf{h} \otimes \mathcal{F}_\mathsf{k})$, define

$$\mathcal{T}z := \int ds\, z(s)$$

and for $z \in L^1_{\text{loc}}(\mathbb{R}_+; \mathsf{h} \otimes \mathcal{F}_\mathsf{k})$, define

$$\mathcal{T}_t z := \mathcal{T}(z_{[0,t[}).$$

Clearly $t \mapsto \mathcal{T}_t z$ is continuous $\mathbb{R}_+ \to \mathsf{h} \otimes \mathcal{F}_\mathsf{k}$.

The next result is also fundamental. It is an integration-by-parts formula for time and Hitsuda-Skorohod integrals of processes of the special type that arise in quantum stochastic calculus. It will give us an Itô product formula for quantum stochastic integrals.

Recall the notation

$$\widehat{\mathsf{k}} = \mathbb{C} \oplus \mathsf{k}, \quad \widehat{c} = \begin{pmatrix} 1 \\ c \end{pmatrix}, \quad \widehat{f}(s) = \widehat{f(s)}$$

for $c \in \mathsf{k}$ and any k-valued function f.

Theorem 3.5. *Let* $f \in L^2(\mathbb{R}_+; \mathsf{k})$, *and let* $z^0 \in L^1(\mathbb{R}_+; \mathsf{h} \otimes \mathcal{F}_\mathsf{k})$ *and* $z^1 \in L^2(\mathbb{R}_+; \mathsf{k} \otimes \mathsf{h} \otimes \mathcal{F}_\mathsf{k})$ *be of the form*

$$z^i_t = x^i_t \otimes \varepsilon(f|_{[t,\infty[}), \text{ for } i = 0, 1,$$

for $x^0 \in L^1_{\text{ad}}(\mathbb{R}_+; \mathsf{h} \otimes \mathcal{F}_\mathsf{k})$ *and* $x^1 \in L^2_{\text{ad}}(\mathbb{R}_+; \mathsf{k} \otimes \mathsf{h} \otimes \mathcal{F}_\mathsf{k})$, *and let* g, w^0 *and* w^1 *be a similar triple. Then*

$$\langle \mathcal{I}z, \mathcal{I}w \rangle = \int dt \left\{ \langle z^1_t, w^1_t \rangle + \langle z_t, \widehat{g}(t) \otimes \mathcal{I}_t w \rangle + \langle \widehat{f}(t) \otimes \mathcal{I}_t z, w_t \rangle \right\} \quad (3.7)$$

where

$$z_t = \begin{pmatrix} z^0_t \\ z^1_t \end{pmatrix}, w_t = \begin{pmatrix} w^0_t \\ w^1_t \end{pmatrix} \text{ and } \mathcal{I}z = \mathcal{T}z^0 + \mathcal{S}z^1.$$

Proof. First note that, since

$$|\langle z_t, \widehat{g}(t) \otimes \mathcal{I}_t w \rangle| = |\langle z^0_t, \mathcal{I}_t w \rangle + \langle z^1_t, g(t) \otimes \mathcal{I}_t w \rangle|$$
$$\leq (\|z^0_t\| + \|z^1_t\| \|g(t)\|) \|\mathcal{I}_t w\|$$

and $\|\mathcal{I}_t(w)\|$ is bounded, by $\|w^0\| + C_g \|w^1\|$, the integral is well-defined.

Ordinary integration by parts gives

$$\langle \mathcal{T}z^0, \mathcal{T}w^0 \rangle = \left(\int dt \int_0^t ds + \int ds \int_0^s dt \right) \langle z^0_s, w^0_t \rangle$$
$$= \int dt\, \langle \mathcal{T}_t z^0, w^0_t \rangle + \int ds\, \langle z^0_s, \mathcal{T}_s w^0 \rangle.$$

The key to unravelling the terms involving Hitsuda-Skorohod integrals is the (a.e.) relation

$$\nabla_t z_s^i = f(t) \otimes z_s^i \quad \text{for } t > s,$$

already used in the proof of Theorem 3.3. Thus

$$\langle Tz^0, Sw^1 \rangle = \int ds \langle z_s^0, Sw^1 \rangle$$

$$= \int ds \int dt \langle \nabla_t z_s^0, w_t^1 \rangle$$

$$= \int ds \int_0^s dt \langle \nabla_t z_s^0, w_t^1 \rangle + \int dt \int_0^t ds \langle f(t) \otimes z_s^0, w_t^1 \rangle$$

$$= \int ds \langle z_s^0, S_s w^1 \rangle + \int dt \langle f(t) \otimes T_t z^0, w_t^1 \rangle.$$

By symmetry,

$$\langle Sz^1, Tw^0 \rangle = \int ds \langle S_s z^1, w_s^0 \rangle + \int dt \langle z_t^1, g(t) \otimes T_t w^0 \rangle.$$

Finally, by Skorohod isometry,

$$\langle Sz^1, Sw^1 \rangle - \langle z^1, w^1 \rangle$$

$$= \int ds \int dt \langle \nabla_s z_t^1, \nabla_t w_s^1 \rangle$$

$$= \int ds \int_0^s dt \langle f(s) \otimes z_t^1, \nabla_t w_s^1 \rangle + \int dt \int_0^t ds \langle \nabla_s z_t^1, g(t) \otimes w_s^1 \rangle$$

$$= \int ds \langle f(s) \otimes S_s z^1, w_s^1 \rangle + \int dt \langle z_t^1, g(t) \otimes S_t w^1 \rangle.$$

Collecting together the nine component terms we have obtained for $\langle \mathcal{I}z, \mathcal{I}w \rangle$ now confirms the identity (3.7), and so the proof is complete.   □

We are now fully prepared to go quantum.

3.3 Brahma, Vishnu and Shiva

First we shall describe the three different kinds of quantum stochastic integration and then, adding in time too, we shall amalgamate them into a single integral. Throughout S is a k-admissible set.

Annihilation integrals are the easiest of the three types of quantum stochastic integral, in view of the eigenrelation

$$a(f_{[t,t+h[})\varepsilon(g) = \int_t^{t+h} ds \langle f(s), g(s) \rangle \varepsilon(g).$$

Definition. (Annihilation integral.) Let F be a measurable $(k \otimes h_1)$-h_2-process, with domain $k \underline{\otimes} \mathcal{D} \underline{\otimes} \mathcal{E}(S)$, such that

$$z : t \mapsto F_t(f(t) \otimes u \otimes \varepsilon(f)) \tag{3.8}$$

is Bochner integrable. Then define

$$A(F) : \mathcal{D} \underline{\otimes} \mathcal{E}(S) \to h_2 \otimes \mathcal{F}_k, \quad A(F)(u \otimes \varepsilon(f)) = \mathcal{T}z.$$

If (3.8) is locally Bochner integrable then define

$$A(F)_t := A(F_{[0,t[}).$$

The following is now an immediate consequence of definitions.

Proposition 3.6. *Under the local Bochner-integrability condition on F, $\{A(F)_t : t \geq 0\}$ defines a continuous h_1-h_2-process with domain $\mathcal{D} \underline{\otimes} \mathcal{E}(S)$.*

In the first example we already see the advantage of *local* Bochner integrability.

Example 3.7 (annihilation process). The 'annihilation processes' defined in (2.9) are annihilation integrals:

$$A_t^{\langle g|} = A(\langle g_{[0,t[}| \otimes I_{h \otimes \mathcal{F}}).$$

As mentioned earlier (on page 219) this may be viewed as a quantum Wiener integral, since the integrand is 'sure'.

Annihilation integrals are also written

$$\int_0^t F_s \, dA(s). \tag{3.9}$$

For creation integrals the following is a key observation.

Lemma 3.8. *For $u \in h$ and $f, g \in K$,*

$$u \otimes a^*(f)\varepsilon(g) = \mathcal{S}(f \otimes u \otimes \varepsilon(g)).$$

Proof. By the adjoint relation $\nabla^* = \mathcal{S}$ and (3.2),

$$
\begin{aligned}
\langle v \otimes \varepsilon(h), u \otimes a^*(f)\varepsilon(g) \rangle &= \langle v, u \rangle \langle \langle f, h \rangle \varepsilon(h), \varepsilon(g) \rangle \\
&= \langle h \otimes v \otimes \varepsilon(h), f \otimes u \otimes \varepsilon(g) \rangle \\
&= \langle \nabla(v \otimes \varepsilon(h)), f \otimes u \otimes \varepsilon(g) \rangle \\
&= \langle v \otimes \varepsilon(h), \mathcal{S}(f \otimes u \otimes \varepsilon(g)) \rangle.
\end{aligned}
$$

$\square$

Definition. (Creation integral.) Let F be a measurable h_1-$(k \otimes h_2)$-process with domain $\mathcal{D} \underline{\otimes} \mathcal{E}(S)$ such that

$$z : t \mapsto F_t(u \otimes \varepsilon(f)) \tag{3.10}$$

is square integrable (i.e. strongly measurable with $t \mapsto \|z_t\|$ being square integrable in the usual sense). Then define

$$A^*(F) : \mathcal{D} \underline{\otimes} \mathcal{E}(S) \to h_2 \otimes \mathcal{F}_k, \quad A^*(F)(u \otimes \varepsilon(f)) = \mathcal{S}z.$$

Again, if (3.10) is locally square integrable, define

$$A^*(F)_t := A^*(F_{[0,t[}).$$

Proposition 3.9. *Under the local square-integrability condition on F, $\{A^*(F)_t : t \geq 0\}$ defines a continuous h_1-h_2-process with domain $\mathcal{D} \underline{\otimes} \mathcal{E}(S)$.*

Proof. By the adaptedness of F,

$$z_t = x(t) \otimes \varepsilon(f|_{[t,\infty[})$$

where

$$\|x(t)\|^2 = \|F_t(u \otimes \varepsilon(f))\|^2 \exp(-\|f_{[t,\infty[}\|^2).$$

The result therefore follows from Corollary 3.4.   □

Definition. (Preservation Integral.) Let F be a measurable $(k \otimes h_1)$-$(k \otimes h_2)$-process with domain $\mathcal{D} \underline{\otimes} \mathcal{E}(S)$ such that

$$z : t \mapsto F_t\big(f(t) \otimes u \otimes \varepsilon(f)\big) \tag{3.11}$$

is square integrable. Then define

$$N(F) : \mathcal{D} \underline{\otimes} \mathcal{E}(S) \to h \otimes \mathcal{F}_k, \quad N(F)\big(u \otimes \varepsilon(f)\big) = \mathcal{S}z.$$

Once more, if (3.11) is locally square-integrable, define

$$N(F)_t := N(F_{[0,t[}).$$

Proposition 3.10. *Under the local square-integrability conditions on F, $\{N(F)_t : t \geq 0\}$ defines a continuous h_1-h_2-process with domain $\mathcal{D} \underline{\otimes} \mathcal{E}(S)$.*

Proof. Again, by the adaptedness of F,

$$z_t = x(t) \otimes \varepsilon(f|_{[t,\infty[}),$$

where

$$\|x(t)\|^2 = \|F_t\big(f(t) \otimes u \otimes \varepsilon(f)\big)\|^2 \exp\big(-\|f_{[t,\infty[}\|^2\big),$$

so the result again follows from Corollary 3.4.   □

Exercise. *Show that creation, preservation and annihilation integral processes are all martingales.*

Along with the quantum stochastic integrals, we need ordinary integrals in time. You know the pattern!

Definition. (Time integral.) Let F be a measurable h_1-h_2-process with domain $\mathcal{D} \overline{\otimes} \mathcal{E}(S)$ such that

$$z : t \mapsto F_t\, u \otimes \varepsilon(f) \tag{3.12}$$

is Bochner integrable. Then define

$$T(F) : \mathcal{D} \overline{\otimes} \mathcal{E}_\mathsf{k} \to \mathsf{h} \otimes \mathcal{F}_\mathsf{k}, \quad T(F)\big(u \otimes \varepsilon(f)\big) = Tz.$$

If (3.12) is locally Bochner integrable, define

$$T(F)_t := T(1_{[0,t[}F).$$

The following is obvious.

Proposition 3.11. *Under the local Bochner-integrability condition on F, $\{T(F)_t : t \geq 0\}$ defines a continuous h_1-h_2-process with domain $\mathcal{D} \overline{\otimes} \mathcal{E}(S)$.*

The notation (3.9) is also used for creation, preservation and time integrals.

Exercise. *Recall the complete martingale defined in (2.7):*

$$M_t = \big|\varepsilon(f|_{[0,s[})\big\rangle\big\langle\varepsilon(g|_{[0,s[})\big| \otimes I_{[s,\infty[}.$$

Show that M is expressible in terms of a sum of quantum stochastic integrals:

$$M_t = M_0 + \int_0^t F_s\, dA^*(s) + \int_0^t G_s\, dN(s) + \int_0^t H_s\, dA(s),$$

in which $G = -M$.

Remark. This generalises to martingales M for which each M_t is an ampliation of a Hilbert-Schmidt operator; the coefficients processes F and H are then essentially Hilbert-Schmidt-valued in the same sense that M is essentially rank one.

Quantum stochastic integrability

We now wish to combine the four into a single integral. This is done via matrices. The following catch-all notation will be often used in the sequel:

$$\Delta := I_\mathsf{H} \otimes P_\mathsf{k} \otimes I_{\mathsf{H}'} \in B(\mathsf{H} \otimes \widehat{\mathsf{k}} \otimes \mathsf{H}') \tag{3.13}$$

where, as with the E^e notation, the Hilbert spaces H and H' will be determined by context.

Definition. (Amalgamated QS integral.) Let L be a measurable $(\widehat{k} \otimes h_1)$-$(\widehat{k} \otimes h_2)$-process with domain $\widehat{k} \underline{\otimes} \mathcal{D} \underline{\otimes} \mathcal{E}(S)$, such that

$$t \mapsto L_t^0\big(\widehat{f}(t) \otimes u \otimes \varepsilon(f)\big) \quad \text{is Bochner integrable and}$$
$$t \mapsto L_t^1\big(\widehat{f}(t) \otimes u \otimes \varepsilon(f)\big) \quad \text{is square integrable,}$$

where

$$L_t^0 := \Delta^{\perp} L_t \text{ and } L_t^1 := \Delta L_t.$$

Then the *QS integral* of L is defined by

$$\Lambda(L) : \mathcal{D} \underline{\otimes} \mathcal{E}(S) \to h \otimes \mathcal{F}_k, \quad \Lambda(L)(u \otimes \varepsilon(f)) = T z^0 + S z^1, \tag{3.14}$$

where

$$z_s^i = L_s^i\big(\widehat{f}(s) \otimes u \otimes \varepsilon(f)\big) \text{ for } i = 0, 1.$$

Writing L in block matrix form:

$$L_t = \begin{bmatrix} K_t & G_t \\ F_t & H_t \end{bmatrix}$$

gives

$$\Lambda(L) = T(K) + A^*(F) + N(H) + A(G).$$

Terminology. A measurable $(\widehat{k} \otimes h_1)$-$(\widehat{k} \otimes h_2)$-process L with domain $\widehat{k} \underline{\otimes} \mathcal{D} \underline{\otimes} \mathcal{E}(S)$ which satisfies the above integrability conditions will be called *quantum stochastically integrable on* $\mathbb{R}_+$; if it satisfies the conditions locally then it will simply be called *QS-integrable*. Note that, by Corollary 3.4, QS-integrability on $\mathbb{R}_+$ implies QS-integrability. Thus QS-integrability for a process L amounts to QS-integrability on $\mathbb{R}_+$ for each process $L_{[0,t[}$ ($t \geq 0$).

Example 3.12. An operator $L \in B(\widehat{k} \otimes h_1; \widehat{k} \otimes h_2)$ may be viewed as a *constant* h_1-h_2-process. It is QS-integrable, and QS integration gives rise to a process $\{\Lambda_t(L) : t \geq 0\}$ that need not be bounded any longer but, like all QS integral processes, *is* continuous.

The identity and inequality in the next result are known as the *fundamental formula!first* and the *Fundamental Estimate* of quantum stochastic calculus.

Theorem 3.13. *Let L be a $(\widehat{k} \otimes h_1)$-$(\widehat{k} \otimes h_2)$-processes with domain $\widehat{k} \underline{\otimes} \mathcal{D} \underline{\otimes} \mathcal{E}(S)$ which is QS-integrable on $\mathbb{R}_+$. Then, for $u \in h_2, v \in \mathcal{D}, g \in K$ and $f \in S$,*

$$\langle u \otimes \varepsilon(g), \Lambda(L)(v \otimes \varepsilon(f)) \rangle$$
$$= \int ds \, \langle \widehat{g}(s) \otimes u \otimes \varepsilon(g), L_s\big(\widehat{f}(s) \otimes v \otimes \varepsilon(f)\big) \rangle \tag{3.15}$$

and

$$\|\Lambda(L)(u \otimes \varepsilon(f))\|$$

$$\leq \int ds \, \|L_s^0(\widehat{f}(s) \otimes u \otimes \varepsilon(f))\| + C_f \Big(\int ds \|L_s^1(\widehat{f}(s) \otimes u \otimes \varepsilon(f))\|^2 \Big)^{1/2}$$

(3.16)

where C_f is given by (3.6).

Proof. Using the adjoint relation $\mathcal{S}^* = \nabla$, the identity (3.2), and the notation (3.14)

$$\langle u \otimes \varepsilon(f), T(z^0) + \mathcal{S}(z^1) \rangle = \int ds \left(\langle u \otimes \varepsilon(f), z_s^0 \rangle + \langle f(s) \otimes u \otimes \varepsilon(f), z_s^1 \rangle \right)$$

$$= \int ds \, \langle \widehat{f}(s) \otimes u \otimes \varepsilon(f), z_s \rangle$$

where

$$z_s = \begin{pmatrix} z_s^0 \\ z_s^1 \end{pmatrix} = L(s)\big(\widehat{g}(s) \otimes v \otimes \varepsilon(g)\big),$$

which proves (3.15). Since

$$\|T z^0 + \mathcal{S} z^1\| \leq \|T z^0\| + \|\mathcal{S} z^1\|$$
$$\leq \|z^0\| + C_f \|z^1\|,$$

by (3.5), (3.16) holds too.   □

Corollary 3.14. *If L has a adjoint process $L^\dagger$, which is QS-integrable on $\mathbb{R}_+$, then*

$$\Lambda(L^\dagger) \subset \Lambda(L)^*.$$

Parity process and Fermi fields

With $\mathsf{h} = \mathsf{k} = \mathbb{C}$ define second quantised operators $J_t = \Gamma(q_t), t \geq 0$, where

$$q_t : f \mapsto -f_{[0,t[} + f_{[t,\infty[}.$$

Thus J is a continuous, unitary and self-adjoint process (in fact a martingale) called the *parity process*.

Exercise. *Using the First Fundamental Formula, show that J satisfies the QS integral equation*

$$J_t = I - 2 \int_0^t J_s \, dN_s.$$

Let $\varphi \in \mathsf{K} = L^2(\mathbb{R}_+)$. Then, for each $f \in \mathsf{K}$, $\overline{\varphi(s)} f(s) J_s \varepsilon(f)$ is (Bochner) integrable in s and $\varphi(s) J_s \varepsilon(f)$ is square integrable in s. Therefore the QS integrals

$$b(\varphi) := \int \overline{\varphi(s)} J_s \, dA_s \quad \text{and} \quad b^*(\varphi) := \int \varphi(s) J_s \, dA_s^*$$

are well-defined on $\mathcal{E}(\mathsf{K})$.

Exercise. *Show that $b(\varphi)$ and $b^*(\varphi)$ are bounded and mutually adjoint operators which satisfy the canonical anticommutation relations (CAR):*

$$\begin{cases} b(\varphi)b^*(\varphi) + b^*(\varphi)b(\varphi) = \|\varphi\|^2, \\ b(\varphi)^2 = 0. \end{cases}$$

[Hint: Work with matrix elements with respect to exponential vectors, and use the Skorohod isometry to obtain an expression for $\langle b^*(\varphi)\varepsilon(f), b^*(\varphi)\varepsilon(g)\rangle$ which may be compared with your expression for $\langle b(\varphi)\varepsilon(f), b(\varphi)\varepsilon(g)\rangle$. Prove boundedness after establishing the CAR.]

Remark. In the language of quantum field theory and operator-valued distributions dA_s corresponds to $a_s \, ds$ where a now represents annihilation for an unsmeared Bose field. Here we are *smearing with an operator process*. This method of transforming Bose fields into Fermi fields may be considered as a *continuous Jordan-Wigner transform*.

3.4 Quantum Itô product formula

Typically a QS process does not leave its exponential domain invariant. This presents an obstruction to composing processes. Whilst it is possible to extend definitions it turns out that one can accommodate this obstruction to a large extent by using the inner product:

$$\langle X_t \xi, Y_t \eta \rangle \text{ for } \xi \in \mathcal{D} \underline{\otimes} \mathcal{E} \text{ and } \eta \in \mathcal{D}' \underline{\otimes} \mathcal{E}',$$

where X is an h_1-h-process with domain $\mathcal{D} \underline{\otimes} \mathcal{E}(S)$ and Y is an h_2-h-process with domain $\mathcal{D}' \underline{\otimes} \mathcal{E}(S')$.

The next result is known as the *Second Fundamental Formula* of quantum stochastic calculus. Recall the Δ notation (3.13).

Theorem 3.15 (Hudson-Parthasarathy). *Let L and M be processes which are QS-integrable on $\mathbb{R}_+$ with domains $\widehat{k} \underline{\otimes} \mathcal{D} \underline{\otimes} \mathcal{E}(S)$ and $\widehat{k} \underline{\otimes} \mathcal{D}' \underline{\otimes} \mathcal{E}(S')$ respectively. Then, for $u \in \mathcal{D}$, $v \in \mathcal{D}'$, $f \in S$ and $g \in S'$,*

$$\langle \Lambda(L)(u \otimes \varepsilon(f)), \Lambda(M)(v \otimes \varepsilon(g)) \rangle =$$

$$\int dt \Big\{ \big\langle \widehat{f}(t) \otimes \Lambda(L)_t(u \otimes \varepsilon(f)), M_t(\widehat{g}(t) \otimes v \otimes \varepsilon(g)) \big\rangle$$

$$+ \big\langle L_t(\widehat{f}(t) \otimes u \otimes \varepsilon(f)), \widehat{g}(t) \otimes \Lambda(M)_t(v \otimes \varepsilon(g)) \big\rangle$$

$$+ \big\langle L_t(\widehat{f}(t) \otimes u \otimes \varepsilon(f)), \Delta M_t(\widehat{g}(t) \otimes v \otimes \varepsilon(g)) \big\rangle \Big\}. \quad (3.17)$$

Proof. Set

$$z_t^0 = \Delta^\perp L_t(\widehat{f}(t) \otimes u \otimes \varepsilon(f)), \quad z_t^1 = \Delta L_t(\widehat{f}(t) \otimes u \otimes e(f)),$$

$$w_t^0 = \Delta^\perp M_t(\widehat{g}(t) \otimes v \otimes \varepsilon(g)), \quad w_t^1 = \Delta M_t(\widehat{g}(t) \otimes v \otimes \varepsilon(g)),$$

and apply Theorem 3.5.   □

In favourable circumstances this takes a more attractive and amenable form.

Corollary 3.16. *Let L and M be bounded QS-integrable processes whose QS integral processes X and Y are bounded. Then*

$$X_t^* Y_t = \int_0^t \left((I_{\widehat{k}} \otimes X_s^*) M_s + L_s^* (I_{\widehat{k}} \otimes Y_s) + L_s^* \Delta M_s \right) d\Lambda_s, \quad t \geq 0, \quad (3.18)$$

provided that the integrand is QS-integrable.

Proof. Rearranging the right-hand side of (3.17)

$$\langle u \otimes \varepsilon(f), X_t^* Y_t (v \otimes \varepsilon(g)) \rangle = \int_0^t ds \left\langle \widehat{f}(s) \otimes u \otimes \varepsilon(f), Z_s \big(\widehat{g}(s) \otimes v \otimes \varepsilon(g) \big) \right\rangle$$

where Z is the integrand process in (3.18). Comparison with (3.15) therefore completes the proof. $\qquad \square$

Example 3.17. If $X_t = \int_0^t F_s \, dA_s$ and $Y_t = \int_0^t G_s \, dA_s^*$, then

$$\int_0^t F_s \, dA_s \int_0^t G_s \, dA_s^* = \int_0^t F_s Y_s \, dA_s + \int_0^t X_s G_s \, dA_s^* + \int_0^t F_s G_s \, ds.$$

The general rule here is that there is a third term "Itô correction" only if the Wick ordering

$$dA^*, \quad dN, \quad dA,$$

is violated, and in this case the correction term is given by the following *quantum Itô table*:

$$
\begin{array}{c|cc}
 & dA_t^* & dN_t \\
\hline
dA_t & dt & dA_t \\
dN_t & dA^* & dN_t
\end{array}
\quad . \qquad (3.19)
$$

Remark. This contains the Itô correction for classical Brownian motion: if $Q_t = A_t + A_t^*$ (one dimension of noise) then, since

$$(dA_t + dA_t^*)^2 = (dA_t)^2 + (dA_t^*)^2 + dA_t \, dA_t^* + dA_t^* \, dA_t,$$

$(dQ_t)^2 = dt.$

Iterated QS integrals

Let $L \in B(\widehat{k}^{\otimes n} \otimes h_1; \widehat{k}^{\otimes n} \otimes h_2)$. As a constant $(\widehat{k} \otimes \widehat{k}^{\otimes(n-1)} \otimes h_1)$-$(\widehat{k} \otimes \widehat{k}^{\otimes(n-1)} \otimes h_2)$-process this is QS-integrable. Also $\Lambda(L)_t$ $(t \geq 0)$ defines a continuous process and so, if $n \geq 2$, this is QS-integrable itself, as a $(\widehat{k} \otimes \widehat{k}^{\otimes(n-2)} \otimes h_1)$ – $(\widehat{k} \otimes \widehat{k}^{\otimes(n-2)} \otimes h_2)$-process. This leads to the following definition.

Definition. For $L \in B(\widehat{\mathsf{k}}^{\otimes n} \otimes \mathsf{h}_1; \widehat{\mathsf{k}}^{\otimes n} \otimes \mathsf{h}_2)$, the *n-fold iterated QS integral process* of L is defined, for $n = 0, 1, 2, \cdots$, recursively by

$$\Lambda_t^0(L) = L \otimes I \text{ and, for } n \geq 1, \ \Lambda_t^n(L) = \int_0^t \Lambda_s^{n-1}(\widetilde{L}) \, d\Lambda(s),$$

where $\widetilde{L}$ is L viewed as a $(\widehat{\mathsf{k}}^{\otimes(n-1)} \otimes \widehat{\mathsf{k}} \otimes \mathsf{h}_1)$-$(\widehat{\mathsf{k}}^{\otimes(n-1)} \otimes \widehat{\mathsf{k}} \otimes \mathsf{h}_2)$-process.

Proposition 3.18. *Let $L \in B(\widehat{\mathsf{k}}^{\otimes n} \otimes \mathsf{h}_1; \widehat{\mathsf{k}}^{\otimes n} \otimes \mathsf{h}_2)$. Then*

$$\langle u \otimes \varepsilon(f), \Lambda_t^n(L)(v \otimes \varepsilon(g)) \rangle = \int_{\Delta_t^n} ds \, \langle \widehat{f}^{\otimes n}(\mathbf{s}) \otimes u, L \left(\widehat{g}^{\otimes n}(\mathbf{s}) \otimes v \right) \rangle \langle \varepsilon(f), \varepsilon(g) \rangle$$

$$(3.20)$$

and

$$\| \Lambda_t^n(L)(u \otimes \varepsilon(f)) \|^2 \leq (C_{f_{[0,t[}})^{2n} \int_{\Delta_t^n} ds \, \| L \left(\widehat{f}^{\otimes n}(\mathbf{s}) \otimes u \right) \|^2 \| \varepsilon(f) \|^2 \quad (3.21)$$

where Δ_t^n denotes the n-simplex $\{ \mathbf{s} \in \mathbb{R}^n : 0 \leq s_1 \leq \cdots \leq s_n \leq t \}$ and $\widehat{f}^{\otimes n}(\mathbf{s}) = \widehat{f}(s_1) \otimes \cdots \otimes \widehat{f}(s_n)$.

Proof. Exercise in iteration of (3.15) and (3.16). □

So far we have not specified the k-admissible set for the exponential domain of the processes—we may in fact choose K itself.

Corollary 3.19. *For a sequence of operators*

$$L := \left(L_n \in B(\widehat{\mathsf{k}}^{\otimes n} \otimes \mathsf{h}_1; \widehat{\mathsf{k}}^{\otimes n} \otimes \mathsf{h}_2) \right)_{n \geq 0}, \quad (3.22)$$

$T > 0$ and vectors $u \in \mathsf{h}$ and $f \in \mathsf{K}$, the series

$$\sum_{n \geq 0} \Lambda_t^n(L_n)(u \otimes \varepsilon(f))$$

converges absolutely and uniformly on $[0, T]$ provided that

$$\sum_{n \geq 0} (n!)^{-1/2} \| L_n \| (\| \widehat{f}_{[0,T]} \| C_{f_{[0,T]}})^n < \infty. \quad (3.23)$$

Warning. Note that $\widehat{f}_{[0,T]}$ denotes $\widehat{f} \, 1_{[0,t[}$, as opposed to $\widehat{f_{[0,T]}}$.

Proof. By symmetry,

$$\int_{\Delta_t^n} ds \| \widehat{f}^{\otimes n}(\mathbf{s}) \|^2 = (n!)^{-1} \| \widehat{f}_{[0,t[} \|^{2n}.$$

The result follows. □

Let $\mathscr{S}$ denote the linear space of sequences (3.22) satisfying

$$\|L_n\| \le C_1 C_2^n \text{ for some constants } C_1 \text{ and } C_2,$$

and let $\mathscr{S}_0$ be the subspace of sequences whose terms are eventually 0. Then, for $L \in \mathscr{S}$, (3.23) holds for all $f \in \mathsf{K}$ and $T > 0$. The resulting continuous h_1-h_2-process will be denoted $\left(\Lambda_t(L)\right)_{t \ge 0}$.

Proposition 3.20. *The map $L \mapsto \Lambda.(L)$ is linear and is injective on $\mathscr{S}$.*

Proof. Linearity may be read from (3.20); injectivity is left as an **exercise**.
□

Further questions. What about multiplicativity? A quantum Itô formula for iterated QS integrals?

Exercise. *Let $L, M \in B(\widehat{\mathsf{k}} \otimes \mathsf{h})$ and recall the Δ notation (3.13). Show that*

$$\Lambda_t^1(L)\Lambda_t^1(M) = \Lambda_t(L * M)$$

where $L * M$ is the sequence defined by

$$(L * M)_1 = L\Delta M, \quad (L * M)_2 = L^1 M^2 + L^2 M^1,$$

and $(L * M)_n = 0$ otherwise, where

$$L^1 = (\pi \overline{\otimes} \mathrm{id}_{B(\mathsf{h})})(I_{\widehat{\mathsf{k}}} \otimes L) \text{ and } L^2 = I_{\widehat{\mathsf{k}}} \otimes L,$$

π is the tensor flip on $B(\mathsf{k} \otimes \mathsf{k})$ and M^1 and M^2 are defined similarly.

Exercise. *Work out the general formula for $L * M$, when $L, M \in \mathscr{S}_0$.*

Notes

Quantum stochastic integrals for finite-dimensional noise were defined in [HP₁], the founding paper of quantum stochastic calculus; see the lecture notes [Hud]. The extension to infinite-dimensional noise was developed in [HP₂] and [MoS]. Fermi fields were realised as quantum stochastic integrals in [HP₃], thereby subsuming fermionic stochastic calculus ([ApH]) into the Hudson-Parthasarathy calculus—this was later generalised in a multidimensional theory incorporating a mixture of Bose and Fermi creation and annihilation processes and $\mathbb{Z}_2$-graded number/exchange processes ([EyH]). There were other contemporary developments, namely an Itô-Clifford stochastic calculus ([BSW₁]), and a finite-temperature/quasi-free stochastic calculus ([BSW₂], [HL₂], [HL₁], [L₁], [LWi], [LMa]). Stochastic calculus in free Fock space was also developed soon afterwards ([KüS]). All three of these have attracted recent attention (see [CaK], [HKK] and [BiS]).

Quantum stochastic integrals may also be viewed as *integral-sum* kernel operators ([Maa]); their theory is described in the lecture notes [L7]. Developments up until the early 1990's were described in the two books [Par] and [Mey]. The direct approach described here, exploiting the gradient and divergence of Malliavin calculus was developed in [L1,2] and adopted in the lecture notes [Bia]; see also [Bel1]. An indirect approach exploiting classical stochastic calculus, through which quantum stochastic integrals are defined *implicitly*, was developed in [AtM]. These latter two approaches were unified and extended in [AtL] where an *adapted gradient operator* D completes a quartet of classical operations: $SDLP$ (P being P_{ad} and L being T), on which the calculus may be founded. For the product formula and injectivity of this 'global' QS integration see [LW4]; in finite dimensions the formula was first stated in [CEH] and proved in [HPu]. Injectivity is closely related to the independence of quantum stochastic integrators: if L is a QS-integrable process for which $A(L)_t = 0$ for each $t \geq 0$ then L is the zero process (see [L3], [Att1] and [LW1] for sufficient conditions for this to hold).

There is also a functional quantum Itô formula ([Vin2]), an interesting Ω-adapted theory ([Belt1]) which is parallel to the identity-adapted theory described in these notes (see [Belt2]), and an important representation theory for martingales ([PSi]) and semimartingales ([Att2]) as quantum stochastic integrals, which is still under active development ([Att3], [Coq2], [Ji], [Paut]); see the lecture notes [Att4].

4 QS Differential Equations

This section contains the basic existence and uniqueness theorem for quantum stochastic differential equations with bounded coefficients and arbitrary noise dimension space. Recall the notations K and $\mathcal{F}_k$, defined in (2.1); also let S be a k-admissible set *consisting of locally essentially bounded functions with essentially finite-dimensional range*, such as $\mathbb{S}_k$.

Exponential noise

Let $k \in L^1_{loc}(\mathbb{R}_+)$, $u, v \in L^2_{loc}(\mathbb{R}_+; k)$ and $W \in L^\infty_{loc}(\mathbb{R}_+; B(k))$, where the latter is defined as on page 219. and, recalling the Fock space operators introduced there, set

$$X_t = \Gamma\big(z_t, u_{[0,t[}, R_t, v_{[0,t[}\big)$$

where

$$z_t = \int_0^t ds\, k(s) \text{ and } (R_t h)(s) = W(s)h_{[0,t[}(s) + h_{[t,\infty[}(s).$$

Now, since $R_t h = R_t h_{[0,t[} + h_{[t,\infty[}$ for $h \in K$, if $f, g \in K$ then

$$X_t \varepsilon(g) = \exp \int_0^t ds \{k(s) + \langle v(s), g(s) \rangle\} \, \varepsilon \big(R_t g_{[0,t[} + u_{[0,t[} + g_{[t,\infty[} \big)$$

so X is adapted, and

$$\langle \varepsilon(f), X_t \varepsilon(g) \rangle = \alpha(t) \langle \varepsilon(f_{[t,\infty[}), \varepsilon(g_{[t,\infty[}) \rangle, \tag{4.1}$$

where

$$\alpha(t) = \exp \int_0^t ds \Big\{ k(s) + \langle v(s), g(s) \rangle + \langle f(s), W(s)g(s) \rangle + \langle f(s), u(s) \rangle \Big\}.$$

Since

$$\|(T_t - T_s)g\|^2 = \int dr \, \|W(r)g_{[s,t[}(r) - g_{[s,t[}(r)\|^2$$
$$\leq (1 + \|W_{[s,t[}\|_\infty^2) \|g_{[s,t[}\|^2,$$

and the exponential map ε is continuous, it follows that X is a continuous process with domain $\mathcal{E}(\mathsf{K})$. Note also that X is a martingale if (and only if) $k = 0$. Now note that, since $\langle \varepsilon(f_{[t,\infty[}), \varepsilon(g_{[t,\infty[}) \rangle = \exp \int_t^\infty ds \, \langle f(s), g(s) \rangle$, the derivative of (4.1) is

$$\Big\{ k(t) + \langle v(t), g(t) \rangle + \langle f(t), (W(t) - I)g(t) \rangle + \langle f(t), u(t) \rangle \Big\} \langle \varepsilon(f), X_t \varepsilon(g) \rangle.$$

Since $X_0 = I$ it follows that

$$\langle \varepsilon(f), (X_t - I) \varepsilon(g) \rangle = \int_0^t ds \, \Big\langle \widehat{f}(s) \otimes \varepsilon(f), (L_s \otimes X_s) \widehat{g}(s) \otimes \varepsilon(g) \Big\rangle,$$

where

$$L_s = \begin{bmatrix} k(s) & \langle v(s)| \\ |u(s)\rangle & W(s) - I_{\mathsf{k}} \end{bmatrix}.$$

Reference to (3.15) reveals that we have 'solved' our second quantum stochastic differential equation:

$$dX_t = L_t \otimes X_t \, d\Lambda_t; \quad X_0 = I. \tag{4.2}$$

In case you are wondering, we solved our first on page 233.

4.1 QSDE's for operator processes

Let L be a bounded $(\widehat{\mathsf{k}} \otimes \mathsf{h}_1)$-$(\widehat{\mathsf{k}} \otimes \mathsf{h}_2)$-process, and let $T \in B(\mathsf{h}_1; \mathsf{h}_2)$. An h_1-h_2-process X with domain $\mathsf{h}_1 \underline{\otimes} \mathcal{E}(S)$ is a *weak solution* of the right quantum stochastic differential equation

$$dX_t = L_t \widehat{X}_t \, d\Lambda(t), \quad X_0 = T \otimes I, \tag{4.3}$$

where $\widehat{X}_t := I_{\widehat{k}} \underline{\otimes} X_t$, if it satisfies

$$
\left\langle u \otimes \varepsilon(f), X_t\big(v \otimes \varepsilon(g)\big)\right\rangle - \langle u, Tv\rangle\langle \varepsilon(f), \varepsilon(g)\rangle =
$$
$$
\int_0^t ds \left\langle \widehat{f}(s) \otimes u \otimes \varepsilon(f), L_s\widehat{X}_t\big(\widehat{g}(s) \otimes v \otimes \varepsilon(g)\big)\right\rangle, \tag{4.4}
$$

for $u \in h_2$, $v \in h_1$, $f \in K$ and $g \in S$. Implicit here is the assumption that the Lebesgue integrals exist, in other words

$$
t \mapsto \left\langle \widehat{f}(t) \otimes u \otimes \varepsilon(f), L_t\widehat{X}_t\big(\widehat{g}(t) \otimes v \otimes \varepsilon(g)\big)\right\rangle
$$

is locally integrable. It is a *strong solution* if furthermore the $(\widehat{k}\otimes h_1)$-$(\widehat{k}\otimes h_2)$-process

$$
t \mapsto L_t\widehat{X}_t
$$

is quantum stochastically integrable. In that case, in view of the First Fundamental Formula (3.15), X satisfies the integral equation

$$
X_t = T \otimes I + \Lambda_t(Z) \text{ where } Z = \big(L_t\widehat{X}_t\big)_{t\geq 0}. \tag{4.5}
$$

Remark. Strong solutions are in particular continuous processes.

Existence and uniqueness

Recall the regularity conditions on processes in Subsection 2.3.

Theorem 4.1. *Let L be a bounded measurable $(\widehat{k} \otimes h_1)$-$(\widehat{k} \otimes h_2)$-process for which*

$$
t \mapsto L_t E_{\widehat{f}(t)} \text{ has a locally uniform bound for each } f \in S.
$$

Then the right QSDE (4.3) has a unique strongly regular strong solution.

Proof. **Exercise** in Picard iteration.   □

For the constant coefficient case, here is a better uniqueness theorem.

Theorem 4.2. *Let $L \in B(\widehat{k} \otimes h_1; \widehat{k} \otimes h_2)$. Then the QSDE (4.3), with $L_t := L \otimes I_{\mathcal{F}_k}$ for each t, has at most one weakly regular weak solution.*

Proof. **Exercise.**   □

Remark. There is a left QSDE too:

$$
dX_t = \widehat{X}_t(I_{\widehat{k}} \otimes L)d\Lambda_t, \; X_0 = T \otimes I, \tag{4.6}
$$

for which existence and uniqueness holds, as above.

Notation. A convenient notation for the solutions to the right and left constant-coefficient QSDE's is $^L X$ and X^L respectively.

Further questions. When is X^L contractive? isometric? unitary?

4.2 QSDE's for mapping processes

Let $\phi : V \to M(\widehat{k}; V)_b$ be a k-*bounded* linear map, for an operator space V. This terminology simply means that ϕ is bounded if $\dim k < \infty$ and is completely bounded otherwise. For such maps there is a k-bounded map

$$\phi^{(k)} : M(\widehat{k}; V)_b \to M(\widehat{k}^{\otimes 2}; V)_b$$

(see Proposition 1.24 for the CB case) and we may iterate this lifting ad infinitum, to form the sequence given by $\phi^0 = \mathrm{id}_V$ and

$$\phi^{n+1} := \phi^{(k^{\otimes n})} \circ \cdots \circ \phi : V \to M(\widehat{k}^{\otimes(n+1)}; V)_b, \quad n \geq 0.$$

Note that the last $\widehat{k}$ to be added into the picture is the left-most one; we need to reverse this. Thus let

$$\phi_n = (\pi \overline{\otimes} \mathrm{id}_V) \circ \phi^n : V \to M(\widehat{k}^{\otimes n}; V)_b, \ n \geq 0, \tag{4.7}$$

were π is the normal automorphism of $B(\widehat{k}^{\otimes n})$ which effects the permutation

$$\pi(T_1 \otimes \cdots \otimes T_n) = T_n \otimes \cdots \otimes T_1.$$

Existence and uniqueness

Let $\phi : V \to M(\widehat{k}; V)_b$ be a linear map defined on an operator space V in $B(h_1; h_2)$. A process k on V satisfies the QSDE

$$dk_t = k_t \circ \phi \, d\Lambda_t; \quad k_0(x) = x \otimes I; \tag{4.8}$$

weakly on $h_1 \underline{\otimes} \mathcal{E}(S)$ if

$$\langle u \otimes \varepsilon(f), \big(k_t(x) - x \otimes I\big) v \otimes \varepsilon(g) \rangle = \\ \int_0^t ds \, \langle u \otimes \varepsilon(f), k_s \big(E^{\widehat{f}(s)} \phi(x) E_{\widehat{g}(s)} \big) v \otimes \varepsilon(g) \rangle \tag{4.9}$$

for all $u \in h_2$, $v \in h_1$, $f \in K$, $g \in S$ and $x \in V$. Recall the iterated QS integrals defined on page 235.

Theorem 4.3. *Let* V *be an operator space in* $B(h_1; h_2)$ *and let* ϕ *be a completely bounded operator* $V \to M(\widehat{k}; V)_b$. *Then, with* ϕ_n *as given by* (4.7),

$$k_t(x)(v \otimes \varepsilon(f)) = \sum_{n \geq 0} \Lambda_t^n(\phi_n(x))(v \otimes \varepsilon(f)) \tag{4.10}$$

defines a continuous strongly regular process which weakly satisfies the QSDE (4.8) *on* $h_1 \underline{\otimes} \mathcal{E}(K)$. *Moreover it is the unique weakly regular process weakly satisfying this equation.*

Proof (Sketch). The sum is well-defined, in view of Corollary 3.19 and the inequalites

$$\|\phi_n\| \le \|\phi^{(\widehat{k}^{\otimes n})}\| \cdots \|\phi\| \le \left(\|\phi\|_{\mathrm{cb}}\right)^n.$$

The resulting process is continuous and it follows from the estimate (3.21) that it is strongly regular.   □

Remarks. The proper hypothesis here is that ϕ has k-*bounded columns*. This means $\phi(\cdot)E_e$ is k-bounded $V \to C(\widehat{k}; V)_b$ for each $e \in k$.

For applications it is necessary to incorporate other kinds of initial conditions. For example, in the construction of Lévy processes on quantum groups, $k_0(x) = x \otimes I$ is replaced by $k_0(x) = \epsilon(x)I$ where ϵ is the counit. Moreover ϕ maps the quantum group into operators on $\widehat{k}$ (see the lectures of Uwe Franz in the second volume of these notes).

Exercise. Complete the proof by showing that k satisfies (4.9), and is unique among weak solution of (4.8).

Notation. This existence and uniqueness result justifies the notation k^ϕ for the solution. In fact k^ϕ satisfies (4.8) in a strong sense. Rather than go into the technicalities of what that might mean in general, we specialise now to *completely bounded processes*.

Definition. A k-bounded process k satisfies the QSDE (4.8) *strongly* if it is measurable, it satisfies (4.8) weakly and, for each $x \in V$, the process

$$t \mapsto k_t^{(\widehat{k})}\big(\phi(x)\big) \text{ is QS-integrable.}$$

Shortly we shall see conditions on ϕ which ensure that the solution process k is completely bounded.

Further questions. When is k^ϕ *-homomorphic? completely positive?

Exercise. Show that the pure number process described in Example 2.5 satisfies the QSDE

$$dk_t = k_t \circ (\psi - \mathrm{id}_V)dN_t, \quad k_0(x) = x \otimes I. \tag{4.11}$$

strongly on $h_1 \underline{\otimes} \mathcal{E}(K)$.

Notes

An existence and uniqueness theorem for the constant-coefficient operator QSDE with finite-dimensional noise space, focusing on the case of unitary solutions, was given in [HP1]. Existence and uniqueness for the constant-coefficient mapping QSDE with finite-dimensional noise, focusing on the case of unital *-homomorphic solutions, was given in [Eva]. These were extended to infinite-dimensional noise in [HP$_2$] and [MoS]. This was simplified somewhat in [Mey] and further analysed in [LW$_1$]. With the introduction of matrix spaces it was possible to obtain solutions *living on a C^*-algebra* (and more generally on an operator space), under natural CB hypotheses ([LW$_3$]). Modified initial conditions are required for the construction of Lévy processes on quantum groups ([Sch]); for their incorporation into the current framework, as opposed to an integral-sum kernel operator approach, see [LSk]. The 'further questions' and the relationship between X^L and $^L X$ are addressed in the next section.

For a nice treatment of stochastic differential equations on infinite-dimensional spaces driven by a Wiener process on a Hilbert space, making extensive use of semigroup theory, see [DaZ].

5 QS Cocycles

In this section we define quantum stochastic cocycles, or Markovian cocycles, for the shift on the Fock space $\mathcal{F}_k$, otherwise known as the CCR flow of index k. In the analysis of these cocycles a central role is played by their *semigroup representation*. We shall see that solutions of QSDE's form cocycles and, in turn, explore the extent to which cocycles have such an infinitesimal description—in terms of an additive cocycle which is an operator linear combination of the fundamental QS processes of creation, preservation and annihilation. A remarkable feature of the unbounded business of solving QSDE's, described in the previous section, is that *unitary* operator-valued cocycles and **-homomorphic*-valued mapping cocycles may be obtained. In fact, the form of the 'stochastic generator' naturally reflects that of the cocycle—just as unitary groups have skew-adjoint generators, contraction semigroups have dissipative generators and *-homomorphic semigroups have *-derivations as generators. These are amongst the reasons why Markovian cocycles are emphasised here.

We begin with classical Brownian motion.

Some Markovian cocycles for Brownian motion

There is an alternative to the Itô approach to continuous-time classical Markov processes which focuses on their *cocycle* structure with respect to the underlying shift. Consider the paradigm case of Brownian motion.

Let $(\phi_t)_{t\in\mathbb{R}}$ be a one-parameter group of completely bounded maps on an ultraweakly closed operator space V. For example

$$V = |h\rangle \text{ and } \phi_t(|v\rangle) = |U_t v\rangle$$

for a strongly continuous unitary group $(U_t)_{t\in\mathbb{R}}$ (Schrödinger evolution) or

$$V = \mathcal{M} \text{ and } \phi_t(x) = e^{t\delta}(x)$$

for a bounded *-derivation δ on a von Neumann algebra $\mathcal{M}$ (simple Heisenberg evolution). Let L^∞ denote the L^∞-space of the canonical Brownian motion

$\{B_t : t \geq 0\}$ on Ω, the path space $\mathcal{C}_0$ defined in (1.28) with $\mathsf{k}^{\mathbb{R}} = \mathbb{R}$, and define the semigroup of shifts on $\mathsf{V} \overline{\otimes} L^\infty = L^\infty(\Omega; \mathsf{V})$, by

$$\sigma_s(f)(\omega) = f(\theta_s \omega)$$

where $(\theta_s)_{s \geq 0}$ is the semigroup of shifts on paths:

$$(\theta_s \omega)(t) = \omega(s+t) - \omega(s).$$

Then

$$\begin{aligned}
\phi_{\omega(s+t)}(x) &= \phi_{\omega(s)} \circ \phi_{[\omega(s+t)-\omega(s)]}(x) \\
&= \phi_{\omega(s)} \circ \phi_{(\theta_s \omega)(t)}(x) \\
&= \phi_{\omega(s)}\big(\sigma_s(\phi_{B_t}(x))(\omega)\big).
\end{aligned} \tag{5.1}$$

Now define a family of CB maps

$$k_t : \mathsf{V} \to \mathsf{V} \overline{\otimes} L^\infty, \quad x \mapsto \phi_{B_t}(x).$$

Let $\widehat{k}_s$ denote the extension of k_s to an operator on $\mathsf{V} \overline{\otimes} L^\infty$, defined by

$$\widehat{k}_s(f)(\omega) = \phi_{\omega(s)}(f(\omega)),$$

for functions f depending only on the path beyond time s. Thus k_s is the restriction of $\widehat{k}_s$ to constant functions. Then (5.1) reads $k_{s+t}(x) = \widehat{k}_s\big(\sigma_s(k_t(x))\big)$, thus the one-parameter family $(k_t)_{t \geq 0}$ satisfies the *cocycle identity*

$$k_{s+t} = \widehat{k}_s \circ \sigma_s \circ k_t. \tag{5.2}$$

If the randomness is averaged out, by defining

$$P_t x = \mathbb{E}[k_t(x)],$$

then the *Markovian semigroup* of the cocycle results:

$$P_{s+t} = P_s P_t, \quad P_0 = \mathrm{id}_{\mathsf{V}}.$$

In the first example above this gives a self-adjoint contraction semigroup:

$$P_t = e^{-\frac{1}{2}tH^2}$$

where H is the Stone generator of the group: $U_t = e^{itH}$. In the second example it gives a CP contraction semigroup: $P_t = e^{-\frac{1}{2}t\delta^2}$.

   Classical probability abounds with examples of such cocycles and associated semigroups. Here we are interested in seeing how they arise in noncommutative probability.

Fock-space shifts

Shifts on the Fock space $\mathcal{F}_k$ are defined by

$$\sigma_t(X) = I_t \otimes (S_t X S_t^*), \quad X \in B(\mathcal{F}_k),$$

where $S_t = \Gamma(s_t)$ is the isometry $\mathcal{F}_k \to \mathcal{F}_{k,[t,\infty[}$ given by

$$S_t \varepsilon(f) = \varepsilon(s_t f), \quad s_t f(s) = f(s-t),$$

and $I_t = I_{\mathcal{F}_{k,[0,t[}}$. They form a normal endomorphism semigroup on $B(\mathcal{F}_k)$. In particular they extend to $B(h_1; h_2) \overline{\otimes} B(\mathcal{F}_k)$, for Hilbert spaces h_1 and h_2, where they map right matrix spaces to right matrix spaces:

$$V \otimes_M B(\mathcal{F}_k) \to V \otimes_M B(\mathcal{F}_{k,[t,\infty[}) \subset V \otimes_M B(\mathcal{F}_k)$$

for any operator space V in $B(h_1; h_2)$. We use the same notation σ_t for the shift on any of these.

The following identity is useful:

$$\langle \varepsilon(f), \sigma_t(X)\varepsilon(g)\rangle$$
$$= \langle \varepsilon(f_{[0,t[}), \varepsilon(g_{[0,t[})\rangle \langle \varepsilon(s_t^*(f|_{[t,\infty[})), X \varepsilon(s_t^*(g|_{[t,\infty[}))\rangle \quad (5.3)$$

Exercise. *Verify the following effect of shifts on exponential operators:*

$$\sigma_t(\Gamma(z, u, T, v)) = I_t \otimes \Gamma(z, s_t u, s_t T s_t^*, s_t v), \quad (5.4)$$

the second tensor component being an operator on $\mathcal{F}_{k,[t,\infty[}$, and the first being the identity operator on $\mathcal{F}_{k,[0,t[}$.

Markovian cocycles for quantum noise

A bounded Hilbert-space operator process X is a *left Markovian cocycle* if it satisfies

$$X_{s+t} = X_s \sigma_s(X_t); \quad X_0 = I.$$

A completely bounded process k on an operator space V in $B(h_1; h_2)$ is a *completely bounded Markovian cocycle* if it satisfies

$$k_{s+t} = \widehat{k}_s \circ \sigma_s \circ k_t, \quad k_0(x) = x \otimes I.$$

where $\widehat{k}_s$ is the right lifting $k_s \otimes_M \mathrm{id}_{B(H)}$ for $H = \mathcal{F}_{k,[s,\infty[}$.

Since

$$\mathcal{F}_{k,[0,s[} \otimes \mathcal{F}_{k,[s,\infty[} = \mathcal{F}_k,$$

the identity $V \otimes_M B(H_1) \otimes_M B(H_2) = V \otimes_M B(H_1 \otimes H_2)$ (which is property 3 following the definition of matrix spaces) ensures that everything fits together properly.

Remark. In fact the Markovian cocycle property can be defined for a wider class of processes; for example processes X^L and k^ϕ need not be bounded (respectively, completely bounded) themselves, however there are good reasons to consider them as Markovian cocycles.

E-semigroup of a Markovian cocycle

Let k be a CB Markovian cocycle on an operator space V in $B(h_1; h_2)$. Then the compositions

$$K_s := \widehat{k}_s \circ \sigma_s \tag{5.5}$$

form a CB semigroup on $V \otimes_M B(\mathcal{F}_k)$. Conversely, if $(K_t)_{t \geq 0}$ is a CB semigroup on $V \otimes_M B(\mathcal{F}_k)$ satisfying

$$K_t(x \otimes b) = k_t(x)(I_{h_1} \otimes \sigma_t(b))$$

for a process k on V, then k is a Markovian cocycle on V.

In particular, normal *-homomorphic Markovian cocycles on von Neumann algebras $\mathcal{M}$ give rise to E-semigroups on $\mathcal{M} \overline{\otimes} B(\mathcal{F}_k)$ (see the lectures of Rajarama Bhat in this volume).

5.1 Semigroup representation

If k is a CB Markovian cocycle on an operator space V then for $d, e \in k$,

$$P_t^{d,e} := E^{\varpi(d_{[0,t[})}k_t(\,\cdot\,)E_{\varpi(e_{[0,t[})} \tag{5.6}$$

where $\varpi(\,\cdot\,)$ denotes the normalised exponential map (1.23), defines a semigroup on V, and $\{P^{d,e} : d, e \in k\}$ is called the set of *associated semigroups* of the cocycle. In turn the associated semigroups determine the cocycle because the cocycle property gives, for (right-continuous) step functions f and g,

$$E^{\varpi(f_{[0,t[})}k_t(\,\cdot\,)E_{\varpi(g_{[0,t[})} = P_{t_1-t_0}^{(0)} \circ \cdots \circ P_{t_{n+1}-t_n}^{(n)}, \tag{5.7}$$

where $\{0 = t_0 \leq t_1 \leq \cdots \leq t_{n+1} = t\}$ contains the discontinuities of both $f_{[0,t[}$ and $g_{[0,t[}$, and $P^{(k)} = P^{d,e}$ where $d = f(t_k)$ and $e = g(t_k)$.

Proposition 5.1. *Let k be a CB process on V. If (5.6) defines a semigroup for each $d, e \in k$ and (5.7) holds for these semigroups, then k is a Markovian cocycle.*

Let k be a completely bounded process on V satisfying the QSDE (4.8) for a CB map ϕ. It follows from (4.9) that, for each $c, d \in k$,

$$t \mapsto E^{\varpi(c_{[0,t[})}k_t(\,\cdot\,)E_{\varpi(d_{[0,t[})}$$

defines a semigroup $P^{c,d}$ on V. Moreover the semigroup is norm continuous, and is completely contractive if k is (since we are using normalised exponential vectors). In turn it is not difficult to verify that (5.7) holds and therefore k is a Markovian cocycle by Proposition 5.1. We call it the Markovian cocycle *generated by* ϕ. Various converse results hold. These are discussed in the following subsections.

Remark. Left Markovian cocycles on a Hilbert space equally have a semigroup representation in terms of associated semigroups.

Markov regularity

A *Markov-regular cocycle* is a Markovian cocycle all of whose associated semi-groups are norm continuous. Here there is a dichotomy. Recall that the CB condition on a Markovian cocycle can be loosened.

Proposition 5.2. *Let k be a Markovian cocycle on an operator space which is bounded with locally uniform bounds. Then either all of the associated semigroups are norm continuous or none of them are.*

Thus, in particular, Markov regularity for a contraction cocycle k is equivalent to the norm continuity of its *Markov semigroup*

$$\left(E^{\varepsilon(0)}k_t(\,\cdot\,)E_{\varepsilon(0)}\right)_{t\geq 0}.\tag{5.8}$$

This observation proves to be rather useful.

Remark. Similar dichotomies hold for pointwise strong and weak continuity and also for bounded operator Markovian cocycles. All these results follow from simple estimates.

Example 5.3 (Pure number/exchange cocycles: multidimensional case). What is the multidimensional analogue of Example 2.5? The QSDE (4.11) generalises easily:

$$dk_t = \widehat{k}_t \circ (\psi - \iota_k)dN_t, \quad k_0(x) = x \otimes I.\tag{5.9}$$

If $\psi : V \to M(k; V)_b$ is k-bounded then this has a strong solution on $h_1 \otimes \mathcal{E}_k$, unique amongst weakly regular weak solutions. What about an explicit form for the Markov-regular cocycle which is its solution (cf. (2.11))?

Exercise. *Using the identification*

$$\mathfrak{h} \otimes \mathcal{F}_k = \bigoplus_{n \geq 0} \left(k^{\otimes n} \otimes \mathfrak{h} \otimes \mathcal{F}^{(n)}_{[0,t[}\right) \otimes \mathcal{F}_{k,[t,\infty[}$$

verify that the process on V defined by

$$k_t(x) = \bigoplus_{n \geq 0} \left(\psi_n(x) \otimes I^{(n)}_{[0,t[}\right) \otimes I_{k,[t,\infty[},$$

where ψ_n is defined as in (4.7) but with ψ and k in place of ϕ and $\widehat{k}$, is weakly regular and satisfies the QSDE (5.9).

Exponential noise, revisited

Consider again the example of pure-noise processes obtained from exponential operators (see Section 4), now with each of the constituent functions k, u, v and W being *constant*. Thus

$$X_t = \Gamma\left(tw, d \otimes 1_{[0,t[}, W \otimes I_{[0,t[} + I \otimes I_{[t,\infty[}, e \otimes 1_{[0,t[}\right)\tag{5.10}$$

for $w \in \mathbb{C}$, $d, e \in k$ and $W \in B(k)$.

Proposition 5.4. *X is a Markovian cocycle.*

Proof. The shifts $s_t : L^2(\mathbb{R}_+; \mathsf{k}) \to L^2([t, \infty[, \mathsf{k})$ satisfy

$$s_t(e \otimes 1_{[0,s[}) = e \otimes 1_{[t,t+s[},$$
$$s_t(W \otimes I_{[0,s[} + I_\mathsf{k} \otimes I_{[s,\infty[})s_t^* = W \otimes I_{[t,t+s[} + I_\mathsf{k} \otimes I_{[t+s,\infty[}.$$

Therefore, by (5.4), adaptedness and (4.1),

$$\langle \varepsilon(f), X_t \sigma_t(X_s)\varepsilon(g) \rangle = AB,$$

where

$$A = \langle \varepsilon(f_{[0,t[}), X_t \varepsilon(g_{[0,t[}) \rangle$$
$$= \exp\left\{ wt + \langle e \otimes 1_{[0,t[}, g \rangle + \langle f, W \otimes I_{[0,t[} g \rangle + 0 + \langle f, d \otimes 1_{[0,t[} \rangle \right\}$$

and B equals

$$\Big\langle \varepsilon(f\big|_{[t,\infty[}),$$
$$\Gamma(ws, d \otimes 1_{[t,t+s[}, W \otimes I_{[t,t+s[} + I_\mathsf{k} \otimes I_{[t+s,\infty[}, e \otimes 1_{[t,t+s[})\varepsilon(g\big|_{[t,\infty[}) \Big\rangle$$
$$= \exp\Big\{ ws + \langle e \otimes 1_{[t,t+s[}, g \rangle + \langle f, W \otimes I_{[t,t+s[}g \rangle$$
$$+ \langle f_{[t+s,\infty[}, g_{[t+s,\infty[} \rangle + \langle f, d \otimes 1_{[t,t+s[} \rangle \Big\}$$

Thus AB has the form of $\langle \varepsilon(f), X_{t+s}\varepsilon(g) \rangle$. Therefore $X_{s+t} = X_s \sigma_s(X_t)$ on $\mathcal{E}(\mathsf{K})$. □

Thus, in particular, the parity process involved in the realisation of Fermi fields as QS integrals (on page 233) is a Markovian cocycle.

Proposition 5.5. *The associated semigroups for the Markovian cocycle X are given by*

$$P_t^{b,c} = \exp t \langle \widehat{b}, L\widehat{c} \rangle$$

where

$$L = \begin{bmatrix} w & \langle e| \\ |d\rangle & W - I_\mathsf{k} \end{bmatrix} \quad and \quad \widehat{c} = \begin{pmatrix} 1 \\ c \end{pmatrix} \in \widehat{\mathsf{k}}.$$

Proof. Using the formula (4.1) once more,

$$\langle \varepsilon(b_{[0,t[}), X_t \varepsilon(c_{[0,t[}) \rangle$$
$$= \exp\left\{ wt + \langle e, c \rangle t + \langle b, Wc \rangle t + \langle b, d \rangle t \right\},$$

so

$$\langle \varpi(b_{[0,t[}), X_t \varpi(c_{[0,t[}) \rangle = \exp t \langle \widehat{b}, L\widehat{c} \rangle.$$

□

We also know (from Section 4) that X satisfies a QSDE (see (4.2)). Here the QSDE is of a special type — its coefficients are constant:

$$dX_t = L \otimes X_t \, d\Lambda_t; \quad X_0 = I$$

where

$$L = \begin{bmatrix} w & \langle e| \\ |d\rangle & W - I_{\mathsf{k}} \end{bmatrix},$$

the identical matrix arising in our representation of the generators of the associated semigroups!

In spite of this example having only the trivial initial space $\mathbb{C}$, which brings about many simplifications (including commutativity!), it has succeeded in revealing a substantial part of the structure of Markovian cocycles. In (5.17) below we shall see how the conditions for contractivity of exponential noise given in the exercise on page 210 generalise to nontrivial initial space.

5.2 Stochastic generation

Solutions of quantum stochastic differential equations form Markovian cocycles. In turn nice Markovian cocycles have an infinitesimal description as a solution of a QSDE. Furthermore, properties of the cocycle are naturally reflected in the structure of their 'stochastic generator'. This applies both to mapping cocycles and to operator cocycles.

We need the following extension of the Δ notation (3.13):

$$\Delta(x) := P_{\mathsf{k}} \otimes x \in B(\widehat{\mathsf{k}}) \otimes_{\mathrm{sp}} \mathsf{V} \tag{5.11}$$

for elements x of an operator space V. Thus $\Delta(1) = \Delta$ when $1 \in \mathsf{V}$.

Completely positive cocycles

The best results here are for CP *contraction* cocycles.

Theorem 5.6. *Let k be a completely positive contraction process on a C^*-algebra $\mathcal{A}$. Then the following are equivalent:*

(i) *k is a Markov-regular cocycle;*
(ii) *$k = k^\phi$ for a completely bounded operator $\phi : \mathcal{A} \to \mathrm{M}(\widehat{\mathsf{k}}; \mathcal{A})_{\mathrm{b}}$.*

We speak of the *stochastic generator* ϕ of the cocycle k. For the next result let $E_{(0)} = E_\chi$ and $E^{(0)} = E^\chi$ where $\chi = \binom{1}{0} \in \widehat{\mathsf{k}}$.

Theorem 5.7. *Let k be a completely positive contraction process on a C^*-algebra $\mathcal{A}$ in $B(\mathsf{h})$, and suppose that k weakly satisfies (4.8) for some bounded operator $\phi : \mathcal{A} \to \mathrm{M}(\widehat{\mathsf{k}}; \mathcal{A})_{\mathrm{b}}$. Then ϕ has the form*

$$\phi(a) = \psi(a) - \Delta(a) + J^* a E^{(0)} + E_{(0)} a J, \tag{5.12}$$

where ψ is completely positive and $J \in C_{\widehat{k}}(\mathcal{A}'')$ and satisfies

$$\phi(1) \leq 0.$$

Remarks. Notice that any map ϕ of the form (5.12) is completely bounded, and so, by uniqueness, $k = k^\phi$. In fact, for an operator ϕ of the form (5.12), k^ϕ is necessarily completely positive and contractive, and so the converse of Theorem 5.7 holds too.

*-homomorphic cocycles

Let $\mathcal{A}$ be a C^*-algebra acting on h. Necessary conditions on a completely bounded map $\theta : \mathcal{A} \to M(\widehat{k}; \mathcal{A})_b$ for the cocycle generated by θ to be *-homomorphic may be obtained quite easily from the quantum Itô product formula.

Proposition 5.8. *Let k be a *-homomorphic Markovian cocycle on a C^*-algebra $\mathcal{A}$ which acts on h, with bounded stochastic generator θ. Then θ is a real map, that is $\theta(a^*) = \theta(a)^*$, and satisfies*

$$\theta(ab) = \theta(a)\iota(b) + \iota(a)\theta(b) + \theta(a)\Delta\theta(b) \tag{5.13}$$

where ι is the ampliation $\iota_{\widehat{k}} : a \mapsto I_{\widehat{k}} \otimes a$.

If we write θ in block matrix form:

$$\theta = \begin{bmatrix} \tau & \delta^\dagger \\ \delta & \rho - \iota_k \end{bmatrix} \tag{5.14}$$

then (5.13), together with reality of θ, reads

$$\rho \text{ is a *-homomorphism } \mathcal{A} \to M(k; \mathcal{A})_b$$
$$\delta \text{ is a } \rho\text{-derivation } \mathcal{A} \to C(k; \mathcal{A})_b$$
$$\tau \text{ is a real map } \mathcal{A} \to \mathcal{A} \text{ satisfying}$$
$$\tau(ab) - \tau(a)b - a\tau(b) = \delta^\dagger(a)\delta(b),$$

where $\delta^\dagger : \mathcal{A} \to R(k; \mathcal{A})_b$ is defined by $\delta^\dagger(a) = \delta(a^*)^{\overset{\cdot}{*}}$, and a ρ-*derivation* is a linear map satisfying the ρ-*Leibniz identity*

$$\delta(ab) = \delta(a)b + \rho(a)\delta(b).$$

The converse is a trickier matter. The next result is quite recent. Recall the definition of k-boundedness (on page 241).

Theorem 5.9. *Let $\theta : \mathcal{A} \to M(\widehat{k}; \mathcal{A})_b$ be a real k-bounded map, satisfying (5.13). Then, in either of the following two cases, the Markovian cocycle generated by θ is *-homomorphic:*

(a) $\theta(a)E_\chi \in |\widehat{\mathsf{k}}\rangle \otimes_{\mathrm{sp}} \mathcal{A}$ for all $a \in \mathcal{A}$, $\chi \in \widehat{\mathsf{k}}$.
(b) $\mathcal{A}$ is a von Neumann algebra and θ is ultraweakly continuous.

Remark. When $\mathcal{A}$ is unital, part (a) is equivalent to $\theta(\mathcal{A}) \subset M(\mathcal{K} \otimes_{\mathrm{sp}} \mathcal{A})$, where M denotes *multiplier algebra* (an important concept in Johan Kustermans' notes in this volume), and $\mathcal{K}$ is the C^*-algebra of compact operators on $\widehat{\mathsf{k}}$.

Exercise. Let j be a *-homomorphic Markov-regular cocycle on a commutative C^*-algebra $\mathcal{A}$. Show that the following family is commutative:

$$\{j_t(a) : a \in \mathcal{A}, t \geq 0\}.$$

Operator cocycles

Contraction cocycles on a Hilbert space satisfy (constant-coefficient) QSDE's, under the assumption of Markov regularity.

Theorem 5.10. *Let X be a Hilbert-space contraction process. Then the following are equivalent:*

(i) X *is a Markov-regular left cocycle;*
(ii) $X = X^L$ *for a bounded operator L.*

Again we refer to the operator L as the *stochastic generator* of the cocycle X.

Corollary 5.11. *Let X be a Hilbert-space contraction process on $\mathbb{C}$. Then the following are equivalent:*

(i) X *is a Markov-regular left cocycle.*
(ii) X *is an exponential noise of the form (5.10).*

In the language of Rajarama Bhat's lectures in this volume, every local cocycle with respect to a CCR flow is an exponential noise. (Strong continuity of the cocycle implies Markov-regularity in this context.)

As with CP contraction cocycles on a C^*-algebra, we may recognise contractivity from the form of the generator. Bearing in mind our discussion of exponential noise, the structure (5.18) and (5.17) below may be compared respectively with (1.32) and the exercise which follows.

Theorem 5.12. *Let X be a bounded Markov-regular left cocycle on a Hilbert space h and suppose that X weakly satisfies a left QSDE of the form (4.6). Then the following are equivalent:*

(i) X *is contractive;*
(ii) L *is bounded and*

$$L^* + L + L^* \Delta L \leq 0; \tag{5.15}$$

(iii) L *is bounded and*

$$L + L^* + L^* \Delta L \leq 0; \tag{5.16}$$

(iv) *There are bounded operators* H, B, M, V *and* W *such that*

$$L = \begin{bmatrix} iH - \frac{1}{2}(M^*M + B^2) & BVS - M^*W \\ M & W - I \end{bmatrix} \tag{5.17}$$

where $H = H^*$, $\|V\|, \|W\| \leq 1$, $B \geq 0$ *and* $S = (1 - W^*W)^{1/2}$.

Furthermore, X *is isometric if and only if equality holds in* (5.15) *if and only if*

$$W \text{ is isometric and } B = 0. \tag{5.18}$$

Finally X *is unitary if and only if*

$$L^* + L + L^*\Delta L = 0 = L + L^* + L\Delta L^*,$$

if and only if W *is unitary and* $B = 0$.

In fact, for a bounded operator L satisfying (5.15), X^L is necessarily contractive as is $^L X$, and its adjoint process is given by

$$(X^L)^* = {}^M X, \text{ where } M = L^*.$$

The next subsection reveals more.

Dual cocycles

Let X be a bounded left operator Markovian cocycle. Then its adjoint process $(Z_t = X_t^*)_{t \geq 0}$ is a *right cocycle*, in other words it satisfies

$$Z_{s+t} = \sigma_s(Z_t)Z_s, \quad Z_0 = I.$$

It is also true that $(R_t X_t R_t)_{t \geq 0}$ defines a right cocycle, where R is the time reversal process defined in Example 2.3. This is most easily verified by exploiting the semigroup representation, since right cocycles have such a representation too, but with the semigroups appearing in the reverse order. Combining these we obtain the *dual cocycle* of X, defined by

$$\widetilde{X}_t := R_t X_t^* R_t.$$

Thus the dual cocycle of a left cocycle is another left cocycle. The stochastic generator of the dual of a Markov-regular left contraction cocycle X is the adjoint of the stochastic generator of X.

Duality plays an important part in the analysis of cocycles. Given that the dual of a contraction (respectively, isometric) cocycle is a contraction (respectively, coisometric) cocycle, the role of duality in Theorem 5.12 is hopefully evident.

Notes

The cocycle viewpoint in classical Markov process theory is promoted in [Pin]. Direct link with the quantum context is investigated in [LSi]. That Markovian cocycles of the above kind satisfy QSDE's was first shown in [HL$_4$], for unitary operator cocycles, and in [Bra], for normal unital *-homomorphic mapping cocycles on a von Neumann algebra, both using the representation of martingales as QS integrals ([PSi]). Subsequently the semigroup representation of cocycles has provided the most effective tool for the natural generalisations of these results ([LP$_2$], [LW$_2$]). The characterisation of the generators of Markov-regular contraction operator cocycles, or rather *contractive solutions of QSDE's* with bounded coefficients, was obtained [Fag$_2$] and [MoP], and refined in [LW$_1$]. The characterisation of the stochastic generators of CP cocycles was obtained in [LP$_2$] for finite-dimensional noise, and extended to infinite dimensions in [LW$_1$], *under the assumption* that the cocycle satisfied a QSDE. Independent work on CP stochastic evolutions may also be found in [Bel$_2$]. That Markov-regular contraction operator cocycles, respectively CP contraction cocycles on a C^*-algebra, *necessarily satisfy* a QSDE was shown in [LW$_2$]. In particular, this made the above assumptions redundant.

The sufficient conditions for a map θ (enjoying the necessary algebraic structure) to stochastically generate a *-homomorphic Markovian cocycle, were obtained in [LW$_{4,5}$] extending the finite-dimensional ([Eva]) and Mohari-Sinha-regular ([MoS]) cases. Our method exploited the algebraic structure of quantum stochastic calculus (cf. [L$_6$]), more specifically a product formula for iterated QS integrals (final exercise of the previous section) and the knowledge that θ is necessarily completely bounded and generates a CP contractive cocycle on the C^*-algebra ([LW$_3$]). The exercise below Theorem 5.9 is from [MoS], and is relevant to the interpretation of Markovian cocycles in terms of classical Markov processes.

An alternative approach to QS cocycles on V amalgamates its associated semigroups into a single semigroup on $B(\widehat{\mathsf{k}}) \otimes_{\mathrm{sp}} \mathsf{V}$. A short proof of Theorem 5.7 may be founded on this approach ([LW$_8$]). This method originated in the paper [AcK] and has been extensively developed in [LW$_6$].

Markovian cocycles were introduced into quantum probability in [Acc], as a tool for perturbing quantum Markov processes, and were further elaborated in the fundamental paper [AFL]. Dual cocycles were introduced in [Jou], for analysing non-regular contraction operator cocycles. For further discussion of the literature see [LW$_2$]. See also the very recent paper [HKK] which points to interesting future developments for interconnections between probability and operator algebras, with stochastic cocycles as a central idea.

6 QS Dilation

For this section *fix a unital C^*-algebra $\mathcal{A}$ acting nondegenerately on a Hilbert space $\mathfrak{h}$*. Let k be a contractive CP Markovian cocycle on $\mathcal{A}$. Then its Markov semigroup

$$P_t := \mathbb{E}_0 \circ k_t = E^{\varepsilon(0)} k_t(\cdot) E_{\varepsilon(0)}$$

is a completely positive contraction semigroup on $\mathcal{A}$, since the conditional expectation $\mathbb{E}_0$ is both contractive and completely positive.

Stochastic dilation problem. Given a CP contraction semigroup P on $\mathcal{A}$, is there a *-homomorphic Markovian cocycle j on $\mathcal{A}$ such that

$$\mathbb{E}_0 \circ j_t = P_t \text{ for } t \geq 0? \tag{6.1}$$

Another name for CP contraction semigroups is quantum dynamical semigroups; these are assumed further to enjoy continuity properties appropriate to the algebra, and usually also to be unital.

CP semigroups

Stinespring's Theorem gives us the form of an individual CP map (Example 1.11). What about a semigroup of such maps? (This question is also posed in Rajarama Bhat's notes in this volume.)

Proposition 6.1 (Evans and Lewis). *Let $\tau \in B(\mathcal{A})$. Then the following are equivalent:*

(i) τ *generates a CP semigroup;*
(ii) $\partial\tau$ *is a nonnegative-definite kernel on $\mathcal{A}$, where*

$$\partial\tau(a,b) = \tau(a^*b) - a^*\tau(b) - \tau(a)^*b + a^*\tau(1)b.$$

Such a semigroup is contractive if and only if its generator satisfies

$$\tau(1) \leq 0.$$

Exercise. *Prove this.*

Example 6.2 (Lindbladians). Let (π, H) be a representation of $\mathcal{A}$, let $D \in B(\mathfrak{h}; \mathsf{H})$ and let $H = H^* \in B(\mathfrak{h})$. Set $\mathcal{L} = \mathcal{L}_{D,\pi,H}$ where

$$\mathcal{L}_{D,\pi,H} : a \mapsto D^*\pi(a)D - \frac{1}{2}\{D^*D, a\} + i[H, a]. \tag{6.2}$$

If $\mathcal{L}(\mathcal{A}) \subset \mathcal{A}$ then $\mathcal{L}$ generates a CP contraction semigroup on $\mathcal{A}$.

Exercise. *Prove this, and generalise it to the noncontractive case.*

For stochastic dilation we need to know that norm-continuous CP semigroups have *completely* bounded generators. The following result is nontrivial. Recall that completely positive maps are completely bounded.

Lemma 6.3 (Christensen). *Let ϕ be an ultraweakly continuous completely positive map on a von Neumann algebra $\mathcal{M}$, satisfying $\|\phi - \mathrm{id}_{\mathcal{M}}\| \leq 10^4$. Then, for any separable Hilbert space h,*

$$\|\phi \overline{\otimes} \mathrm{id}_{B(\mathsf{h})} - \mathrm{id}_{\mathcal{M} \overline{\otimes} B(\mathsf{h})}\| \leq 10^4 \|\phi - \mathrm{id}_{\mathcal{M}}\|^{1/4}.$$

To obtain the result we want, the fact that the bidual of a C^*-algebra is naturally a von Neumann algebra (more correctly a W^*-algebra) may be used.

Corollary 6.4. *Let τ be the generator of a norm-continuous completely positive semigroup on a C^*-algebra $\mathcal{C}$. Then τ is completely bounded.*

Proof. Let h be an infinite dimensional separable Hilbert space. A semigroup of CP maps on $\mathcal{C}^{**} \overline{\otimes} B(\mathsf{h})$ is defined by $(e^{t\tau^{**}} \overline{\otimes} \mathrm{id}_{B(\mathsf{h})})_{t\geq 0}$. By the lemma this is norm continuous. Its generator, which extends $\tau^{**} \otimes \mathrm{id}_{B(\mathsf{h})}$, is thus bounded. Hence τ is completely bounded. $\qquad\qquad\square$

6.1 Stochastic dilation

Given the block matrix form of the stochastic generator of a $*$-homomorphic Markov-regular cocycle, namely (5.14), the generator of its Markov semigroup is the top-left component of the matrix. The following result is key for QS dilation.

Proposition 6.5. *Let $\tau \in B(\mathcal{A})$ be the generator of a CP contraction semigroup on $\mathcal{A}$. Then there is a triple $(\mathsf{k}, \rho, \delta)$ consisting of a Hilbert space k, a representation $\rho : \mathcal{A} \to B(\mathsf{k}) \overline{\otimes} \mathcal{A}''$ and a ρ-derivation $\delta : \mathcal{A} \to |\mathsf{k}\rangle \overline{\otimes} \mathcal{A}''$ satisfying*

$$\tau(a^*b) - \tau(a)^*b - a^*\tau(b) = \delta(a)^*\delta(b). \tag{6.3}$$

If p denotes the orthogonal projection onto $\overline{\mathrm{Lin}}\,\delta(\mathcal{A})\mathfrak{h}$, then

$$p \in (B(\mathsf{k}) \overline{\otimes} \mathcal{A}'') \cap \rho(\mathcal{A})'.$$

Moreover, if $\mathcal{A}$ is a von Neumann algebra and τ is ultraweakly continuous then so are δ and ρ.

Proof. Let $\gamma_1 : \mathcal{A} \to B(\mathfrak{h}; \mathsf{K}_1)$ be a minimal Kolmogorov map for the nonnegative-definite kernel $\partial\tau : \mathcal{A} \times \mathcal{A} \to \mathcal{A} \subset B(\mathfrak{h})$ given by

$$\partial\tau(a,b) = \tau(a^*b) - a^*\tau(b) - \tau(a)^*b + a^*\tau(1)b.$$

The identity

$$\big(\gamma_1(ua) - \gamma_1(u)a\big)^* \big(\gamma_1(ub) - \gamma_1(u)b\big) = \gamma_1(a)^*\gamma_1(b)$$

holds for all $a, b \in \mathcal{A}$ and isometric u in $\mathcal{A}$. (***Exercise.*** Verify this.) Therefore, in view of the minimality of γ_1, there is a unique isometry $\pi_1(u)$ on K_1 satisfying

$$\pi_1(u)\gamma_1(a) = \gamma_1(ua) - \gamma_1(u)a.$$

The map $u \mapsto \pi_1(u)$ extends uniquely to a unital representation of $\mathcal{A}$ on K_1. Now define

$$\mathsf{K} = \mathsf{K}_1 \oplus \overline{\operatorname{Ran}} \tau(1), \quad \pi(a) = \begin{bmatrix} \pi_1(a) & 0 \\ 0 & 0 \end{bmatrix} \text{ and } \gamma(a) = \begin{bmatrix} \gamma_1(a) \\ (-\tau(1))^{1/2}a \end{bmatrix}.$$

Then $\gamma : \mathcal{A} \to B(\mathfrak{h}; \mathsf{K})$ is a minimal Kolmogorov map for the nonnegative-definite kernel $(a, b) \mapsto \tau(a^*b) - \tau(a)^*b - a^*\tau(b)$:

$$\tau(a^*b) - \tau(a)^*b - a^*\tau(b) = \gamma(a)^*\gamma(b); \quad \overline{\operatorname{Lin}}\,\gamma(\mathcal{A})\mathfrak{h} = \mathsf{K}.$$

Moreover $\pi : \mathcal{A} \to B(\mathsf{K})$ is a representation of $\mathcal{A}$ (nonunital unless $\tau(1) = 0$) such that γ is a π-derivation:

$$\gamma(ab) = \gamma(a)b - \pi(a)\gamma(b).$$

In view of the identity

$$\big(\gamma(a)u'\big)^*\big(\gamma(b)u'\big) = \gamma(a)^*\gamma(b),$$

which holds for all $a, b \in \mathcal{A}$ and isometric u' in $\mathcal{A}'$, the minimality of γ implies that there is a unique isometry $\pi'(u')$ on K satisfying

$$\pi'(u')\gamma(a) = \gamma(a)u'.$$

Again $u' \mapsto \pi'(u')$ extends uniquely to a representation of $\mathcal{A}'$ on K. Now π' is normal and unital, and also

$$\pi'(\mathcal{A}) \subset \pi(\mathcal{A})'. \tag{6.4}$$

By the structure of normal representations of von Neumann algebras (a good reference is [Tak]) it follows that there is a Hilbert space k and an isometry $V : \mathsf{K} \to \mathsf{k} \otimes \mathfrak{h}$ such that

$$\pi'(a') = V^*\big(I_\mathsf{k} \otimes a'\big)V \text{ for } a' \in \mathcal{A}', \text{ and } p := VV^* \in \big(I_\mathsf{k} \otimes \mathcal{A}'\big)' = B(\mathsf{k})\overline{\otimes}\mathcal{A}''.$$

Now define representations ρ and ρ' of $\mathcal{A}$ and $\mathcal{A}'$ respectively, and a ρ-derivation δ, by

$$\rho(a) = V\pi(a)V^*, \quad \delta(a) = V\gamma(a) \text{ and } \rho'(a') = \big(I_\mathsf{k} \otimes a'\big)p.$$

Since $\rho'(a') = V\pi'(a')V^*$ it follows from (6.4) that $\rho'(\mathcal{A}') \subset \rho(\mathcal{A})'$; in particular $p \in \rho(\mathcal{A})'$. The remaining properties of ρ and δ follow.

If $\mathcal{A}$ is a von Neumann algebra and τ is ultraweakly continuous then the ultraweak continuity of γ_1, π and γ are easily checked; the ultraweak continuity of ρ and δ follows.   □

Remark. The representation $(\rho, \mathsf{k} \otimes \mathfrak{h})$ is typically nonunital.

Stochastic dilation on a von Neumann algebra

Combining the previous proposition with Theorem 5.9 and Corollary 6.4 gives the following dilation theorem.

Theorem 6.6. *Suppose that $\mathcal{A}$ be a von Neumann algebra. Let P be a completely positive contraction semigroup with bounded and ultraweakly continuous generator τ. Then, with $(\mathsf{k}, \rho, \delta)$ as in the previous theorem,*

$$\begin{bmatrix} \tau & \delta^\dagger \\ \delta & \rho - \iota_\mathsf{k} \end{bmatrix}$$

*generates a *-homomorphic Markovian cocycle j which dilates P, in the sense* (6.1).

Remark. The E-semigroup $\left(J_s := \widehat{j}_s \circ \sigma_s \right)_{s \geq 0}$ on $\mathcal{A} \overline{\otimes} B(\mathcal{F}_\mathsf{k})$ therefore also dilates P.

Stochastic dilation on a C^*-algebra

In order to achieve stochastic dilation on a C^*-algebra we need to appeal to a little Hilbert C^*-module theory (for which [Lan] is recommended).

Theorem 6.7. *Suppose that $\mathcal{A}$ is separable. Let P be a norm-continuous completely positive contraction semigroup with generator τ. Then there is a separable Hilbert space k and a completely bounded map $\theta : \mathcal{A} \to M(\mathcal{K}(\mathsf{k}) \otimes_{\mathrm{sp}} \mathcal{A})$ such that the Markovian cocycle j generated by θ is *-homomorphic and dilates the semigroup P.*

For the notation here see the remark following Theorem 5.9.

Proof. Let $(\pi, \mathsf{H}, \gamma)$ be as in the first part of the proof of Proposition 6.5. Set $F = \overline{\mathrm{Lin}}\,\gamma(\mathcal{A})\mathcal{A} \subset B(\mathfrak{h}; \mathsf{H})$. Then, since $\gamma(a)^*\gamma(b) = \tau(a^*b) - \tau(a)^*b - a^*\tau(b) \in \mathcal{A}$ for $a, b \in \mathcal{A}$, F is a Hilbert C^*-module with $\mathcal{A}$-valued inner product given by $\langle f_1, f_2 \rangle := f_1^* f_2$. By the separability of $\mathcal{A}$, F is countably generated and so, by Kasparov's Embedding Theorem, there is an adjointable isometry $\phi : F \to |\mathsf{k}\rangle \otimes_{\mathrm{sp}} \mathcal{A}$, for some separable Hilbert space k. By the nondegeneracy of $\mathcal{A}$ on $\mathfrak{h}$ and the minimality of γ, the map $fu \mapsto \phi(f)u$ $(f \in F, u \in \mathfrak{h})$ extends uniquely to an isometry $V : \mathsf{H} \to \mathsf{k} \otimes \mathfrak{h}$. Now, letting $\phi : \mathcal{A} \to B(\widehat{\mathsf{k}} \otimes \mathfrak{h})$ be the map with block matrix form (5.14):

$$\theta = \begin{bmatrix} \tau & \delta^\dagger \\ \delta & \rho - \iota_\mathsf{k} \end{bmatrix}$$

where $\rho = V\pi(\cdot)V^*$, $\delta = \phi \circ \gamma = V\gamma(\cdot)$ and ι_k is the ampliation $a \mapsto I_\mathsf{k} \otimes a$, it is easily checked that θ satisfies (the given equivalent of) (5.13). Now $\mathrm{Ran}\,\delta \subset \mathrm{Ran}\,\phi \subset |\mathsf{k}\rangle \otimes_{\mathrm{sp}} \mathcal{A}$, and so also $\mathrm{Ran}\,\delta^\dagger \subset \langle\mathsf{k}| \otimes_{\mathrm{sp}} \mathcal{A}$, and

$$\rho(a)E_e = V\pi(a)V^*E_e = \phi\big(\pi(a)\phi^*(|e\rangle \otimes 1_{\mathcal{A}})\big) \subset \text{Ran}\,\phi$$

for $a \in \mathcal{A}$ and $e \in \mathsf{k}$. It follows that θ satisfies condition (a) of Theorem 5.9. Again Corollary 6.4 implies that θ is CB — the complete boundedness of δ following from the identity $\delta(a)^*\delta(b) = \tau(a^*b) - \tau(a)^*b - a^*\tau(b)$. The result therefore follows from Theorem 5.9. □

Remark. A smarter proof is obtained by directly appealing to the Hilbert-C^*-module-theoretic Kolmogorov map ([Mur₂]).

6.2 Decomposition via perturbation

To analyse the structure of the von Neumann algebraic stochastic dilation, we need to know what bounded ρ-derivations look like. It turns out that they are all *inner*.

Theorem 6.8 (Christensen-Evans). *Let* (π, H) *be a representation of* $\mathcal{A}$ *and let* $\gamma : \mathcal{A} \to B(\mathfrak{h}; \mathsf{H})$ *be a bounded π-derivation. Then there is an element* $g \in \overline{\text{Lin}}^{\text{uw}}\gamma(\mathcal{A})\mathcal{A}$ *such that* $\gamma = \delta_{g,\pi}$:

$$a \mapsto ga - \pi(a)g.$$

This may be used to obtain a Lindbladian structure for the generators of CP contraction semigroups which is well suited to quantum stochastic dilation.

Theorem 6.9. *Let* $\tau \in B(\mathcal{A})$ *be the generator of a CP contraction semigroup on* $\mathcal{A}$. *Then there is a quadruple* (k, ρ, d, h) *consisting of a Hilbert space* k, *a* *-homomorphism* $\rho : \mathcal{A} \to B(\mathsf{k})\overline{\otimes}\mathcal{A}''$, *and elements* $d \in |\mathsf{k}\rangle\overline{\otimes}\mathcal{A}''$ *and* $h = h^* \in \mathcal{A}''$ *such that* $\tau = \mathcal{L}_{d,\rho,h}$:

$$a \mapsto d^*\rho(a)d - \frac{1}{2}\{d^*d, a\} + i[h, a]. \tag{6.5}$$

When $\mathcal{A}$ *is a von Neumann algebra and* τ *is ultraweakly continuous* ρ *may be chosen to be normal.*

Proof. Let $(\mathsf{k}, \rho, \delta)$ be a triple as in Proposition 6.5, with ρ chosen normal if $\mathcal{A}$ is a von Neumann algebra and τ is ultraweakly continuous. Using Theorem 6.8 let $d \in \overline{\text{Lin}}^{\text{uw}}\delta(\mathcal{A})\mathcal{A} \subset |\mathsf{k}\rangle\overline{\otimes}\mathcal{A}''$ be such that $\delta = \delta_{d,\rho}$. Then, setting $\mathcal{L} = \mathcal{L}_{d,\rho,0}$ (see (6.2)),

$$\mathcal{L}(a^*b) - \mathcal{L}(a)^*b - a^*\mathcal{L}(b) = \delta(a)^*\delta(b).$$

Since τ also satisfies this identity it follows that τ differs from $\mathcal{L}$ only by a derivation η say, in $B(\mathcal{A})$. Applying Theorem 6.8 once more shows that $\tau - \mathcal{L} = \delta_{ih} : a \mapsto [ih, a]$ where $h \in \overline{\text{Lin}}^{\text{uw}}\eta(\mathcal{A})\mathcal{A} \subset \mathcal{A}''$. Reality (of τ and $\mathcal{L}$) implies that h may be chosen to be self-adjoint. It follows that

$$\tau = \mathcal{L}_{d,\rho,0} + \delta_{ih} = \mathcal{L}_{d,\rho,h}.$$

□

With this Lindbladian structure for the semigroup generator one can express the QS dilation described earlier as a perturbation of the Markovian cocycle generated by the preservation-only part of the cocycle generator, at least in the von Neumann algebra case. For this purpose the following perturbation theorem is needed. Recall Theorem 4.1, and the Δ-notation (5.11).

Theorem 6.10. *Suppose that $\mathcal{A}$ is a von Neumann algebra. Let j be a Markov-regular normal *-homomorphic cocycle on $\mathcal{A}$ with stochastic generator $\theta : \mathcal{A} \to B(\widehat{\mathsf{k}})\overline{\otimes}\mathcal{A}$, and let W be the unique solution of the QSDE*

$$dW_t = (\mathrm{id}\,\overline{\otimes}j_t)(l)W_t d\Lambda_t, \quad W_0 = I, \tag{6.6}$$

where $l \in B(\widehat{\mathsf{k}})\overline{\otimes}\mathcal{A}$ satisfies $l + l^* + l^*\Delta l \le 0$. Then

(a) *W is contractive;*
(b) *$k_t := W_t^* j_t(\cdot)W_t$ defines a (CP contractive) Markov-regular cocycle;*
(c) *the stochastic generator of k is given by*

$$\phi(a) =$$
$$\theta(a) + \iota(a)l + l^*\iota(a) + l^*\Delta(a)l + l^*\Delta\theta(a) + \theta(a)\Delta l + l^*\Delta\theta(a)\Delta l. \tag{6.7}$$

This result was established for the purpose of obtaining a process-wise Stinespring decomposition for CP contraction cocycles. It is included here since it may be used for proving the following result, which reveals some of the structure of the von Neumann algebraic QS dilation of Theorem 6.6.

Theorem 6.11. *Suppose that $\mathcal{A}$ is a von Neumann algebra. For a normal representation $\rho : \mathcal{A} \to B(\mathsf{k})\overline{\otimes}\mathcal{A}$, and elements $d \in |\mathsf{k})\overline{\otimes}\mathcal{A}$ and $h = h^* \in \mathcal{A}$ set*

$$\theta^0 = \begin{bmatrix} 0 & 0 \\ 0 & \rho - \iota_\mathsf{k} \end{bmatrix}, \quad l = \begin{bmatrix} -(ih + \tfrac{1}{2}d^*d) & d^* \\ -qd & q - 1 \end{bmatrix} \text{ and } \theta = \begin{bmatrix} \mathcal{L} & \delta^\dagger \\ \delta & \rho - \iota_\mathsf{k} \end{bmatrix}$$

where

$$q = \rho(1), \quad \mathcal{L} = \mathcal{L}_{d,\rho,h} \text{ and } \delta = \delta_{d,\rho}.$$

If j^0 and j are the (*-homomorphic) Markovian cocycles generated by θ^0 and θ respectively, and W is the contractive solution of the QSDE

$$dW_t = (\mathrm{id}\,\underline{\otimes}j_t^0)(l)W_t d\Lambda_t, \quad W_0 = I,$$

then W is partial isometry-valued and

$$j_t = W_t^* j_t^0(\cdot)W_t, \quad t \ge 0. \tag{6.8}$$

Remark. In Example 5.3 we saw the explicit form taken by cocycles with pure number/exchange generators like j^0.

Notes

Quantum stochastic dilation of norm-continuous quantum dynamical semigroups on $B(\mathfrak{h})$ was achieved in the original paper [HP$_1$] for a single Lindbladian generator, and in [HP$_2$] for the general case ($\mathfrak{h}$ separable). Its extension to a general von Neumann algebra was carried out in [GoS], and simplified in [GLSW] where it was also extended to the dilation of *cocycles*. Extension to separable unital C^*-algebras was done in [GPS], and simplified in [LW$_5$]. The innerness result for bounded π-derivations was proved in [ChE] as the key step in establishing their generalisation (Theorem 6.9) of the Lindblad, Gorini-Kossakowski-Sudarshan characterisation of generators of norm-continuous, normal, unital CP semigroups on $B(\mathsf{h})$ ([Lin], [GKS]). The perturbation theorem was proved in [GLW], extending earlier results in [EvH] and [DaS]. The decomposition given in Theorem 6.11 was obtained for the case of one-dimensional noise and the von Neumann algebra $B(\mathfrak{h})$ in [EvH].

Minimality for QS dilations is discussed in the lectures of Rajarama Bhat in this volume, for ultraweakly continuous unital CP semigroups on $B(\mathsf{h})$ with bounded generator. In that case, as explained there, *the minimal dilation* may be realised as a QS dilation (6.8) in which j^0 is simply ampliation and W is unitary-valued. On the other hand, for CP contraction semigroups on a general von Neumann algebra recent research shows that QS dilations typically *cannot* be minimal. This suggests that a deeper QS analysis may be called for, founded on Hilbert modules.

The final theorem is used in the recent result that the product system of a quantum stochastic E-semigroup on a von Neumann algebra is necessarily 'exponential' ([BhL]). Since these are product systems of Hilbert W^*-modules, a full explanation of this result would take us into their theory, which is sadly beyond this course.

Afterword

These notes constitute an introduction to quantum stochastic analysis from a current perspective. They are 'introductory' in the sense that they build the theory from scratch and therefore, due not least to limitations of space, cover only a fraction of the subject. Many other topics in quantum stochastic analysis could not reasonably be covered in these notes. For a further idea of the scope, in particular applications in areas such as quantum optics, quantum measurement theory and quantum filtering theory, Mathematical Reviews may be consulted (see [QSC]). The second volume of the lecture notes of the Grenoble Summer School ([QP$_{12}$]) contains an extensive bibliography which might also be useful.

A very specific sense in which they are introductory is that the treatment of QSDE's with unbounded operator coefficients and, correspondingly, Markovian cocycles whose Markov semigroup is not norm continuous (i.e.

non-Markov-regular cocycles), has been entirely omitted. There is now an extensive literature on the former, much of it by Franco Fagnola and Alexander Chebotarev; see the lecture notes [Fag₃]. The earliest results were obtained in [App], [Vin₁], [Fag₁], [Moh], [Fag₂] and [MoP]. For the latter I am aware only of the following papers: [Jou], [AJL], [Fag₂], [AcM] and [LW₇]. Below there is but a taste of this work; for proofs see [Fag₃], [Mey] and [LW₇], and for recent work on the (more difficult) *right* QSDE see [FW₁,₂].

Recall the review of c_0-semigroup (pre-)generators and dissipative operators (given on page 200). For the left QSDE

$$dX_t = \widehat{X}_t(F \underline{\otimes} I)\, d\Lambda_t, \quad X_0 = I, \tag{*}$$

in which F is an operator on $\widehat{k} \otimes \mathfrak{h}$ with dense domain $\widehat{D} \underline{\otimes} \mathcal{D}$, having block-matrix form $\left[\begin{smallmatrix} K & M \\ L & C-I \end{smallmatrix}\right]$, consider the conditions

(A) $2\mathrm{Re}\,\langle \xi, F\xi \rangle + \|\Delta F\xi\|^2 \le 0$, for all $\xi \in \widehat{D} \otimes \mathcal{D}$;
(B) the operator K is a pregenerator of a c_0-semigroup on $\mathfrak{h}$.

Exercise. *Show that condition (A) is necessary for the left QSDE (*) to have a strong contractive solution on $\mathcal{D} \underline{\otimes} \mathcal{E}_D$ (cf. Theorem 5.12).*

The conditions are not wholly independent since, defining

$$K_d^c := K + E^c L + M E_d + E^c C E_d - \tfrac{1}{2}\|c\|^2 - \tfrac{1}{2}\|d\|^2 \quad (c \in k, d \in D),$$

condition (A) implies that each operator K_d^c is dissipative.

Weak and strong solutions are defined as for the case where F is bounded except that now solutions have domains of the form $\mathcal{D} \underline{\otimes} \mathcal{E}(S)$. For the results below recall Proposition 2.1.

Theorem (Mohari and Parthasarathy). *If condition* (B) *holds then there is at most one contractive weak solution of the left QSDE (*) on $\mathcal{D} \underline{\otimes} \mathcal{E}_D$.*

Here is an application of uniqueness. If (*) has a unique contractive weak solution X on $\mathcal{D} \underline{\otimes} \mathcal{E}_D$ then it is necessarily a left Markovian cocycle. This is proved by verifying that, for each $t \ge 0$, the contraction process defined by

$$X_s^t := \begin{cases} X_s & s \le t \\ X_t \sigma_t(X_{s-t}) & s > t \end{cases}$$

also satisfies (*) on $\mathcal{D} \underline{\otimes} \mathcal{E}_D$.

Exercise. *Check this by viewing $(X_{t+r}^t)_{r\ge 0}$ as an $\mathfrak{h} \otimes \mathcal{F}_{k,[0,t[}$ process and using the explicit action of shifts on exponential vectors (5.3).*

Remark. *If a contractive weak solution of the left QSDE is strongly measurable then it is in fact a strong solution since the integrability condition is trivially satisfied. In particular, there is no distinction between weak and strong contractive solutions of the left QSDE when $\mathfrak{h}$ and k are both separable.*

Theorem (Fagnola). *Existence of a strong contractive solution of the left QSDE* (*) *on* $\mathcal{D}\underline{\otimes}\mathcal{E}_D$ *is assured if conditions* (A) *and* (B) *hold and furthermore*

(C) *the Hilbert spaces* $\mathfrak{h}$ *and* k *are separable.*

Thus, under conditions (A), (B) and (C), F stochastically generates a contractive left Markovian cocycle and its Markov semigroup has generator $\overline{K}$.

Here is a recent variant on this result, obtained by the global semigroup methods mentioned in the notes to Section 5.

Theorem. *Existence of a strong contractive solution of* (*) *on* $\mathcal{D}\underline{\otimes}\mathcal{E}_D$ *is assured if condition* (A) *holds and furthermore*

(B)$'$ $K_{0,d}$ *is a pregenerator of a* c_0-*semigroup on* $\mathfrak{h}$, *for each* $d \in$ T,

where T *is any subset of* k *containing* 0 *which linearly spans* D.

Remark. In fact, from these hypotheses it follows that K_d^c is a pregenerator of a c_0-contraction semigroup for each $c \in \mathsf{k}$ and $d \in D$.

Exercise. *What are all these contraction semigroups being generated here?*

Question. To what extent do non-Markov-regular contraction cocycles satisfy QSDE's? Jean-Lin Journé gave an example to show that a strongly continuous contraction cocycle need not do so. Here are sufficient conditions (which are also necessary) in terms of its associated semigroups. It is not hard to verify that if a contraction cocycle is strongly continuous then its associated semigroups are c_0-contraction semigroups (i.e. they are strongly continuous too).

Theorem. *Let* X *be a strongly continuous left Markovian cocycle on* $\mathfrak{h}$ *with associated semigroup generators* $\{\mathcal{G}_{c,d} : c, d \in \mathsf{k}\}$, *and let* T *be a subset of* k *containing* 0. *If*

$$\mathcal{D} := \bigcap_{d\in T} \mathrm{Dom}\, \mathcal{G}_{0,d} \ \text{ and }\ D := \mathrm{Lin}\, T$$

are dense in $\mathfrak{h}$ *and* k *respectively, then* X *satisfies the left QSDE* (*) *on* $\mathcal{D}\underline{\otimes}\mathcal{E}_D$ *for some operator* F *with domain* $\hat{D}\otimes\mathcal{D}$.

Remark. Under the conditions of the theorem, $\mathrm{Dom}\, \mathcal{G}_{c,d} \supset \mathcal{D}$ for all $c \in \mathsf{k}$ and $d \in D$ (cf. the remark following the previous theorem).

Acknowledgements

I am grateful to Michael Schürmann and Uwe Franz for the opportunity to give a course of lectures in the Greifswald Spring School on Quantum Independent Increment Processes in March 2003; also to the students of the School, particularly Roman Rozhin, Adam Skalski and Lisa Steiner, whose keen participation made lecturing a pleasure. The notes for that course have been revised and expanded for publication. I am grateful to Nils Gebhardt, Rolf Gohm, Orawan Sanhan, and A.K. Vijayarajan for generously combing for misprints. Special thanks go to Alex Belton, Robin Hudson, Adam Skalski, Nick Weatherall and Stephen Wills whose thoughtful contributions have improved these notes considerably. Thanks to Steve also for agreeing to the inclusion of some unpublished work (in the Afterword). Finally, I am indebted to Cathie Shipley for typing the original lecture notes, and for her good-humoured assistance in producing these, and to Uwe Franz for his patience and great editorial work.

Special notations (from Section 2 onwards)

k   A fixed Hilbert space, the *noise dimension space*
K $L^2(\mathbb{R}_+; k)$

Special notations for Section 6

$\mathfrak{h}$ Another fixed Hilbert space, the *system space*
$\mathcal{A}$ A fixed C^*-algebra acting nondegenerately on $\mathfrak{h}$

General notations and conventions

$\mathrm{Map}(S; T)$ The set of all functions from a set S to a set T
$\mathrm{Ran}\, f$ Range=Image of a function f
$F_{[r,t[}$ Function $s \mapsto \begin{cases} F(s) & \text{if } s \in [r,t[\\ 0 & \text{otherwise} \end{cases}$,
       for a vector-valued function F defined on (part of) the real line
$\delta^\dagger$ Map $x \mapsto \delta(x^*))^*$,
       for a linear map δ between involutive spaces, see page 250
$\mathrm{Lin}\, S$ Linear span, for a subset S of a vector space
$\mathrm{Dom}\, T$ Domain of an operator T, see page 197
$\overline{\mathrm{Lin}\, S}$ Closure of $\mathrm{Lin}\, S$, for a subset S of a normed space
$\overline{\mathrm{Ran}\, f}$ Closure of $\mathrm{Ran}\, f$, for a normed space-valued function f
$\overline{\mathscr{S}}^{\mathrm{uw}}$ Ultraweak closure, for a subset $\mathscr{S}$ of $B(\mathsf{H}; \mathsf{H}')$
$\mathscr{S}'$ Commutant: $\{A \in B(\mathsf{H}) : \forall_{X \in \mathscr{S}} AX = XA\}$,
       for a subset $\mathscr{S}$ of $B(\mathsf{H})$
$\mathscr{S}''$ Double commutant $(\mathscr{S}')'$.
$\langle\,,\,\rangle$ Inner products are linear in their *second* argument
$\widehat{\mathsf{h}}$ $\mathbb{C} \oplus \mathsf{h}$, for a Hilbert space h
$\widehat{c}$ $\binom{1}{c} \in \widehat{\mathsf{h}}$, for $c \in \mathsf{h}$
P_k Orthogonal projection in $B(\widehat{\mathsf{k}})$ with range $\{\binom{0}{c} : c \in \mathsf{k}\}$
Δ $I_\mathsf{H} \otimes P_\mathsf{k} \otimes I_{\mathsf{H}'} \in B(\mathsf{H} \otimes \widehat{\mathsf{k}} \otimes \mathsf{H}')$,
       with H and H' determined by context
$\Delta(x)$ $P_\mathsf{k} \otimes x$, for $x \in B(\mathsf{h}; \mathsf{h}')$, for Hilbert spaces h, h'

$\langle u|$ Dirac "bra-": element of $H^* = B(H; \mathbb{C})$ given by $v \mapsto \langle u, v \rangle$

$|u\rangle$ Dirac "-ket": element of $B(\mathbb{C}; H)$ given by $\lambda \mapsto \lambda u$

E_u $I_h \otimes |u\rangle \otimes I_{h'}$, where h and h' are determined by context

E^u $(E_u)^* = I_h \otimes \langle u| \otimes I_{h'}$

$M_{n,m}$, $M_{n,m}(\mathbb{C})$,
            with norm given by its usual identification with $B(\mathbb{C}^m; \mathbb{C}^n)$

$L(U; V)$ Vector space of linear maps between vector spaces U and V

$B(X; Y)$ Normed space of bounded operators,
            between normed spaces X and Y

$CB(V; W)$ Operator space of completely bounded operators in $B(V; W)$

$M(k, h; V)_b$ h-k-matrix space over V, see page 195

$\otimes$ Algebraic tensor product

$\otimes_{sp}$ Spatial tensor product, see page 194

$\overline{\otimes}$ Ultraweak tensor product, see page 194

$\otimes_M$ Matrix-space tensor product, see (1.16)

ι_h Ampliation maps $T \mapsto I_h \otimes T$, or $T \otimes I_h$,
            depending on context

$\mathbb{P}$ and $\mathbb{P}_b$ Classes of (bounded) processes, see pages 216 to 220

References

[Acc]   L. Accardi, On the quantum Feynman-Kac formula, *Rend. Sem. Mat. Fis. Milano* **48** (1978), 135–180 (1980).

[AcK]   L. Accardi and S.V. Kozyrev, On the structure of Markov flows, *Chaos Solitions Fractals* **12** (2001) no. 14-15, 2639–2655.

[AFL]   L. Accardi, A. Frigerio and J.T. Lewis, Quantum stochastic processes, *Publ. Res. Inst. Math. Sci.* **18** (1982) no. 1, 97–133.

[AJL]   L. Accardi, J.-L. Journé and J.M. Lindsay, On multidimensional Markovian cocycles, *in* [QP$_4$], pp. 59–67.

[AcM]   L. Accardi and A. Mohari, On the structure of classical and quantum flows, *J. Funct. Anal.* **135** (1996), 421–455.

[App]   D.B. Applebaum, Unitary evolutions and horizontal lifts in quantum stochastic calculus, *Comm. Math. Phys.* **140** (1991) no. 1, 63–80.

[ApH]   D.B. Applebaum and R.L. Hudson, Fermion Itô's formula and stochastic evolutions, *Comm. Math. Phys.* **96** (1984) no. 4, 473–496.

[Arv]   W. Arveson, Continuous analogues of Fock space, *Mem. Amer. Math. Soc* **80** (1989), no. 409.

[Att$_1$]   S. Attal, Problèmes d'unicité dans les représentations d'operateurs sur l'espace de Fock, *in* [Sém$_{26}$], pp. 619–632.

[Att$_2$]   S. Attal, An algebra of non-commutative bounded semimartingales, Square and angle quantum brackets, *J. Funct. Anal.* **124** (1994) no. 2, 292–332.

[Att$_3$]   S. Attal, The structure of the quantum semimartingale algebras, *J. Operator Theory* **46** (2001) no. 2, 391–410.

[Att$_4$]   S. Attal, Extensions of quantum stochastic calculus, *in* [QP$_{11}$], pp. 1–37.

[AtB]   S. Attal and A. Bernard, Orthogonal spaces associated with exponentials of indicator functions on Fock space, *J. London Math. Soc.* (2) **66** (2002) no. 2, 487–498.

[AtL]   S. Attal and J.M. Lindsay, Quantum stochastic calculus with maximal operator domains, *Ann. Probab.* **32** (2004) no. 1A, 488–529.

[AtM]   S. Attal and P.-A. Meyer, Interprétation probabiliste et extension des intégrales stochastiques non commutatives, *in* [Sém27], pp. 312–327.

[Bel1]   V.P. Belavkin, A quantum nonadapted Itô formula and stochastic analysis in Fock scale, *J. Funct. Anal.* **102** (1991), 414–447.

[Bel2]   V.P. Belavkin, Quantum stochastic positive evolutions: characterization, construction, dilation, *Comm. Math. Phys.* **184** (1997) no. 3, 533-566.

[Belt1]   A.C.R. Belton, Quantum Ω-semimartingales and stochastic evolutions, *J. Funct. Anal.* **187** (2001) no. 1, 94–109.

[Belt2]   A.C.R. Belton, An isomorphism of quantum semimartingale algebras, *Quart. J. Math.* **55** (2004) no. 2, 135–165.

[Bha]   B.V.R. Bhat, Cocycles of CCR flows, *Mem. Amer. Math. Soc.* **149** (2001), no. 709.

[BhL]   B.V.R. Bhat and J.M. Lindsay, Regular quantum stochastic cocycles have exponential product systems, *in* [QP18] (to appear).

[Bia]   Ph. Biane, Calcul stochastique non-commutatif, *in*, "Lectures on Probability Theory: Ecole d'Eté de Probabilités de Saint-Flour XXIII — 1993," P. Bernard, ed., Lecture Notes in Mathematics **1608**, Springer, Berlin 1995.

[BiS]   Ph. Biane and R. Speicher, Stochastic calculus with respect to free Brownian motion and analysis in Wigner space, *Probab. Theory Related Fields* **112** (1998) no. 3, 373–409.

[Bra]   W.S. Bradshaw, Stochastic cocycles as a characterisation of quantum flows, *Bull. Sci. Math.* (2) **116** (1992), 1–34.

[BR1]   O. Bratteli and D.W. Robinson, "Operator Algebras and Quantum Statistical Mechanics I, C^*- and W^*-algebras, Symmetry Groups, Decomposition of States," Texts and Monographs in Physics, Springer-Verlag, New York–Heidelberg 1979.

[BR2]   O. Bratteli and D.W. Robinson, "Operator Algebras and Quantum Statistical Mechanics II, Equilibrium states. Models in quantum statistical mechanics," Texts and Monographs in Physics, Springer-Verlag, New York–Heidelberg 1981.

[BSW1]   C. Barnett, I.F. Wilde and R.F. Streater, The Itô-Clifford integral, *J. Funct. Anal.* **48** (1982) no. 2, 172–212.

[BSW2]   C. Barnett, I.F. Wilde and R.F. Streater, Quasifree quantum stochastic integrals for the CAR and CCR, *J. Funct. Anal.* **52** (1983) no. 1, 19–47.

[CaK]   E.A. Carlen and P. Krée, On martingale inequalities in non-commutative stochastic analysis, *J. Funct. Anal.* **158** (1998) no. 2, 475–508.

[CEH]   P.B. Cohen, T.M.W. Eyre and R.L. Hudson, Higher order Itô product formula and generators of evolutions and flows, *Internat. J. Theoret. Phys.* **34** (1995) no. 8, 1481–1486.

[Chr]   E. Christensen, Generators of semigroups of completely positive maps, *Comm. Math. Phys.* **62** (1978) no. 2, 167–171.

[ChE]   E. Christensen and D.E. Evans, Cohomology of operator algebras and quantum dynamical semigroups, *J. London Math. Soc.* **20** (1979) no. 2, 358–368.

[CoH]   A.M. Cockroft and R.L. Hudson, Quantum mechanical Wiener processes, *J. Multivariate Anal.* **7** (1977) no. 1, 107–124.

[Coo]   J.M. Cook, The mathematics of second quantization, *Trans. Amer. Math. Soc.* **74** (1953), 222–245.

[Coq$_1$] A. Coquio, Why there are only three quantum noises, *Probab. Theory Related Fields* **118** (2000) no. 3, 349–364.

[Coq$_2$] A. Coquio, Stochastic integral representation of unbounded operators in Fock spaces, *Prépublication de l'Institut Fourier, Grenoble*, **517** (2002).

[DaZ] G.I. Da Prato and J. Zabczyk, "Stochastic Equations in Infinite Dimensions," Encyclopedia of Mathematics and its Applications **45**, Cambridge Universite Press, Cambridge 1992.

[DaS] P.K. Das and K.B. Sinha, Quantum flows with infinite degrees of freedom and their perturbations, *in* [QP$_7$], pp. 109–123.

[Dav] E.B. Davies, "One-Parameter Semigroups," London Mathematical Society Monographs **15**, Academic Press, London 1980.

[Dix$_1$] J. Dixmier, "C^*-algebras," (Transl. from 2nd French edn. by F. Jellett, with corr. and enlarged biblio.) North-Holland Mathematical Library **15**, North-Holland Publishing, Amsterdam 1977.

[Dix$_2$] J. Dixmier, "von Neumann Algebras," With a preface by E.C. Lance, (Transl. from 2nd French edn. by F. Jellett) North-Holland Mathematical Library **27**, North-Holland, Amsterdam 1981.

[EfR] E.G. Effros and Z.-J. Ruan, "Operator Spaces," London Mathematical Society Monographs, New Series **23**, Oxford University Press, 2000.

[Eme] M. Emery, Classical probability theory – an outline of stochastic integrals and diffusions, *in* [QP$_{11}$], pp. 87–121.

[Eva] M.P. Evans, Existence of quantum diffusions, *Probab. Theory Related Fields* **81** (1989) no. 4, 473–483.

[EvH] M.P. Evans and R.L. Hudson, Perturbations of quantum diffusions, *J. London Math. Soc.* (2) **41** (1990) no. 2, 373–384.

[EvL] D.E. Evans and J.T. Lewis, Dilations of irreversible evolutions in algebraic quantum theory, *Comm. Dublin Inst. Adv. Studies, Ser. A (Theor. Phys.)* **24**, 1977.

[EvK] D.E. Evans and Y. Kawahigashi, "Quantum Symmetries on Operator Algebras," Oxford Mathematical Monographs. Oxford University Press, New York 1998.

[EyH] T.M.W. Eyre and R.L. Hudson, Representations of Lie superalgebras and generalised Boson-Fermion equivalence in quantum stochastic calculus, *Comm. Math. Phys.* **186** (1997), 87–94.

[Fag$_1$] F. Fagnola, On quantum stochastic differential equations with unbounded coefficients, *Probab. Theory Related Fields* **86** (1990) no. 4, 501–516.

[Fag$_2$] F. Fagnola, Characterization of isometric and unitary weakly differentiable cocycles in Fock space, *in* [QP$_8$], pp. 143–164.

[Fag$_3$] F. Fagnola, Quantum stochastic differential equations, *in* [QP$_{11}$], pp. 123–170.

[FW$_1$] F. Fagnola and S.J. Wills, Mild solutions of quantum stochastic differential equations, *Electron. Comm. Probab.* **5** (2000), 158–171.

[FW$_2$] F. Fagnola and S.J. Wills, Solving quantum stochastic differential equations with unbounded coefficients, *J. Funct. Anal.* **198** (2003) no. 2, 279–310.

[Fan] M. Fannes, Canonical commutation and anticommutation relations, *in* [QP$_{11}$], pp. 171–198.

[FeL] J.F. Feinstein and J.M. Lindsay, Exponentials of indicators of intervals are not total, *Unpublished note* (2000).

[Foc] V. Fock, Konfigurationsraum und Zweite Quantelung, *Zeitschrift für Physik* **75** (1932), 622–647.

[GKS]   V. Gorini, A Kossakowski and E.C.G. Sudarshan, Completely positive dynamical semigroups of N-level systems, *J. Mathematical Phys.* **17** (1976) no. 5, 821–825.

[GLSW] D. Goswami, J.M. Lindsay, K.B. Sinha and S.J. Wills, Dilation of Markovian cocycles on a von Neumann algebra, *Pacific J. Math.* **211** (2003) no. 2, 221–247.

[GLW]   D. Goswami, J.M. Lindsay and S.J. Wills, A stochastic Stinespring theorem, *Math. Ann.* **319** (2001) no. 4, 647–673.

[GPS]   D. Goswami, A.K. Pal and K.B. Sinha, Stochastic dilation of a quantum dynamical semigroup on a separable unital C^*-algebra, *Infin. Dimens. Anal. Quantum Probab. Relat. Top.* **3** (2000) no. 1, 177-184.

[GoS]   D. Goswami and K.B. Sinha, Hilbert modules and stochastic dilation of a quantum dynamical semigroup on a von Neumann algebra, *Comm. Math. Phys.* **205** (1999) no. 2, 377–405.

[Gui]   A. Guichardet, "Symmetric Hilbert Spaces and Related Topics. Infinitely divisible positive definite functions. Continuous products and tensor products. Gaussian and Poissonian stochastic processes," Lecture Notes in Mathematics **261**, Springer-Verlag, Berlin 1972.

[HHKKR] J. Hellmich, R. Honneger, C. Köstler, B. Kümmerer and A. Rieckers, Couplings to classical and non-classical squeezed white noise as stationary Markov processes, *Publ. Res. Inst. Math. Sci.* **38** (2002) no. 1, 1–31.

[HKK]   J. Hellmich, C. Köstler and B. Kümmerer, Noncommutative continuous Bernoulli shifts, *Preprint, Queen's University, Kingston* (2004).

[Hid]   T. Hida, "Brownian Motion," (Transl. from Japanese edn. by author and T.P. Speed, with an added chapter) Applications of Mathematics **11**, Springer-Verlag, Berlin 1980.

[HiP]   E. Hille and R.S. Phillips, "Functional Analysis and Semi-groups," Third printing of revised 1957 edn., American Mathematical Society Colloquium Publications, Vol. XXXI, AMS, Providence, R. I. 1974.

[Hud]   R.L. Hudson, An introduction to quantum stochastic calculus and some of its applications, *in* [QP$_{11}$], pp. 221–271.

[HL$_1$]   R.L. Hudson and J.M. Lindsay, A noncommutative martingale representation theorem for non-Fock quantum Brownian motion, *J. Funct. Anal.* **61** (1985) no. 2, 202–221.

[HL$_2$]   R.L. Hudson and J.M. Lindsay, Uses of non-Fock quantum Brownian motion and a quantum martingale representation theorem, *in* [QP$_2$], pp. 276–305.

[HL$_3$]   R.L. Hudson and J.M. Lindsay, The classical limit of reduced quantum stochastic evolutions, *Ann. Inst. H. Poincaré Phys. Théor.* **43** (1985) no. 2, 133–145.

[HL$_4$]   R.L. Hudson and J.M. Lindsay, On characterizing quantum stochastic evolutions, *Math. Proc. Cambridge Philos. Soc.* **102** (1987) no. 2, 363–369.

[HP$_1$]   R.L. Hudson and K.R. Parthasarathy, Quantum Itô's formula and stochastic evolutions, *Comm. Math. Phys.* **93** (1984) no. 3, 301–323.

[HP$_2$]   R.L. Hudson and K.R. Parthasarathy, Stochastic dilations of uniformly continuous completely positive semigroups, *Acta Appl. Math.* **2** (1984) no. 3-4, 353–378.

[HP$_3$]   R.L. Hudson and K.R. Parthasarathy, Unification of fermion and boson stochastic calculus, *Comm. Math. Phys.* **104** (1986) no. 3, 457–470.

[HPu]   R.L. Hudson and S. Pulmannová, Chaotic expansion of elements of the universal enveloping algebra of a Lie algebra associated with a quantum stochastic calculus, *Proc. London Math. Soc.* (3) **77** (1998) no. 2, 462–480.

[Ji]   U.C. Ji, Stochastic integral representation theorem for quantum semimartingales, *J. Funct. Anal.* **201** (2003) no. 1, 1–29.

[Jou]   J.-L. Journé, Structure des cocycles markoviens sur l'espace de Fock, *Probab. Theory Related Fields* **75** (1987) no. 2, 291–316.

[KR$_{1,2}$] R. Kadison and J.R. Ringrose, "Fundamentals of the theory of operator algebras I & II," (Reprint of 1983 edn., resp. corrected reprint of 1986 edn.) Graduate Studies in Mathematics **15** & **16**, AMS, Providence, RI 1997.

[KR$_{3,4}$] R. Kadison and J.R. Ringrose, "Fundamentals of the theory of operator algebras III & IV," Birkhäuser, Boston, MA 1991 & 1992.

[KüS]   B. Kümmerer and R. Specicher, Stochastic integration on the Cuntz algebra O_∞, *J. Funct. Anal.* **103** (1992) no. 2, 372–408.

[Lan]   E.C. Lance, "Hilbert C^*-modules. A toolkit for operator algebraists," London Mathematical Society Lecture Note Series **210**, CUP, Cambridge 1995.

[Lin]   G. Lindblad, On the generators of quantum dynamical semigroups, *Comm. Math. Phys.* **48** (1976) no. 2, 119–130.

[L$_1$]   J.M. Lindsay, Fermion martingales, *Probab. Theory Relat. Fields* **71** (1986) no. 2, 307–320.

[L$_2$]   J.M. Lindsay, On set convolutions and integral-sum kernel operators, *in* "Probability Theory and Mathematical Statistics II," Proceedings, Fifth Vilnius Conference, 1989 *B. Grigelionis, Yu V. Prohorov, V.V. Sazanov and V. Stratulevicius, eds.*, VSP, Utrecht 1990, pp. 105–123.

[L$_3$]   J.M. Lindsay, Independence for quantum stochastic integrators, *in* [QP$_6$], pp. 325–332; Addendum, *in* [QP$_{10}$], pp. 363.

[L$_4$]   J.M. Lindsay, Quantum and non-causal stochastic calculus, *Probab. Theory Related Fields* **97** (1993) no. 1-2, 65–80.

[L$_5$]   J.M. Lindsay, Gaussian hypercontractivity revisited, *J. Funct. Anal.* **92** (1990) no. 2, 313–324.

[L$_6$]   J.M. Lindsay, On the algebraic structure of quantum stochastic calculus, *Tatra Mt. Math. Publ.* **10** (1997), 281-290.

[L$_7$]   J.M. Lindsay, Integral-sum kernel operators, *in* [QP$_{12}$], pp. 1–21.

[LMa]   J.M. Lindsay and H. Maassen, Stochastic calculus for quantum Brownian motion of nonminimal variance—an approach using integral-sum kernel operators, *in* Mark Kac Seminar on Probability and Physics, CWI Syllabi **32** Math. Centrum, Centrum Wisk. Inform., Amsterdam 1992, pp. 97–167.

[LMe]   J.M. Lindsay and P.-A. Meyer, Fermionic hypercontractivity, *in* [QP$_7$], pp. 211–220.

[LP$_1$] J.M. Lindsay and K.R. Parthasarathy, Cohomology of power sets with applications in quantum probability, *Comm. Math. Phys.* **124** (1989) no. 3, 337–364.

[LP$_2$] J.M. Lindsay and K.R. Parthasarathy, On the generators of quantum stochastic flows, *J. Funct. Anal.* **158** (1998) no. 2, 521–549.

[LSa]   J.M. Lindsay and O. Sanhan, A note on matrix spaces, *Preprint* (2005).

[LSi]   J.M. Lindsay and K.B. Sinha, Feynman-Kac representation of some noncommutative elliptic operators, *J. Funct. Anal.* **147** (1997) no. 2, 400–419.

[LSk]   J.M. Lindsay and A.G. Skalski, Quantum stochastic convolution cocycles, *Ann. Inst. H. Poincaré Probab. Statist.* Special issue, in memory of Paul-André Meyer (to appear).

[LWi]   J.M. Lindsay and I.F. Wilde, On non-Fock boson stochastic integrals, *J. Funct. Anal.* **65** (1986) no. 1, 76–82.

[LW$_1$]   J.M. Lindsay and S.J. Wills, Existence, positivity, and contractivity for quantum stochastic flows with infinite dimensional noise, *Probab. Theory Related Fields* **116** (2000) no. 4, 505–543.

[LW$_2$]   J.M. Lindsay and S.J. Wills, Markovian cocycles on operator algebras, adapted to a Fock filtration, *J. Funct. Anal.* **178** (2000) no. 2, 269–305.

[LW$_3$]   J.M. Lindsay and S.J. Wills, Existence of Feller cocycles on a C^*-algebra, *Bull. London Math. Soc.* **33** (2001) no. 5, 613–621.

[LW$_4$]   J.M. Lindsay and S.J. Wills, Homomorphic Feller cocycles on a C^*-algebra, *J. London Math. Soc.* (2) **68** (2003) no. 1, 255–272.

[LW$_5$]   J.M. Lindsay and S.J. Wills, Multiplicativity via a hat trick, *in* [QP$_{15}$], pp. 181–193.

[LW$_6$]   J.M. Lindsay and S.J. Wills, Markovian cocycles and semigroups on operator spaces, *Preprint* (2004).

[LW$_7$]   J.M. Lindsay and S.J. Wills, Operator Markovian cocycles via asociated semigroups, *Preprint* (2004).

[LW$_8$]   J.M. Lindsay and S.J. Wills, On the generators of completely positive Markovian cocycles, *Preprint* (2004).

[Maa]   H. Maassen, Quantum Markov processes on Fock space described by integral kernels, *in* [QP$_2$], pp. 361–374.

[Mey]   P.-A. Meyer, "Quantum Probability for Probabilists," 2nd Edn., Lecture Notes in Mathematics **1538**, Springer-Verlag, Berlin 1995.

[Moh]   A. Mohari, Quantum stochastic differential equations with unbounded coefficients and dilations of Feller's minimal solution, *Sankhyā Ser. A* **53** (1991) no. 3, 255–287.

[MoP]   A. Mohari and K.R. Parthasarathy, A quantum probabilistic analogue of Feller's condition for the existence of unitary Markovian cocycles in Fock spaces, *in* "Statistics and Probability: A Raghu Raj Bahadur Festschrift," *J.K. Ghosh, S.K. Mitra, K.R. Parthasarathy and B.L.S. Prakasa Rao, eds.*, Wiley Eastern, New Delhi 1993, pp. 475–497.

[MoS]   A. Mohari and K.B. Sinha, Quantum stochastic flows with infinite degrees of freedom and countable state Markov processes, *Sankhyā Ser. A* **52** (1990) no. 1, 43–57.

[Mur$_1$] G.J. Murphy, "C^*-algebras and operator theory," Academic Press, Boston 1990.

[Mur$_2$] G.J. Murphy, Positive definite kernels and Hilbert C^*-modules, *Proc. Edinburgh Math. Soc.* (2) **40** (1997), no. 2, 367–374.

[Par]   K.R. Parthasarathy, "An Introduction to Quantum Stochastic Calculus," Monographs in Mathematics **85**, Birkhäuser Verlag, Basel 1992.

[PSi]   K.R. Parthasarathy and K.B. Sinha, Stochastic integral representation of bounded quantum martingales in Fock space, *J. Funct. Anal.* **67** (1986) no. 1, 126–151.

[PSu]   K.R. Parthasarathy and V.S. Sunder, Exponentials of indicator function are total in the boson Fock space $\Gamma(L^2[0,1])$, *in* [QP$_{10}$], pp. 281–284.

[Pau]   V. Paulsen, "Completely bounded maps and operator algebras," Cambridge Studies in Advanced Mathematics **78**, CUP, Cambridge 2002.

[Paut]   Y. Pautrat, Stochastic integral representations of second quantization operators, *J. Funct. Anal.* **208** (2004) no. 1, 163–193.

[Ped]   G.K. Pedersen, "C^*-algebras and their Automorphism Groups," London Mathematical Society Monographs **14**, Academic Press, London 1979.

[Pet]   D. Petz, "An invitation to the algebra of canonical commutation relations," Leuven Notes in Mathematical and Theoretical Physics, Series A: Mathematical Physics **2**, Leuven University Press, Leuven 1990.

[Pin]   M.A. Pinsky, Stochastic integral representation of multiplicative operator functionals of a Wiener process, *Trans. Amer. Math. Soc.* **167** (1972), 89–104.

[Pis]   G. Pisier, "Introduction to Operator Space Theory," London Mathematical Society Lecture Note Series **294**, CUP, Cambridge 2003.

[QP$_2$]   "Quantum Probability and Applications II," Proceedings, Heidelberg 1984. *L. Accardi & W. von Waldenfels, eds.*, Lecture Notes in Mathematics **1136**, Springer-Verlag, Berlin 1985.

[QP$_4$]   "Quantum Probability and Applications IV," Proceedings, Rome Symposium 1986-1987, *L. Accardi & W. von Waldenfels, eds.*, Lecture Notes in Mathematics **1396**, Springer-Verlag, Heidelberg 1989.

[QP$_6$]   "Quantum Probability and Related Topics VI," *L. Accardi, ed.*, World Scientific, Singapore 1991.

[QP$_7$]   "Quantum Probability and Related Topics VII," *L. Accardi, ed.*, World Scientific, Singapore 1992.

[QP$_8$]   "Quantum Probability and Related Topics VIII," *L. Accardi ed.*, World Scientific, Singapore 1993.

[QP$_{10}$]   "Quantum Probability and Communications X," *R.L. Hudson & J.M. Lindsay, eds.*, World Scientific, Singapore 1998.

[QP$_{11}$]   "Quantum Probability Communications XI," Summer School, Grenoble 1998, *S. Attal & J.M. Lindsay, eds.*, World Scientific, Singapore 2003.

[QP$_{12}$]   "Quantum Probability Communications XII," Summer School, Grenoble 1998, *S. Attal & J.M. Lindsay, eds.*, World Scientific, Singapore 2003.

[QP$_{15}$]   "Quantum Probability and White Noise Analysis XV," Proceedings, Burg/Spreewald 2001, *W. Freudenberg, ed.*, World Scientific, Singapore 2003.

[QP$_{18}$]   "Quantum Probability and White Noise Analysis XVIII," Proceedings, Greifswald, 2003, *U. Franz & M. Schürmann, eds.*, World Scientific, Singapore 2005 (to appear).

[QSC]   Quantum stochastic calculus, 2000 Mathematics Subject Classification 81S25, *Mathematical Reviews*.

[RS$_1$]   M. Reed and B. Simon, "Methods of Modern Mathematical Physics I: Functional Analysis," 2nd Edn., Academic Press, New York–London 1980.

[RS$_2$]   M. Reed and B. Simon, "Methods of Modern Mathematical Physics II: Fourier analysis, Self-Adjointness," Academic Press New York 1975.

[Sak]   S. Sakai, "C^*-algebras and W^*-algebras," (Reprint of 1971 edn.) Classics in Mathematics, Springer, Berlin 1998.

[Sau]   J.-L. Sauvageot, A survey of operator algebras, *in* [QP$_{12}$], pp. 173–194.

[Sch]   M. Schürmann, "White Noise on Bialgebras," Lecture Notes in Mathematics **1544**, Springer-Verlag, Berlin 1993.

[Seg]   I.E. Segal, Tensor algebras over Hilbert spaces, *Ann. of Math.* (2) **63** (1956), 160–175.

[Sém$_{26}$]   "Séminaire de Probabilités XXVI," *J. Azéma, P.-A. Meyer & M. Yor, eds.*, Lecture Notes in Mathematics **1526**, Springer, Berlin 1992.

[Sém27] "Séminaire de Probabilités XXVII," *J. Azéma, P.-A. Meyer & M. Yor, eds.*, Lecture Notes in Mathematics **1557**, Springer, Berlin 1993.

[Sen] I.R. Senitzky, Dissipation in quantum mechanics. The harmonic oscillator I, *Phys. Rev.* (2) **119** 1960 670–679.

[Ske] M. Skeide, Indicator functions of intervals are totalising in the symmetric Fock space $\Gamma(L^2(\mathbb{R}_+))$, *in* "Trends in Contemporary Infinite Dimensional Analysis and Quantum Probability," Volume in honour of Takeyuki Hida, *L. Accardi, H.-H. Kuo, N. Obata, K. Saito, Si Si & L. Streit, eds.*, Instituto Italiano di Cultura (ISEAS), Kyoto 2000.

[Spe] R. Speicher, Free calculus, *in* [QP$_{12}$], pp. 209–235.

[StZ] S. Strătilă and L. Zsidó, "Lectures on von Neumann algebras," (Revision of 1975 original; transl. from the Romanian by S. Teleman) Abacus Press, Tunbridge Wells 1979.

[Sun] V. Sunder, "An Invitation to von Neumann Algebras," Universitext, Springer-Verlag, New York 1987.

[Tak] M. Takesaki, "Theory of Operator Algebras I," (Reprint of 1979 edn.) Encyclopaedia of Mathematical Sciences **124**, Operator Algebras and Non-commutative Geometry **5**, Springer-Verlag, New York 2002.

[Vin$_1$] G.F. Vincent-Smith, Unitary quantum stochastic evolutions, *Proc. London Math. Soc.* (3) **63** (1991) no. 2, 401–425.

[Vin$_2$] G.F. Vincent-Smith, The Itô formula for quantum semimartingales, *Proc. London Math. Soc.* (3) **75** (1997) no. 3, 671–720.

[VDN] D.V. Voiculescu, K.J. Dykema and A. Nica, "Free Random variables," CRM Monograph Series **1**, American Mathematical Society, Providence R.I. 1992.

[Weg] N.E. Wegge-Olsen, "K-theory and C^*-algebras. A friendly approach," Oxford Science Publications, Clarendon Press, New York 1993.

Dilations, Cocycles and Product Systems

B. V. Rajarama Bhat

Indian Statistical Institute
Bangalore, India
bhat@isibang.ac.in

Notation: Throughout these lectures $\mathcal{B}(\mathcal{H})$ will denote the von Neumann algebra of all bounded operators on a Hilbert space $\mathcal{H}$. All our Hilbert spaces will be complex with an inner product $\langle \cdot, \cdot \rangle$, which is anti-linear in the first variable. Usually we restrict ourselves to separable Hilbert spaces.

1 Dilation theory basics

We begin with the most basic theorem in dilation theory, which shows that contractions on Hilbert spaces are corners of isometries.

Theorem 1.1. *(Sz. Nagy's dilation theorem): Suppose that $T \in \mathcal{B}(\mathcal{H})$ with $\|T\| \leq 1$ for some Hilbert space $\mathcal{H}$. Then there exists a Hilbert space $\mathcal{K}$ containing $\mathcal{H}$ with an isometry $V \in \mathcal{B}(\mathcal{K})$ such that,*

$$T^n = P_{\mathcal{H}} V^n|_{\mathcal{H}} \quad \forall n \geq 0. \tag{1.1}$$

Moreover, if $\overline{\mathrm{span}}\{V^n u : n \geq 0, u \in \mathcal{H}\} = \mathcal{K}$ the pair $(\mathcal{K}, V)$ is unique up to unitary equivalence in the sense that if $(\mathcal{K}', V')$ is another such pair, then there exists a unitary $U : \mathcal{K} \to \mathcal{K}'$, such that $Uu = u$ for $u \in \mathcal{H}$, and $V' = UVU^*$.

If we have an operator V as in Theorem 1.1, then it is called a (power) dilation of T. One can also construct *unitary* dilations for contractions but we will not talk about them!

Our main tool for proving all dilation theorems will be the Kolmogorov map construction, which shows when we can embed a set inside a Hilbert space (see J. M. Lindsay [Li] or K. R. Parthasarathy [Pa]). We shall use only complex-valued kernels.

Definition 1.2. *Let $\mathcal{M}$ be a set. A map $K : \mathcal{M} \times \mathcal{M} \to \mathbb{C}$ is called a positive definite kernel if*

$$\sum_{i,j} \bar{c}_i c_j K(x_i, x_j) \geq 0$$

for all $c_1, c_2, \ldots, c_n \in \mathbb{C}$, $x_1, x_2, \ldots, x_n \in \mathcal{M}$ and $n \geq 1$.

In other words K on $\mathcal{M} \times \mathcal{M}$ is a positive definite kernel if the matrix $[K(x_i, x_j)]$ is positive definite for all choices of a finite number of points $x_1, x_2, \ldots, x_n$ from $\mathcal{M}$.

Theorem 1.3. *(Kolmogorov map / GNS Construction): Suppose K is a positive definite kernel on a set $\mathcal{M}$. Then there exists a Hilbert space $\mathcal{K}$ with a mapping $\lambda : \mathcal{M} \to \mathcal{K}$ such that*

$$\langle \lambda(x), \lambda(y) \rangle = K(x, y)$$

for all x, y in $\mathcal{M}$. Moreover, if

$$\overline{span}\{\lambda(x) : x \in \mathcal{M}\} = \mathcal{K},$$

then the pair $(\mathcal{K}, \lambda)$ is unique up to unitary equivalence, that is, if $(\mathcal{K}', \lambda')$ is another such pair, then there exists a unitary operator $U : \mathcal{K} \to \mathcal{K}'$ such that $U\lambda(x) = \lambda'(x)$ for all $x \in \mathcal{M}$.

Sketch of a Proof of Theorem 1.1: Suppose we had a dilation $(\mathcal{K}, V)$ as above. Then we see that for vectors u, v in $\mathcal{H}$,

$$\langle V^m u, V^n v \rangle = \begin{cases} \langle u, T^{n-m} v \rangle & m \leq n \\ \langle u, (T^*)^{m-n} v \rangle & n < m. \end{cases}$$

This suggests that we take $\mathcal{M}$ as the set $\{(m, u) : m \geq 0, u \in \mathcal{H}\}$, and define $K : \mathcal{M} \times \mathcal{M} \to \mathbb{C}$ by

$$K((m, u), (n, v)) = \begin{cases} \langle u, T^{n-m} v \rangle & m \leq n \\ \langle u, (T^*)^{m-n} v \rangle & n < m. \end{cases}$$

Next note that positive definiteness of K is equivalent to the block operator matrix $[A_{ij}]$, defined by

$$A_{ij} = \begin{cases} T^{j-i} & 0 \leq i \leq j \leq n \\ (T^*)^{i-j} & 0 \leq i < j \leq n, \end{cases}$$

being positive for all orders n. And this can be proved through simple matrix manipulations and induction. So we can apply the GNS construction to have a

Hilbert space $\mathcal{K}$ with a Kolmogorov map $\lambda : \mathcal{M} \to \mathcal{K}$ for the kernel K. We see that by identifying u in $\mathcal{H}$ with $\lambda(0, u)$ in $\mathcal{K}$ we have an isometric embedding of $\mathcal{H}$ in $\mathcal{K}$. Further define V on $\mathcal{K}$ by setting $V\lambda(m, u) = \lambda(m + 1, u)$ and extending linearly to get an isometry. Now it is not difficult to see that V is indeed a dilation of T. $\square$

Theorem 1.4. *(Stinespring's theorem): Let $\mathcal{A}$ be a unital C^*-algebra and let $\tau : \mathcal{A} \to \mathcal{B}(\mathcal{H})$ be a contractive completely positive map, for some Hilbert space $\mathcal{H}$. Then there exists a Hilbert space $\mathcal{K}$ containing $\mathcal{H}$, with a $*$-homomorphism $\pi : \mathcal{A} \to \mathcal{B}(\mathcal{K})$ and an isometry $V : \mathcal{H} \to \mathcal{K}$ such that*

$$\tau(X) = V^*\pi(X)V$$

for all $X \in \mathcal{A}$.

We have already seen Stinespring's theorem and its proof in earlier lectures [Li]. It shows that contractive completely positive maps are compressions of $*$-homomorphisms. One of the disadvantages with Stinespring's theorem is that when we have a completely positive map τ from $\mathcal{A}$ into itself, we also would like to consider powers τ^n of τ, but we can't talk of π^n due to domain problems, though each τ^n is a contractive completely positive map in its own right. In other words we are looking for a 'power dilation'. Contrast this with the Sz. Nagy dilation and the following more general theorem from multi-variable operator theory.

Theorem 1.5. *(Bunce, Frazho, Popescu): Let $(T_1, \ldots, T_n)$ be an n-tuple of bounded operators on a Hilbert space $\mathcal{H}$, for some $n \geq 1$, such that $\sum T_i T_i^* \leq I$. Then there exists a Hilbert space $\mathcal{K}$ containing $\mathcal{H}$ with an n-tuple $(V_1, \ldots, V_n)$ of isometries such that:*

(i) $V_i^* V_j = \delta_{ij} I$, $1 \leq i, j \leq n$.
(ii) $V_i^* u = T_i^* u$, *for* $u \in \mathcal{H}$ *and* $1 \leq i \leq n$;

Moreover if $\overline{\mathrm{span}}\{V_{i_1} \ldots V_{i_k} u : 1 \leq i_r \leq n, \;\; \forall r, k \geq 1, u \in \mathcal{H}\} = \mathcal{K}$ then this tuple is unique up to unitary equivalence.

Note that here (i) says that the V_i's are isometries with orthogonal ranges and (ii) says that the V_i's leave $\mathcal{H}^\perp$ invariant and that they form a dilation of the T_i's in the sense that:

$$T_{i_1} \ldots T_{i_k} = P_{\mathcal{H}} V_{i_1} \ldots V_{i_k}|_{\mathcal{H}}$$

for all tuples $i_1, \ldots, i_k$. Instead of looking at the n-tuple $(T_1, \ldots T_n)$ we may consider the completely positive map $\tau : \mathcal{B}(\mathcal{H}) \to \mathcal{B}(\mathcal{H})$ defined by

$$\tau(X) = \sum T_i X T_i^* \quad \forall X \in \mathcal{B}(\mathcal{H}).$$

Then the condition $\sum T_i T_i^* \leq I$ means precisely that τ is contractive. In a similar way that the V_i's being isometries with orthogonal ranges means that the map $\theta : \mathcal{B}(\mathcal{K}) \to \mathcal{B}(\mathcal{K})$ defined by

$$\theta(X) = \sum V_i X V_i^* \quad \forall X \in \mathcal{B}(\mathcal{K}),$$

is a $*$-endomorphism of $\mathcal{B}(\mathcal{K})$, that is, $\theta : \mathcal{B}(\mathcal{K}) \to \mathcal{B}(\mathcal{K})$ is a linear map satisfying $\theta(X^*) = \theta(X)^*$ and $\theta(XY) = \theta(X)\theta(Y)$ for all X, Y in $\mathcal{B}(\mathcal{K})$. Furthermore, (ii) implies that θ is a dilation of τ in the sense that:

$$\tau^n(X) = P_{\mathcal{H}}\theta^n(X)P_{\mathcal{H}}$$

where X in $\mathcal{B}(\mathcal{H})$ is identified with $P_{\mathcal{H}}XP_{\mathcal{H}}$ in $\mathcal{B}(\mathcal{K})$ to talk of $\theta^n(X)$. So these completely positive maps have $*$-endomorphic dilations. Below we are looking for similar results for general quantum dynamical semigroups.

Definition 1.6. *Let $\mathcal{A} \subset \mathcal{B}(\mathcal{H})$ be a unital C^*-algebra. Let $\tau = \{\tau_t : t \geq 0\}$ be a contractive quantum dynamical semigroup (semigroup of completely positive maps) on $\mathcal{A}$. A (subordinated) weak Markov flow with expectation semigroup τ is a triple $(\mathcal{K}, F, j)$, where*

(i) *$\mathcal{K}$ is a Hilbert space containing $\mathcal{H}$;*
(ii) *$F = \{F_t : t \geq 0\}$ is an increasing family of projections on $\mathcal{K}$ with $F(0)$ being the projection onto $\mathcal{H}$;*
(iii) *$j = \{j_t : t \geq 0\}$ is a family of $*$-homomorphisms, $j_t : \mathcal{A} \to \mathcal{B}(\mathcal{K})$, with $j_0(X) = XF(0)$;*
(iv) *$F(s)j_t(X)F(s) = j_s(\tau_{t-s}(X))$ for $0 \leq s \leq t$, $X \in \mathcal{A}$.*

It is said to be minimal *if*

$$\overline{\operatorname{span}}\{j_{t_1}(X_1)\ldots j_{t_n}(X_n)u : t_i \geq 0, X_i \in \mathcal{A}, u \in \mathcal{H}\} = \mathcal{K}.$$

In a weak Markov flow $(\mathcal{K}, F, j)$, $\mathcal{K}$ is known as the *dilation space*, the family of projections F is known as the *filtration* and the family of $*$-homomorphisms j is known as *the weak Markov process*. The word 'weak' here refers to the fact that the j_t's are non-unital $*$-homomorphisms. The property (iv) is known as the Markov property. The idea of this kind of dilation has been around for some time. For the formulation used here and for references on other variations see [BP1-2]. The main advantage of weak Markov flows is the following existence and uniqueness theorem and the fact that practically any definition of Markov dilation one considers almost always contains a weak dilation as a component.

Theorem 1.7. *Given a contractive quantum dynamical semigroup τ there always exists a minimal weak Markov flow $(\mathcal{K}, F, j)$ with τ as its expectation semigroup. Moreover, such a triple is unique up to unitary equivalence.*

The proof is once again through a GNS-type construction. This is possible as inner products between vectors of the form $j_{t_1}(X_1)\ldots j_{t_n}(X_n)u$ are completely determined by τ and we can show positive definiteness of the associated kernel.

The next step in dilation theory is to construct a semigroup of endomorphisms. Let $(\mathcal{K}, F, j)$ be a minimal weak Markov flow for a quantum dynamical semigroup τ as above. Let $\mathcal{B}$ be the C^*-subalgebra of $\mathcal{B}(\mathcal{K})$ generated by $\{j_t(X) : X \in \mathcal{A}, t \geq 0\}$. Define $\theta_t : \mathcal{B} \to \mathcal{B}$, by setting

$$\theta_t(j_s(X)) = j_{s+t}(X),$$

and extending $*$-homomorphically. One has to check that θ_t is well-defined and has $*$-homomorphic extension [Bh1-2]. But once we know that such endomorphisms of $\mathcal{B}$ exist, it is easy to verify that $\{\theta_t : t \geq 0\}$ is a semigroup of $*$-endomorphisms of $\mathcal{B}$. If we are in the von Neumann algebra setup, that is if $\mathcal{A}$ is a von Neumann algebra, each τ_t is normal and $t \mapsto \tau_t(X)$ is ultraweakly continuous, then we get a semigroup of normal $*$-endomorphisms on the von Neumann algebra generated by $\{j_t(X)\}$ (see [BS], Section 12).

Once we have a semigroup of $*$-endomorphisms θ as above we may actually forget about the weak Markov flow and consider the triple $(\mathcal{K}, \mathcal{B}, \theta)$ as a dilation of $(\mathcal{H}, \mathcal{A}, \tau)$. Note that as $P := F(0)$ is the projection onto $\mathcal{H}$, identifying $X \in \mathcal{B}(\mathcal{H})$ with $PXP \in \mathcal{B}(\mathcal{K})$, we actually have

$$P\theta_t(X)P = \tau_t(X) \quad \forall X \in \mathcal{A}, t \geq 0.$$

In other words we have a semigroup of $*$-endomorphisms as a dilation of a quantum dynamical semigroup. The minimality here depends upon whether we want to consider the C^*-algebra setup or the von Neumann algebra setup. But in either case, there is a suitable notion of minimality and there is a unique minimal dilation.

2 E_0-semigroups and product systems

Let $\mathcal{H}, \mathcal{P}$ be two non-zero, complex, separable Hilbert spaces. Let $W : \mathcal{H} \otimes \mathcal{P} \to \mathcal{H}$ be an isometry. [*] Now consider the map ψ on $\mathcal{B}(\mathcal{H}) \to \mathcal{B}(\mathcal{H})$ defined by

$$\psi(X) = W(X \otimes I_\mathcal{P})W^*, \quad X \in \mathcal{B}(\mathcal{H}). \tag{2.1}$$

We easily verify that ψ has the following properties:

(i) ψ is linear;
(ii) $\psi(XY) = \psi(X)\psi(Y)$, for all $X, Y \in \mathcal{B}(\mathcal{H})$;
(iii) $\psi(X^*) = \psi(X)^*$, for all $X \in \mathcal{B}(\mathcal{H})$;
(iv) ψ is normal (ultraweakly continuous).

---

[*] Note that if $\mathcal{H}$ is finite dimensional, we can't have such an isometry unless $\mathcal{P}$ is one-dimensional, as we will be having $\dim(\mathcal{H} \otimes \mathcal{P}) > \dim(\mathcal{H})$. However there are no such constraints if $\mathcal{H}$ is infinite dimensional.

A mapping ψ of $\mathcal{B}(\mathcal{H})$ which satisfies properties (i)-(iv) is said to be a normal $*$-endomorphism of $\mathcal{B}(\mathcal{H})$. It is not hard to see that any normal $*$-endomorphism of $\mathcal{B}(\mathcal{H})$ necessarily has the form (2.1). Indeed if ψ is a given $*$-endomorphism, choose a unit vector $a \in \mathcal{H}$, take $\mathcal{P}$ as the range of the projection $\psi(|a\rangle\langle a|)$ and consider $W : \mathcal{H} \otimes \mathcal{P} \to \mathcal{H}$ defined by

$$W(x \otimes \psi(|a\rangle\langle a|)y) = \psi(|x\rangle\langle a|)y.$$

Then we see that W is an isometry with $W^*z = \sum_i e_i \otimes \psi(|a\rangle\langle e_i|)z$ for any orthonormal basis $\{e_i\}$ of $\mathcal{H}$ and (2.1) is satisfied.

Here is another way of expressing ψ. Let $\{e_i : i \geq 1\}$ be an orthonormal basis of $\mathcal{P}$. Define $V_i \in \mathcal{B}(\mathcal{H})$ by setting $V_i x = W(x \otimes e_i)$, $\forall x \in \mathcal{H}$. We leave it to the reader to verify that the V_i's are isometries with orthogonal ranges, that is,

$$V_i^* V_j = \delta_{ij} I \quad \forall i, j \tag{2.2}$$

and

$$\psi(X) = \sum_i V_i X V_i^*, \tag{2.3}$$

for $X \in \mathcal{B}(\mathcal{H})$. Here if the number of terms is infinite, that is, if $\mathcal{P}$ is infinite dimensional, the series converges in the strong operator topology. Note that each V_i is an element of the space $\mathcal{E}$ of intertwiners:

$$\mathcal{E} := \{Y \in \mathcal{B}(\mathcal{H}) : \psi(X)Y = YX \; \forall X \in \mathcal{B}(\mathcal{H})\}.$$

If we take any two elements Y, Z in $\mathcal{E}$, we see that Y^*Z commutes with every $X \in \mathcal{H}$, and so is a scalar. It is another little exercise to show that taking $\langle Y, Z \rangle I = Y^*Z$ makes $\mathcal{E}$ into a Hilbert space, and that $\{V_i : i \geq 1\}$ forms an orthonormal basis for this Hilbert space. This of course shows that $\mathcal{P}$ and $\mathcal{E}$ are isomorphic as Hilbert spaces.

We are interested in one parameter semigroups of $*$-endomorphisms, but before moving further let us note that: $\psi^n(X) = W_n(X \otimes I_{\mathcal{P}^{\otimes n}})W_n^*$, where $W_n : \mathcal{H} \otimes \mathcal{P}^{\otimes n} \to \mathcal{H}$ are isometries inductively defined as $W_1 = W$ and $W_{n+1} = W(W_n \otimes I_{\mathcal{P}})$. In other words we need the Hilbert space $\mathcal{P}^{\otimes n}$ to describe the n-th power of ψ. Here we have a discrete product system of Hilbert spaces as: $\mathcal{P}^{\otimes(m+n)} = \mathcal{P}^{\otimes m} \otimes \mathcal{P}^{\otimes n}$.

Definition 2.1. *An E-semigroup θ on $\mathcal{B}(\mathcal{H})$ is a family, $\theta = \{\theta_t : t \geq 0\}$, of linear maps of $\mathcal{B}(\mathcal{H})$ such that:*

(i) *For every t, $\theta_t : \mathcal{B}(\mathcal{H}) \to \mathcal{B}(\mathcal{H})$ is a normal $*$-endomorphism;*
(ii) *θ is a semigroup, that is, $\theta_0(X) = X$, and $\theta_{s+t}(X) = \theta_s(\theta_t(X))$ for all $X \in \mathcal{B}(\mathcal{H})$ and $s, t \geq 0$.*
(iii) *$t \mapsto \theta_t(X)$ is continuous in the weak (or equivalently strong) operator topology, for every X in $\mathcal{B}(\mathcal{H})$.*

If further $\theta_t(I) = I$ for all t, then θ is said to be an E_0-semigroup.

Example 2.2. *Let $V = \{V_t, t \geq 0\}$ be a one parameter semigroup of isometries on $\mathcal{H}$, that is, (i) $V_t^* V_t = I$, for all t; (ii) $V_0 = I$, $V_{s+t} = V_s V_t$, for all s, t; (iii) $t \mapsto V_t$ is continuous in the strong operator topology. Take $\theta_t(X) = V_t X V_t^*$, then θ is an E-semigroup. It is an E_0-semigroup iff V_t is unitary for every t. For instance, let $\mathcal{H} = L^2(\mathbf{R}_+)$ and let V_t be defined by*

$$V_t f(x) = \begin{cases} f(x - t) & t \leq x < \infty \\ 0 & 0 < x < t, \end{cases}$$

for $f \in L^2(\mathbf{R}^+)$, then we have a semigroup of isometries. On the other hand if we take $\mathcal{H} = L^2(\mathbf{R})$, and let V_t be defined by

$$V_t f(x) = f(x - t) \quad -\infty < x < \infty$$

for $f \in L^2(\mathbf{R})$ we have a semigroup of unitaries.

Example 2.3. *(Fock-space shift): Take $\mathcal{H}$ as the symmetric Fock space $\Gamma_\mathcal{K} = \Gamma(L^2(\mathbf{R}_+, \mathcal{K}))$, where $\mathcal{K}$ is some other Hilbert space. Then*

$$\sigma_t(X) = I_t \otimes (S_t X S_t^*)$$

where S_t is the second quantization of the shift on $L^2(\mathbf{R}_+, \mathcal{K})$, as described in [Li], is an E_0-semigroup.

We wish to study and classify E_0-semigroups. For instance, we wish to say that Examples 2.2 and 2.3 are really distinct. The first step in this direction is to attach a 'product system of Hilbert spaces' with every E_0-semigroup. This was first done in [Ar1].

A *product system* E is a 'measurable' family of Hilbert spaces, $E = \{\mathcal{E}_t : t \geq 0\}$, with a collection of unitaries, $U_{s,t} : \mathcal{E}_s \otimes \mathcal{E}_t \to \mathcal{E}_{s+t}$, such that we have associativity in the sense that: $U_{s_1+s_2,s_3}(U_{s_1,s_2} \otimes I_{s_3}) = U_{s_1,s_2+s_3}(I_{s_1} \otimes U_{s_2,s_3})$ as unitary maps from $\mathcal{E}_{s_1} \otimes \mathcal{E}_{s_2} \otimes \mathcal{E}_{s_3}$ to $\mathcal{E}_{s_1+s_2+s_3}$ for any s_1, s_2, s_3. (See [Ar1], or [Lie] for the measurability details). There is a natural notion of isomorphism of product systems.

Let θ be an E_0-semigroup on $\mathcal{B}(\mathcal{H})$ of some Hilbert space $\mathcal{H}$. Arveson's idea was to look at the space of intertwining operators. Thus define

$$\mathcal{E}_t = \{Y \in \mathcal{B}(\mathcal{H}) : \theta_t(X)Y = YX \ \forall X \in \mathcal{B}(\mathcal{H})\}.$$

We have already seen that this is a Hilbert space, with the inner product $\langle Y, Z \rangle I = Y^* Z$. Now define $U_{s,t}(\mathcal{E}_s \otimes \mathcal{E}_t) \to \mathcal{E}_{s+t}$ by $U_{s,t}(Y_1 \otimes Y_2) = Y_1 Y_2$. It is easy to verify that $U_{s,t}$ is an isometry. A little bit of extra work shows that it is actually a unitary. Some further technicalities of measurability have to be taken care of, see [Ar1].

We have another construction of the product system as follows. Fix a unit vector $a \in \mathcal{H}$. Note that by the *-endomorphism property, $\theta_t(|a\rangle\langle a|)$ is a projection. Take $\mathcal{P}_t = $ range $\theta_t(|a\rangle\langle a|)$. Of course this depends upon a but we are suppressing it in the notation. Define $W_t : \mathcal{H} \otimes \mathcal{P}_t \to \mathcal{H}$ by

$$W_t(x \otimes \theta_t(|a\rangle\langle a|)y) = \theta_t(|x\rangle\langle a|)y.$$

Once again we verify that W_t is a unitary. (If θ were non-unital, W_t would only be an isometry onto the range $\theta_t(I)$). Further, let $V_{s,t}$ be W_t restricted to domain $\mathcal{P}_s \otimes \mathcal{P}_t$ and range $\mathcal{P}_{s+t}$. Then it is a unitary and $(\mathcal{P}_t, V_{s,t})$ forms a product system. We might think that the two product systems we have got now from θ must be isomorphic. It is almost true, actually they are *opposites* of each other, in the sense that if instead of $V_{s,t}$, if we had taken $V_{t,s}T_{s,t}$, where $T_{s,t} : \mathcal{P}_s \otimes \mathcal{P}_t \to \mathcal{P}_t \otimes \mathcal{P}_s$ is the twist unitary operator: $T_{s,t}(x \otimes y) = y \otimes x$, then we would have got an isomorphism. † The reason we want to stay with $V_{s,t}$ is that it is 'natural' and is the appropriate one when one works later on with product systems of Hilbert C^*-modules.

It is not hard to see that the product system $\mathcal{P}_t$ associated with Example 2.2, is the trivial product system where each $\mathcal{P}_t$ is isomorphic to $\mathbb{C}$ and we have the unitary $V_{s,t}(x \otimes y) = xy$, with respect to this isomorphism.

For Example 2.3, the product system is $\mathcal{H}_t = \Gamma(L^2([0,t),\mathcal{K}))$, where the product system structure comes from the isomorphism $\mathcal{H}_{s+t} = \mathcal{H}_s \otimes \mathcal{H}_t$, by once again making use of second quantization of the shift to identify $\mathcal{H}_t$ with $\Gamma(L^2([s, s+t),\mathcal{K}))$. In fact if we take a as the vacuum vector $\varepsilon(0)$ and compute $\mathcal{P}_t$ and $V_{s,t}$, we see that $V_{s,t}(\varepsilon(f) \otimes \varepsilon(g)) = \varepsilon(f + S_s g)$. Here ε stands for exponential vector.

We also need some notions for comparing E_0-semigroups in order to classify them.

Definition 2.4. *Let θ and θ' be E_0-semigroups acting on $\mathcal{B}(\mathcal{H})$, and $\mathcal{B}(\mathcal{H}')$ respectively. Then θ and θ' are said to be* conjugate *if there exists a unitary $M : \mathcal{H} \to \mathcal{H}'$, such that*

$$\theta'_t(X) = M\theta_t(M^*XM)M^* \tag{2.4}$$

for all $X \in \mathcal{B}(\mathcal{H}'), t \geq 0$.

This is the notion of unitary equivalence, the main point being that the same unitary works for all t. There is a weaker notion of equivalence which is useful and for that we need to talk about *cocycles*.

Definition 2.5. *A strongly continuous family of operators $G = \{G_t \in \mathcal{B}(\mathcal{H}) : t \geq 0\}$ is said to be a* (left) cocycle *for an E_0-semigroup θ of $\mathcal{B}(\mathcal{H})$ if*

$$G_{s+t} = G_s\theta_s(G_t) \quad \forall s,t \geq 0, \ G_0 = I.$$

A cocycle G is said to be local *if G_t is in the commutant $(\theta_t(\mathcal{B}(\mathcal{H})))'$ for all t, that is, if G_t commutes with $\theta_t(Z)$ for all Z. It is said to be* positive *(respectively* unitary, isometric, contractive*) if each G_t is positive (respectively unitary, isometric, contractive).*

---

† In general the opposite product system of a product system need not be isomorphic to the original product system [Tsi], but there is no such problem for these examples.

Definition 2.6. *Let θ, θ' be E_0-semigroups acting on $\mathcal{B}(\mathcal{H})$ and $\mathcal{B}(\mathcal{H}')$ respectively. Then θ and θ' are said to be* cocycle conjugate *if there exists a third E_0-semigroup θ'' on $\mathcal{B}(\mathcal{H})$ and a unitary cocycle $U = \{U_t : t \geq 0\}$ of θ such that:*

(i) $\theta''_t(X) = U_t \theta_t(X) U_t^*$, $\forall X \in \mathcal{B}(\mathcal{H}), t \geq 0$;
(ii) θ'' *is conjugate to* θ'.

Theorem 2.7. *(Arveson [Ar1]): Two E_0 semigroups on $\mathcal{B}(\mathcal{H})$ and $\mathcal{B}(\mathcal{H}')$ respectively with $\mathcal{H} \cong \mathcal{H}'$, are cocycle conjugate if and only if they have isomorphic product systems.*

Arveson has also shown that every product system arises as a product system of some E_0-semigroup. (No simple proof of this result is known). So now we have reduced the problem of classifying E_0-semigroups up to cocycle conjugacy to that of classifying product systems of Hilbert spaces up to unitary isomorphism. How do we see that the Fock space product systems for different noise spaces $\mathcal{K}$ and $\mathcal{K}'$ are non-isomorphic if they have different dimensions? In other words can we recover the dimension of the noise space from the product system? This requires the notion of units and index.

Definition 2.8. *Consider a product system $(\mathcal{E}_t, U_{s,t})$. A 'measurable' family $u = \{u_t : t \geq 0\}$ of non-zero vectors with $u_t \in \mathcal{E}_t$ is said to be a* unit *for the system, if $u_{s+t} = U_{s,t}(u_s \otimes u_t)$ for all s, t.*

Any unit u for the trivial product system has the form $u_t = e^{tq}$ for some $q \in \mathbb{C}$. The units for the Fock space product system are given by: $u_t = e^{qt}\varepsilon(x\chi_{[0,t)})$ for some $(q, x) \in \mathbb{C} \times \mathcal{K}$.

Suppose that u and v are two units of a product system. We have $\langle u_{s+t}, v_{s+t}\rangle = \langle u_s, v_s\rangle\langle u_t, v_t\rangle$ for all s, t. Then by measurability, it follows that $\langle u_t, v_t\rangle = e^{t\gamma_{u,v}}$ for some complex constant $\gamma_{u,v}$ depending only on the units. The function $\gamma : \mathcal{U} \otimes \mathcal{U} \to \mathbb{C}$, obtained this way is known as the *covariance function*. It is not hard to see that it is a conditionally positive definite kernel, that is,

$$\sum \bar{c}_i c_j \gamma(u_i, u_j) \geq 0$$

for any choice of $u_1, \ldots, u_n$ in $\mathcal{U}$ and $c_1, \ldots, c_n$ in $\mathbb{C}$ with $\sum c_i = 0$, where $\mathcal{U}$ is the collection of all units. There is a GNS construction for conditionally positive definite kernels and this gives us a 'minimal' Hilbert space $\mathcal{K}_\gamma$. It can be shown that we will only get a separable Hilbert space. The dimension of this space is an invariant of the product system (it depends only on the isomorphism class of the product system). This number (possibly infinity) is known as the Arveson index or the numerical index of the product system. For the trivial product system $\mathcal{K}_\gamma = \{0\}$ and for the Fock product system described above, $\mathcal{K}_\gamma$ is isomorphic to $\mathcal{K}$. Therefore the Arveson indices for these product systems are 0 and $\dim \mathcal{K}$ respectively.

Definition 2.9. *A product system $E = \{\mathcal{E}_t\}$ is called* spatial *if it has a unit. It is said to be* divisible *if the units generate the product system, that is, $\mathcal{E}_t = \overline{span}\{u_{t_1}^1 \otimes u_{t_2}^2 \otimes \cdots \otimes u_{t_n}^n : u^1, \ldots, u^n$ are units and $t_1 + \cdots + t_n = t\}$.*

The examples of product systems we have seen so far, namely trivial and Fock space product systems are spatial and in fact they are divisible and they are the only divisible product systems [Ar1]. The divisible product systems are also known as Type I product systems. The product systems which are spatial but not divisible (they have units but not sufficiently many to generate the product system) are known as Type II product systems. The product systems which are non-spatial are Type III (they have no units). Initially, Type II, III product systems were hard to come by, there were just some stray examples constructed by R. T. Powers. But now Tsirelson [Tsi] and Liebscher [Lie] have plenty of examples. We still don't know how to completely classify product systems. It seems to be a very hard problem. We will come back to this in the fourth Lecture.

3 Domination and minimality

Let us recall our dilation theorem for quantum dynamical semigroups on the von Neumann algebra of all bounded operators on a Hilbert space. Here we consider only ultraweakly continuous semigroups of normal completely positive maps. We also assume that the completely positive semigroup which we want to dilate is unital. Suppose that $\tau = \{\tau_t : t \geq 0\}$ is a quantum dynamical semigroup of $\mathcal{B}(\mathcal{H}_0)$. If $\mathcal{H}$ is a Hilbert space containing $\mathcal{H}_0$ as a closed subspace and if $\theta = \{\theta_t : t \geq 0\}$ is an E_0-semigroup of $\mathcal{B}(\mathcal{H})$ such that

$$\tau_t(X) = P\theta_t(X)P, \quad t \geq 0, \quad X \in \mathcal{B}(\mathcal{H}_0) = P\mathcal{B}(\mathcal{H})P \subseteq \mathcal{B}(\mathcal{H}) \quad (3.1)$$

where P is the orthogonal projection of $\mathcal{H}$ onto $\mathcal{H}_0$, then θ is called a *dilation* of τ and τ is called a *compression* of θ.

The dilation θ is said to be *minimal* if the subspace generated by its action on $\mathcal{H}_0$ is $\mathcal{H}$, that is, if the subspace $\widehat{\mathcal{H}}$ defined by

$$\overline{span}\{\theta_{r_1}(X_1) \cdots \theta_{r_n}(X_n)u : r_i \geq 0, X_i \in \mathcal{B}(\mathcal{H}_0), u \in \mathcal{H}_0, 1 \leq i \leq n, n \geq 0\}$$

is the whole of $\mathcal{H}$. We know that a minimal dilation exists and that it is unique up to unitary isomorphisms. We will denote it by $\widehat{\tau}$. Given any dilation θ compressing it to $\mathcal{B}(\widehat{\mathcal{H}})$ we get the minimal dilation $\widehat{\tau}$.

A dilation θ is said to be *primary* if $\overline{span}\{\theta_t(P)x : x \in \mathcal{H}, t \geq 0\}$ is $\mathcal{H}$. Note that in our present case τ is unital, this forces $\theta_t(P)$ to be an increasing family of projections. So here the dilation is primary if $\theta_t(P)$ increases to the identity operator on $\mathcal{H}$. Clearly this is a necessary condition for minimality and usually this property can be checked easily. Unfortunately, this is not a sufficient condition. In the following we will *assume that the dilation θ is primary.*

The problem of deciding whether a given dilation is minimal or not seems to be extremely hard in most situations. Here we want to develop a scheme for testing minimality. The key to this is an analysis of 'domination'. A quantum dynamical semigroup $\alpha = \{\alpha_t : t \geq 0\}$ is said to be dominated by a quantum dynamical semigroup $\tau = \{\tau_t : t \geq 0\}$, if $\tau_t - \alpha_t$ is completely positive for every t. We denote this by $\alpha \leq \tau$. Let $\mathcal{D}_\tau$ denote the set of all (not necessarily unital) quantum dynamical semigroups dominated by τ. Then $\mathcal{D}_\tau$ is a partially ordered set with partial order $\leq$. Something special happens when we have domination by $*$-homomorphisms.

Theorem 3.1. *Let $\mathcal{A}$ and $\mathcal{B}$ be unital C^*-algebras and let α and β be linear maps from $\mathcal{A}$ to $\mathcal{B}$, where α is a unital $*$-homomorphism and β is completely positive. Suppose that $\alpha - \beta$ is positive. Then $\beta(X) = \alpha(X)\beta(1) = \beta(1)\alpha(X)$, for every $X \in \mathcal{A}$. (In particular $\beta(1)$ commutes with the range of α.)*

Proof: Let $\mathcal{B}$ be a unital subalgebra of $\mathcal{B}(\mathcal{H})$ for some Hilbert space $\mathcal{H}$. Then Stinespring's theorem applied to β provides us with a Hilbert space $\mathcal{K}$, an isometry $V : \mathcal{H} \to \mathcal{K}$ and a representation $\gamma : \mathcal{A} \to \mathcal{B}(\mathcal{K})$, such that $\beta(X) = V^*\gamma(X)V$. As $\alpha - \beta$ is positive, for any $X \in \mathcal{A}$ we have $V^*\gamma(X^*X)V \leq \alpha(X^*X)$. Hence for any $z \in \mathbb{C}, H \in \mathcal{A}, u \in \mathcal{H}$

$$\|\gamma(e^{zH})Vu\|^2 \leq \|\alpha(e^{zH})u\|^2.$$

Taking $u = \alpha(e^{-zH})v, v \in \mathcal{H}$, we have

$$|\langle v, \gamma(e^{zH})V\alpha(e^{-zH})v\rangle| \leq \|v\|\|\gamma(e^{zH})Vu\| \leq \|v\|^2.$$

Therefore the entire function $z \mapsto \langle v, \gamma(e^{zH})V\alpha(e^{-zH})v\rangle$ is bounded. Hence by Liouville's theorem it is constant. So we get

$$\gamma(e^{zH})V\alpha(e^{-zH}) = \gamma(1)V\alpha(1) = \gamma(1)V,$$

or $\gamma(e^{zH})V = \gamma(1)V\alpha(e^{zH})$ for all $z \in \mathbb{C}, H \in \mathcal{A}$. This clearly implies that $\gamma(X)V = \gamma(1)V\alpha(X)$ and hence $\beta(X) = V^*\gamma(X)V = V^*\gamma(1)V\alpha(X) = \beta(1)\alpha(X)$ for all $X \in \mathcal{A}$. By taking adjoints the proof is complete. $\square$

An immediate consequence of this is the following theorem.

Theorem 3.2. *Let θ be an E_0-semigroup of $\mathcal{B}(\mathcal{H})$, and let ψ be a quantum dynamical semigroup of $\mathcal{B}(\mathcal{H})$. Then the following are equivalent:*

(i) *θ dominates ψ;*
(ii) *$\theta_t - \psi_t$ is positive for every $t \geq 0$;*
(iii) *$\psi_t(Z) = G_t\theta_t(Z) \; \forall Z \in \mathcal{B}(\mathcal{H})$, for some positive, contractive, local cocycle of θ;*
(iv) *ψ is absorbing for θ, that is,*

$$\psi_t(Z)\theta_t(W) = \psi_t(ZW) \quad \forall Z, W \in \mathcal{B}(\mathcal{H}), t \geq 0.$$

(Only continuity of the map $t \mapsto G_t$ needs some work.)

Theorem 3.3. *Let $(\mathcal{H}, \theta)$ be a primary dilation of $(\mathcal{H}_0, \tau)$ and let $(\widehat{\mathcal{H}}, \widehat{\tau})$ be the minimal dilation as above. Then one can show that compression by projection $P = P_{\mathcal{H}_0}$ maps $\mathcal{D}_\theta$ surjectively to $\mathcal{D}_\tau$. This compression map is injective if and only if θ is the minimal dilation of τ.*

This structure can be pictorially represented as follows:

$$
\begin{array}{ccc}
\widetilde{\alpha} \leq \widetilde{\tau} \leq \theta & & \\
\downarrow \quad \downarrow \;\nearrow & & \\
\widehat{\alpha} \leq \widehat{\tau} & & \\
\downarrow \quad \downarrow & & \\
\alpha \leq \tau & &
\end{array}
$$

where as before $\leq$ denotes domination and arrows indicate compressions by appropriate projections. (Here α and τ act on $\mathcal{B}(\mathcal{H}_0)$, $\widehat{\alpha}$ and $\widehat{\tau}$ act on $\mathcal{B}(\widehat{\mathcal{H}})$ and $\widetilde{\alpha}$, $\widetilde{\tau}$ and θ act on $\mathcal{B}(\mathcal{H})$). $\widetilde{\tau}$ is called the *induced semigroup* of the dilation, it is the 'smallest' E-semigroup dominated by θ which compresses to τ. The primary dilation θ is the minimal dilation if and only if $\widetilde{\tau} = \theta$. This gives us a completely algebraic characterization of minimality.

Next, we want to apply this criterion to flows (E_0-semigroups) coming from quantum stochastic calculus. For this we also need to know all the positive (or at least projection) local cocycles of the Fock-space shift. Actually such a cocycle G_t is nothing but a positive contractive Markovian cocycle, or 'exponential noise', as obtained in [Li].

Let $\mathcal{K}$ be a complex separable Hilbert space. Then Hudson-Parthasarathy quantum stochastic differential equations can be written on the space $\widetilde{\mathcal{H}} = \mathcal{H}_0 \otimes \Gamma(L^2(\mathbf{R}_+, \mathcal{K}))$. Here $\mathcal{H}_0$ (identified with $\mathcal{H}_0 \otimes \varepsilon(0)$) is known as the initial space and $\mathcal{H} = \Gamma(L^2(\mathbf{R}_+, \mathcal{K}))$ as the noise space. On $\widetilde{\mathcal{H}}$ we have the E_0-semigroup $\widetilde{\sigma} = id \otimes \sigma$, where $\sigma = \sigma^{\mathcal{K}}$ is the CCR flow (Fock-space shift) on $\mathcal{H}$. By a result of Arveson we know that $\widetilde{\sigma}$ is cocycle conjugate to σ.

Let τ be a unital quantum dynamical semigroup on $\mathcal{B}(\mathcal{H}_0)$ with bounded generator. Then we know that its generator has a very special form (cf. [Li]). Namely if we denote the generator by $\mathcal{L}$ so that $\tau_t(X) = e^{t\mathcal{L}}(X), X \in \mathcal{B}(\mathcal{H}_0), t \geq 0$, then there exists a family of bounded operators $\{L_k \in \mathcal{B}(\mathcal{H}_0) : k \geq 1\}$ and a self-adjoint operator $H \in \mathcal{B}(\mathcal{H}_0)$ such that $\sum_{k \geq 1} L_k^* L_k$ is strongly convergent, and

$$
\mathcal{L}(X) = i[H, X] - \frac{1}{2} \sum_{k \geq 1} (L_k^* L_k X + X L_k^* L_k - 2 L_k^* X L_k) \tag{3.2}
$$

for every $X \in \mathcal{B}(\mathcal{H}_0)$. Now to obtain a dilation of τ using quantum stochastic calculus we fix one such representation of $\mathcal{L}$ and consider a Hilbert space $\mathcal{K}$ with $\dim \mathcal{K}$ equal to the number of L_k's. To avoid trivialities we assume that this number is non-zero.

Let $\{S_j^i : i,j \geq 1\}$ be bounded operators on $\mathcal{H}_0$ such that $\sum_{i,j \geq 1} S_j^i \otimes |e_i\rangle\langle e_j|$ is a unitary operator in $\mathcal{H}_0 \otimes \mathcal{K}$. Define

$$
L_j^i = \begin{cases}
S_j^i - \delta_{ij} & \text{if } i,j \geq 1; \\
L_i & \text{if } i \geq 1, j = 0; \\
-\sum_{k \geq 1} L_k^* S_j^k & \text{if } j \geq 1, i = 0; \\
-(iH + \frac{1}{2}\sum_{k \geq 1} L_k^* L_k) & \text{if } i = j = 0.
\end{cases}
$$

Then by Theorem 27.8 of [Pa] there exists a unique unitary operator valued adapted process $U = \{U_t : t \geq 0\}$ on $\mathcal{B}(\widetilde{\mathcal{H}})$ satisfying the quantum stochastic differential equation

$$
dU = \left(\sum_{i,j \geq 0} L_j^i d\Lambda_i^j\right) U, \quad U_0 = I \tag{3.3}
$$

on $\mathcal{M}$. Here $d\Lambda_i^j$ refers to differentials of the fundamental processes of time, creation, conservation and annihilation, with respect to an orthonormal basis for $\mathcal{K}$.

We then have an E_0-semigroup η of $\mathcal{B}(\widetilde{\mathcal{H}})$ defined by

$$
\eta_t(Z) = U_t^* \widetilde{\sigma}_t(Z) U_t \qquad \text{for } Z \in \mathcal{B}(\widetilde{\mathcal{H}}). \tag{3.4}
$$

Conventionally the family $j = \{j_t : t \geq 0\}$ of representations of $\mathcal{B}(\mathcal{H}_0)$ defined by

$$
j_t(X) = \eta_t(X \otimes 1_{\mathcal{H}}) = U_t^*(X \otimes 1_{\mathcal{H}})U_t, \quad X \in \mathcal{B}(\mathcal{H}_0) \tag{3.5}
$$

is known as the Evans-Hudson flow (or EH flow) associated with the Hudson-Parthasarathy cocycle U. By a small modification of the terminology we refer to the E_0-semigroup η as the Evans-Hudson flow.

As we identify $\mathcal{H}_0$ with $\mathcal{H}_0 \otimes \varepsilon(0)$, $X \in \mathcal{B}(\mathcal{H}_0)$ is to be identified with $X \otimes |\varepsilon(0)\rangle\langle\varepsilon(0)|$ in $\mathcal{B}(\widetilde{\mathcal{H}})$. So for $X \in \mathcal{B}(\mathcal{H}_0)$, by $\eta_t(X)$ we mean $\eta_t(X \otimes |\varepsilon(0)\rangle\langle\varepsilon(0)|)$, which is not the same as $j_t(X)$. However adaptedness of U_t gives

$$
\langle f\varepsilon(0), \eta_t(X)g\varepsilon(0)\rangle = \langle f\varepsilon(0), j_t(X)g\varepsilon(0)\rangle \quad f, g \in \mathcal{H}_0, X \in \mathcal{B}(\mathcal{H}_0)
$$

that is, compressions of $\eta_t(X), j_t(X)$ to $\mathcal{B}(\mathcal{H}_0)$ are the same. Then by standard computation (see Corollary 27.9 of [Pa]) we deduce that

$$
\langle f\varepsilon(0), \eta_t(X)g\varepsilon(0)\rangle = \langle f\varepsilon(0), Xg\varepsilon(0)\rangle + \int_0^t \langle f\varepsilon(0), \eta_s(\mathcal{L}(X))g\varepsilon(0)\rangle ds.
$$

As $\mathcal{L}$ is the generator of τ, η is a dilation of τ. The problem is to determine the minimality of this dilation.

Theorem 3.4. *The Evans-Hudson flow* $\eta = \{\eta_t : \eta_t(\cdot) = U_t^*(\tilde{\sigma}_t(\cdot))U_t, t \geq 0\}$ *coming from the Hudson-Parthasarathy cocycle* $\{U_t\}$ *as above is a minimal dilation of* τ *if and only if* $\{L_i : i \geq 0\}$ *are linearly independent in the* l^2-*sense, where* L_0 *is taken as the identity operator.*

This is proved making use of Theorem 3.3 and details can be found in [Bh3]. Here we give only a brief sketch. In order to apply Theorem 3.3, at first we need to determine the quantum dynamical semigroups dominated by η. In view of Theorem 3.2 we know them if we know the positive contractive local cocycles of η. It is easy to see that these cocycles are necessarily of the form $U_t^*(1 \otimes G_t)U_t$, where $\{G_t\}$ is a positive, contractive, local cocycle of the CCR flow $\sigma^\mathcal{K}$. For CCR flows such local cocycles can be completely parametrized and they have been computed quite explicitly in the Section 7 of [Bh3]. (You may also find them in Lindsay [Li] where they are seen to satisfy quantum stochastic differential equations). The rest of the proof requires only an application of the first fundamental formula of quantum stochastic calculus to check the injectivity of the compression map by computing the compressions of dominated semigroups.   □

This shows that minimal dilation of unital quantum dynamical semigroups on $\mathcal{B}(\mathcal{H}_0)$, with bounded generators can be realized through Hudson-Parthasarathy quantum stochastic calculus. Moreover, since the minimal dilation is unique it shows that the minimal dilation of such quantum dynamical semigroups automatically satisfies a quantum stochastic differential equations.

4 Product systems: Recent developments

This talk will be in two parts. The first part is about the current state of affairs in product systems of Hilbert spaces and the second part is about product systems of Hilbert C^*-modules.

I Exotic product systems

As described earlier, product systems may be divided into three groups called type I, II and III. Further classification comes from the index. So for example a type II product system of index n will be called a type II_n product system. Note that there is no index for type III product system as the index is defined through units and they have no units.[‡]

We know all type I product systems. They are either trivial or Fock. In other words we have exactly one type I_n product system for $n \in \{\infty, 0, 1, \ldots\}$. How to construct type II product systems? Here is a brief sketch of a type II_0 example by B. Tsirelson.

[‡] Arveson sometimes takes index of any type III product system as c, the cardinality of the continuum, as it makes the formula: 'index of tensor product of product systems equals the sum of indices' correct in all situations! (Of course, this is only a convention and is of little help in classification.)

We construct a product system $\{E_t\}$, where each Hilbert space is an L^2 space. That is, $E_t = L^2(\Omega_t, \mathcal{F}_t, P_t)$ for some probability space $(\Omega_t, \mathcal{F}_t, P_t)$. Here Ω_t is the set of compact subsets of $[0, t]$. This set is a metric space with the Hausdorff metric: $d(A, B) = \inf\{\epsilon > 0 : A \subset N_\epsilon(B), B \subset N_\epsilon(A)\}$, where N_ϵ denotes 'ϵ-neighborhood'. This makes Ω_t into a compact metric space with empty set as an isolated point. The σ-field $\mathcal{F}_t$ is nothing but the Borel σ-field of this topology. The probability measure P_t comes from Brownian motion. Start a standard Brownian motion B_t^a at a point a different from 0. Let Z_t be the set of zeros of this Brownian motion in the interval $[0, t]$. That is,

$$Z_t(\omega) = \{s : 0 \le s \le t, B_s^a(\omega) = 0\}.$$

Note that as the Brownian paths are continuous, Z_t is a compact subset of $[0, t]$ (It could be empty). In other words, Z is a function from the space $C^a[0, \infty)$ of continuous paths starting at a to Ω_t. The measure P_t is the induced measure, that is,

$$P_t(\mathcal{C}) = P(Z_t \in \mathcal{C})$$

for any $\mathcal{C}$ in $\mathcal{F}_t$ ($\mathcal{C}$ is a collection of compact subsets of $[0, t]$.)

Now we have to describe the product system structure. Note that Ω_t is isomorphic to $\Omega_{[s, s+t]}$ (compact subsets of $[s, s+t]$) by the shift. Using this isomorphism in the second component, $\Omega_s \times \Omega_t$ is essentially Ω_{s+t}. It is not an exact equality as there are problems if the compact subsets under consideration contain the point $\{s\}$. Tsirelson argues that this can be ignored as the probability that Brownian motion hits 0 at s is 0 for any fixed s. This way, $(\Omega_{s+t}, \mathcal{F}_{s+t})$ is 'essentially' the product space. It would have been easy if P_{s+t} is also the product measure. Tsirelson notices that the measure P_{s+t} is equivalent to the product measure, in the sense that they have same zero-sets and furthermore the L^2-space of a measure depends only on the measure type, that is, L^2 spaces of equivalent measures are naturally isomorphic. The isomorphism is as follows: Suppose that $(\Omega, \mathcal{F})$ is a measurable space and μ, ν are two equivalent measures on it. Then $U : L^2(\Omega, \mathcal{F}, \mu) \to L^2(\Omega, \mathcal{F}, \nu)$ defined by

$$Uf = f\sqrt{\frac{d\mu}{d\nu}}$$

where $\frac{d\mu}{d\nu}$ is the Radon-Nikodym derivative, is a unitary. (We leave it to you to ponder the sense in which it is natural).

We don't need all the nice properties of Brownian moton for this to work. Similar constructions are possible with more general Markov processes. V. Liebscher [Lie] shows that all we need is a 'stationary factorizing measure type' on $(\Omega_1, \mathcal{F}_1)$, In the converse direction he shows that given a product sub-system of a product system one can construct 'random sets' or stationary factorizing measure types.

Tsirelson [Tsi] has also found several type III examples. The construction here is quite different from the type II case. First we see that Fock space examples come from the facts that:

$$\Gamma(\mathcal{H} \oplus \mathcal{K}) \equiv \Gamma(\mathcal{H}) \otimes \Gamma(\mathcal{K}),$$

$$L^2[0, s) \oplus L^2[s, s+t) = L^2[0, s+t).$$

In other words taking Fock space is a kind of exponentiation. It takes direct sums to tensor products. So a sum system on exponentiation gives a product system. Tsirelson's idea is to replace this kind of exact sum systems by almost sum systems, or quasi-sum systems.

A Hilbert space G is a quasi-direct sum of two subspaces $\mathcal{G}_1, \mathcal{G}_2$ if there exists a linear map $A : \mathcal{G}_1 \oplus \mathcal{G}_2 \to \mathcal{G}$ such that $A(\mathcal{G}_1 \oplus 0) = \mathcal{G}_1$, $A(0 \oplus \mathcal{G}_2) = \mathcal{G}_2$, A is 1-1, onto with bounded inverse and $I - (A^*A)^{\frac{1}{2}}$ is Hilbert-Schmidt.

Now the notion of quasi-sum systems should be apparent. Surprisingly one needs sum systems of real Hilbert spaces here to build product systems. It is not clear as to whether one can also construct type II product systems by this procedure. In [BSr] it is shown that under some assumptions only type I and type III systems arise this way.

Finally R.T. Powers has recently constructed several type II examples by dilating quantum dynamical semigroups with unbounded generators. It is not yet clear whether they are different from Tsirelson's examples.

II. Product systems of Hilbert C^*-modules.

This is a different approach to studying dilations of quantum dynamical semigroups on C^*-algebras and leads to a lot of interesting mathematics. Hilbert C^*-modules generalize the notion of Hilbert space, where now the inner product takes values in a C^*-algebra. The book of E. C. Lance [La] is a basic reference for the subject.

Suppose that $\mathcal{A}$ and $\mathcal{B}$ are unital C^*-algebras and $\tau : \mathcal{A} \to \mathcal{B}$ is a unital completely positive map. Then there exists a Hilbert $\mathcal{A} - \mathcal{B}$ module $\mathcal{E}$ (inner products are taking values in $\mathcal{B}$ and there is a left action of $\mathcal{A}$ on $\mathcal{E}$), with a unit vector ξ in $\mathcal{E}$, such that $\tau(a) = \langle \xi, a\xi \rangle$ for all $a \in \mathcal{A}$. This generalization of the GNS construction from states to unital completely positive maps was proved by Paschke in [Pa]. Unlike Stinespring's theorem, the construction here has functorial properties for compositions of completely positive maps. Making use of this and some inductive limit arguments in [BS] we show the following: Given a unital quantum dynamical semigroup $\{\tau_t\}$ on a unital C^*-algebra $\mathcal{B}$, there exists a product system $\{\mathcal{E}_t\}$ of Hilbert C^*-modules over $\mathcal{B}$, with a unital unit $\{\xi_t\}$ such that $\tau_t(b) = \langle \xi_t, b\xi_t \rangle$ for all $b \in \mathcal{B}$. In a sense this is a dilation of the quantum dynamical semigroup. Another inductive limit argument is required to obtain a semigroup of $*$-endomorphisms dilation.

Now it becomes entirely natural to try and understand product systems of Hilbert C^*-modules. How should we classify them? Well, we do not yet have a good answer to this, but what is clear is that there is a class of product systems which can be called as type I, they are the so-called time-ordered Fock modules (or exponential product systems). Their units, positive cocycles, unitary cocycles and so on, can be determined and have structures exactly analogous to our familiar Fock space product systems of Hilbert spaces. Unital quantum dynamical semigroups with bounded generators give rise to these product systems [BBLS]. Skeide [Sk] has a comprehensive treatment of these connections between Hilbert C^*-modules and dilation theory. Recently Muhly and Solel found a sort of dual approach [MS] where one gets product systems of modules on the commutant $\mathcal{B}'$. The connections between these dilations and quantum stochastic calculus are currently being explored [BL].

Exercises

1. State and prove a natural generalization of the Sz. Nagy dilation to one parameter semigroups of contractions.

2. Suppose that $(A_1, A_2, \ldots, A_n)$ is a n-tuple of positive operators on a Hilbert space $\mathcal{H}$ such that $A_1 + A_2 + \cdots + A_n = I$. Show that there is a Hilbert space $\mathcal{K}$ containing $\mathcal{H}$ with an n-tuple $(P_1, P_2, \ldots, P_n)$ of mutually orthogonal projections such that A_i is the compression of P_i to $\mathcal{H}$. (Hint: Think of unital completely positive maps on $\mathbb{C}^n$.) A generalization of this result which shows that positive-operator-valued measures 'dilate' to projection-valued measures is known as Naimarks theorem.

3. Suppose tht $\tau : \mathcal{B}(\mathcal{H}) \to \mathcal{B}(\mathcal{H})$ is a completely positive map of the form $\tau(X) = LXL^*$ for some $L \in \mathcal{B}(\mathcal{H})$, and $\beta : \mathcal{B}(\mathcal{H}) \to \mathcal{B}(\mathcal{H})$ is a completely positive map dominated by τ. Show that $\beta(X) = aLXL^*$ for some $0 \le a \le 1$.

4. Suppose that θ is an E_0-semigroup on $\mathcal{B}(\mathcal{H})$ and that P is the orthogonal projection of $\mathcal{H}$ onto a subspace $\mathcal{H}_0$. Show that the compression τ of θ by P is a unital quantum dynamical semigroup of $\mathcal{B}(\mathcal{H}_0)$ if and only if $\theta_t(P) \ge P$ for all P.

5. Suppose that τ is a contractive quantum dynamical semigroup on a unital C^*-algebra $\mathcal{A}$. Show that $\hat{\tau}$ defined on $\mathcal{A} \oplus \mathbb{C}$, by $\hat{\tau}_t(a \oplus z) = \tau_t(a) + z(1 - \tau_t(1)) \oplus z$ is a unital quantum dynamical semigroup. Show that if τ is a semigroup of $*$-endomorphisms then so is $\hat{\tau}$. (This 'unitization' trick is quite useful in dilation theory as it often helps us to extend results from the unital case to the contractive case.)

6. Suppose that θ is an E_0-semigroup of $\mathcal{B}(\mathcal{H})$ (with $\mathcal{H}$ infinite dimensional). Show that $\widetilde{\theta}$ defined by $\widetilde{\theta}_t(X \otimes Y) = X \otimes \theta_t(Y)$ on $\mathcal{B}(\mathcal{K} \otimes \mathcal{H})$ is cocycle conjugate to θ, for any Hilbert space $\mathcal{K}$.

Main References:

- For quantum dynamical semigroups: [Da], [Pa].
- For E-semigroups and product systems: [Ar1], [Ar6], [Ar7].
- For dilations and Markov processes: [Ku1-2], [BP1-2], [Bh1], [Go].
- Domination and minimality of Hudson-Parthasarathy flows: [Bh3].
- Product systems of Hilbert modules etc: [BS], [BBLS], [Sk], [MS].

References

[AFL] Accardi, L., Frigerio, A., Lewis, J.T. : Quantum Stochastic Processes, Publ. Res. Inst. Math. Sci, **18** (1982), 97-133.

[AL]   Alicki, R., Lendi, K. : *Quantum Dynamical Semigroups and Applications*, Springer Letcure Notes in Phys. **286** (1987), Berlin.

[Ar1] Arveson, W. : *Continuous Analogues of Fock Space*, Mem. Amer. Math. Soc.,**409** (1989).

[Ar2] – : An addition formula for the index of semigroups of endomorphisms of $\mathcal{B}(\mathcal{H})$., Pacific J. Math., **137** (1989), 19-36.

[Ar3] – : The index of a quantum dynamical semigroup, J. Funct. Anal., **146**, (1997) 557-588.

[Ar4] – : On the index and dilations of completely positive semigroups, Internat. J. Math., **10** (1999), 791-823.

[Ar5] – : Continuous analogues of Fock space IV : essential states, Acta Math., **164** (1990), 265-300.

[Ar6] – : Four lectures on noncommutative dynamics, in *Advances in Quantum Dynamics*, Contemporary Mathematics, **335**, Ameircan Mathematical Society (2003), 1-55.

[Ar7] – : Noncommutative Dynamics and E-Semigroups, Springer Monographs in Mathematics, Springer-Verlag (2003).

[Be]   Belavkin, V. P.: A reconstruction theorem for a quantum random process, (English translation) Theoret. and Math. Phys. **62**, no. 3, (1985) 275-289.

[BBLS] Barreto, S.D., Bhat, B.V. Rajarama, Liebscher V., Skeide M.: Type I product systems of Hilbert modules, J. Funct. Anal., **212** (2004), 121-181.

[Bh1] Bhat, B. V. Rajarama : An index theory for quantum dynamical semigroups, Trans. of Amer. Math. Soc., **348** (1996) 561-583.

[Bh2] – : Minimal dilations of quantum dynamical semigroups to semigroups of endomorphisms of C^*-algebras, J. Ramanujan Math. Soc., **14** (1999) 109-124.

[Bh3] – : *Cocycles of CCR flows,* Memoirs of the American Mathematical Society, **149**, no. 709 (2001),

[BL]   Bhat, B.V.Rajarama, Lindsay, J.M.: Regular quantum stochastic cocycles have exponential product systems, in *Quantum Probabiloity and Infinite Dimensional Analysis*, Proceedings, Greifswald, World Scientific (to appear).

[BP1] Bhat, B.V.Rajarama, Parthasarathy, K.R. : Markov dilations of nonconservative dynamical semigroups and a quantum boundary theory, Ann. Inst. Henri Poincaré, Probabilités et Statistiques, **31**, no. 4 (1995) 601-651.

[BP2] – : Kolmogorov's existence theorem for Markov processes in C^*-algebras, Proc. Ind. Acad. Sci. Math. Sci., **103** (1994), 253-262.

[BS]   Bhat, B. V. Rajarama, Skeide, M. : Tensor product systems of Hilbert modules and dilations of completely positive semigroups, Infin. Dimens. Anal. Quantum Probab. Relat. Top., Vol. **3**, Number 4, 519-575(2000).

[BSr] Bhat, B. V. Rajarama, Srinivasan, R. : On product systems arising from sum system, preprint (2003).

[CE]   Christensen, E. and Evans, D., Cohomology of operator algebras and quantum dynamical semigroups, J. London Math. Soc. **20** (1979), 358-368.

[Da]   Davies, E.B. : *Quantum Theory of Open Systems,* Acad. Press (1976) New York.

[EL]   Evans, D.E., Lewis, J.T. : *Dilations of Irreducible Evolutions in Algebraic Quantum Theory*, Comm. Dublin Inst. Adv. Stud., Ser. A no.**24** (1977).

[Em]   Emch, G. G.: Minimal dilations of CP flows, C^* algebras and applications to physics, Springer Lecture Notes in Math. **650** (1978), 156-159.

[GKS]   Gorini, V., Kossakowski, A., Sudarshan, E.C.G. : Completely positive dynamical semigroups of n-level systems, J. Math. Phys. **17** (1976), 821-825.

[Go]   Gohm, R.: *Noncommutative Stationary Processes*, Lecture Notes in Math. **1839**, Springer-Verlag, Berlin, 2004.

[Gu]   Guichardet, A.: *Symmetric Hilbert spaces and related topics*, Springer Lecture Notes in Math. **261** (1972) Berlin.

[HP]   Hudson, R.L., Parthasarathy, K.R. : Quantum Ito's formula and stochastic evolutions, Commun. Math. Phys. **93**, (1984) 301-323.

[Ku1]   Kümmerer, B. : Markov dilations on W^*-algebras, J. Funct. Anal. **63**,(1985) 139-177.

[Ku2]   – : Survey on a theory of non-commutative stationary Markov processes, Quantum Prob. and Appl.-III, Springer Lecture Notes in Math. **1303**, (1987) 154-182.

[La]   Lance, E. C. : Hilbert C^*-modules, Cambridge University Press, 1995.

[Li]   Lindsay, J. M. : Quantum stochastic analysis – an introduction, this volume.

[Lie]   Liebscher, V. : Random sets and invariants for (Type II) continuous tensor product systems of Hilbert spaces, preprint 2002.

[Lin]   Lindblad, G. : On the generators of quantum dynamical semigroups, Comm. Math. Phys. **48** (1976), 119-130.

[MS]   Muhly, P.S., Solel, B. : Quantum Markov processes (correspondences and dilations), Internat. J. Math. **13** (2002), 863-906.

[Pa]   Parthasarathy, K.R. : *An Introduction to Quantum Stochastic Calculus*, Monographs in Math., Birkhäuser Verlag (1991) Basil.

[Po]   Popescu, G.: Isometric dilations for infinite sequences of noncommuting operators, Trans. Amer. Math. Soc., 316(1989), 523-536.

[Po1]   Powers, R.T. : An index theory for semigroups of endomorphisms of $\mathcal{B}(\mathcal{H})$ and type II_1 factors, Canad. J. Math., **40** (1988), 86-114.

[Po2]   – : A non-spatial continuous semigroup of *-endomorphisms of $\mathcal{B}(\mathcal{H})$, Publ. Res. Inst. Math. Sci. **23** (1987), 1053-1069.

[Po3]   – : New examples of continuous spatial semigroups of endomorphisms of $\mathcal{B}(\mathcal{H})$, Internat. J. Math. **10** (1999), 215-288.

[Po4]   – : Possible classification of continuous spatial semigroups of *- endomorphisms of $\mathcal{B}(\mathcal{H})$, Proceedings of Symposia in Pure Math., Amer. Math. Soc., vol. **59** (1996) 161-173

[Sa]   Sauvageot, J-L.: Markov quantum semigroups admit covariant Markov C^* dilations, Comm. Math. Phys. **106** (1986), 91-103.

[Sk]   Skeide, M.: Hilbert modules and applications in quantum probability, Habilitationsschrift, Cottbus, 2001.

[St]   Stinespring, W.F. : Positive functions on C^* algebras,. Proc. Amer. Math Soc. **6** (1955), 211-216.

[SzF]   Sz.-Nagy, B., Foias, C. : *Harmonic Analysis of Operators on Hilbert Space*, North-Holland (1970) Amsterdam.

[Tsi]   Tsirelson, B.: Non-isomorphic product systems, preprint 2002 (arXiv: math.FA/0210457 v1 30 Oct 2002).

Index

Printing and Binding: Strauss GmbH, Mörlenbach

Vol. 1780: J. H. Bruinier, Borcherds Products on O(2,1) and Chern Classes of Heegner Divisors (2002)

Vol. 1781: E. Bolthausen, E. Perkins, A. van der Vaart, Lectures on Probability Theory and Statistics. Ecole d' Eté de Probabilités de Saint-Flour XXIX-1999. Editor: P. Bernard (2002)

Vol. 1782: C.-H. Chu, A. T.-M. Lau, Harmonic Functions on Groups and Fourier Algebras (2002)

Vol. 1783: L. Grüne, Asymptotic Behavior of Dynamical and Control Systems under Perturbation and Discretization (2002)

Vol. 1784: L.H. Eliasson, S. B. Kuksin, S. Marmi, J.-C. Yoccoz, Dynamical Systems and Small Divisors. Cetraro, Italy 1998. Editors: S. Marmi, J.-C. Yoccoz (2002)

Vol. 1785: J. Arias de Reyna, Pointwise Convergence of Fourier Series (2002)

Vol. 1786: S. D. Cutkosky, Monomialization of Morphisms from 3-Folds to Surfaces (2002)

Vol. 1787: S. Caenepeel, G. Militaru, S. Zhu, Frobenius and Separable Functors for Generalized Module Categories and Nonlinear Equations (2002)

Vol. 1788: A. Vasil'ev, Moduli of Families of Curves for Conformal and Quasiconformal Mappings (2002)

Vol. 1789: Y. Sommerhäuser, Yetter-Drinfel'd Hopf algebras over groups of prime order (2002)

Vol. 1790: X. Zhan, Matrix Inequalities (2002)

Vol. 1791: M. Knebusch, D. Zhang, Manis Valuations and Prüfer Extensions I: A new Chapter in Commutative Algebra (2002)

Vol. 1792: D. D. Ang, R. Gorenflo, V. K. Le, D. D. Trong, Moment Theory and Some Inverse Problems in Potential Theory and Heat Conduction (2002)

Vol. 1793: J. Cortés Monforte, Geometric, Control and Numerical Aspects of Nonholonomic Systems (2002)

Vol. 1794: N. Pytheas Fogg, Substitution in Dynamics, Arithmetics and Combinatorics. Editors: V. Berthé, S. Ferenczi, C. Mauduit, A. Siegel (2002)

Vol. 1795: H. Li, Filtered-Graded Transfer in Using Noncommutative Gröbner Bases (2002)

Vol. 1796: J.M. Melenk, hp-Finite Element Methods for Singular Perturbations (2002)

Vol. 1797: B. Schmidt, Characters and Cyclotomic Fields in Finite Geometry (2002)

Vol. 1798: W.M. Oliva, Geometric Mechanics (2002)

Vol. 1799: H. Pajot, Analytic Capacity, Rectifiability, Menger Curvature and the Cauchy Integral (2002)

Vol. 1800: O. Gabber, L. Ramero, Almost Ring Theory (2003)

Vol. 1801: J. Azéma, M. Émery, M. Ledoux, M. Yor (Eds.), Séminaire de Probabilités XXXVI (2003)

Vol. 1802: V. Capasso, E. Merzbach, B.G. Ivanoff, M. Dozzi, R. Dalang, T. Mountford, Topics in Spatial Stochastic Processes. Martina Franca, Italy 2001. Editor: E. Merzbach (2003)

Vol. 1803: G. Dolzmann, Variational Methods for Crystalline Microstructure - Analysis and Computation (2003)

Vol. 1804: I. Cherednik, Ya. Markov, R. Howe, G. Lusztig, Iwahori-Hecke Algebras and their Representation Theory. Martina Franca, Italy 1999. Editors: V. Baldoni, D. Barbasch (2003)

Vol. 1805: F. Cao, Geometric Curve Evolution and Image Processing (2003)

Vol. 1806: H. Broer, I. Hoveijn. G. Lunther, G. Vegter, Bifurcations in Hamiltonian Systems. Computing Singularities by Gröbner Bases (2003)

Vol. 1807: V. D. Milman, G. Schechtman (Eds.), Geometric Aspects of Functional Analysis. Israel Seminar 2000-2002 (2003)

Vol. 1808: W. Schindler, Measures with Symmetry Properties (2003)

Vol. 1809: O. Steinbach, Stability Estimates for Hybrid Coupled Domain Decomposition Methods (2003)

Vol. 1810: J. Wengenroth, Derived Functors in Functional Analysis (2003)

Vol. 1811: J. Stevens, Deformations of Singularities (2003)

Vol. 1812: L. Ambrosio, K. Deckelnick, G. Dziuk, M. Mimura, V. A. Solonnikov, H. M. Soner, Mathematical Aspects of Evolving Interfaces. Madeira, Funchal, Portugal 2000. Editors: P. Colli, J. F. Rodrigues (2003)

Vol. 1813: L. Ambrosio, L. A. Caffarelli, Y. Brenier, G. Buttazzo, C. Villani, Optimal Transportation and its Applications. Martina Franca, Italy 2001. Editors: L. A. Caffarelli, S. Salsa (2003)

Vol. 1814: P. Bank, F. Baudoin, H. Föllmer, L.C.G. Rogers, M. Soner, N. Touzi, Paris-Princeton Lectures on Mathematical Finance 2002 (2003)

Vol. 1815: A. M. Vershik (Ed.), Asymptotic Combinatorics with Applications to Mathematical Physics. St. Petersburg, Russia 2001 (2003)

Vol. 1816: S. Albeverio, W. Schachermayer, M. Talagrand, Lectures on Probability Theory and Statistics. Ecole d'Eté de Probabilités de Saint-Flour XXX-2000. Editor: P. Bernard (2003)

Vol. 1817: E. Koelink, W. Van Assche(Eds.), Orthogonal Polynomials and Special Functions. Leuven 2002 (2003)

Vol. 1818: M. Bildhauer, Convex Variational Problems with Linear, nearly Linear and/or Anisotropic Growth Conditions (2003)

Vol. 1819: D. Masser, Yu. V. Nesterenko, H. P. Schlickewei, W. M. Schmidt, M. Waldschmidt, Diophantine Approximation. Cetraro, Italy 2000. Editors: F. Amoroso, U. Zannier (2003)

Vol. 1820: F. Hiai, H. Kosaki, Means of Hilbert Space Operators (2003)

Vol. 1821: S. Teufel, Adiabatic Perturbation Theory in Quantum Dynamics (2003)

Vol. 1822: S.-N. Chow, R. Conti, R. Johnson, J. Mallet-Paret, R. Nussbaum, Dynamical Systems. Cetraro, Italy 2000. Editors: J. W. Macki, P. Zecca (2003)

Vol. 1823: A. M. Anile, W. Allegretto, C. Ringhofer, Mathematical Problems in Semiconductor Physics. Cetraro, Italy 1998. Editor: A. M. Anile (2003)

Vol. 1824: J. A. Navarro González, J. B. Sancho de Salas, C^∞ - Differentiable Spaces (2003)

Vol. 1825: J. H. Bramble, A. Cohen, W. Dahmen, Multiscale Problems and Methods in Numerical Simulations, Martina Franca, Italy 2001. Editor: C. Canuto (2003)

Vol. 1826: K. Dohmen, Improved Bonferroni Inequalities via Abstract Tubes. Inequalities and Identities of Inclusion-Exclusion Type. VIII, 113 p, 2003.

Vol. 1827: K. M. Pilgrim, Combinations of Complex Dynamical Systems. IX, 118 p, 2003.

Vol. 1828: D. J. Green, Gröbner Bases and the Computation of Group Cohomology. XII, 138 p, 2003.

Vol. 1829: E. Altman, B. Gaujal, A. Hordijk, Discrete-Event Control of Stochastic Networks: Multimodularity and Regularity. XIV, 313 p, 2003.

Vol. 1830: M. I. Gil', Operator Functions and Localization of Spectra. XIV, 256 p, 2003.

Vol. 1831: A. Connes, J. Cuntz, E. Guentner, N. Higson, J. E. Kaminker, Noncommutative Geometry, Martina Franca, Italy 2002. Editors: S. Doplicher, L. Longo (2004)

Vol. 1832: J. Azéma, M. Émery, M. Ledoux, M. Yor (Eds.), Séminaire de Probabilités XXXVII (2003)

Vol. 1833: D.-Q. Jiang, M. Qian, M.-P. Qian, Mathematical Theory of Nonequilibrium Steady States. On the Frontier of Probability and Dynamical Systems. IX, 280 p, 2004.

Vol. 1834: Yo. Yomdin, G. Comte, Tame Geometry with Application in Smooth Analysis. VIII, 186 p, 2004.

Vol. 1835: O.T. Izhboldin, B. Kahn, N.A. Karpenko, A. Vishik, Geometric Methods in the Algebraic Theory of Quadratic Forms. Summer School, Lens, 2000. Editor: J.-P. Tignol (2004)

Vol. 1836: C. Năstăsescu, F. Van Oystaeyen, Methods of Graded Rings. XIII, 304 p, 2004.

Vol. 1837: S. Tavaré, O. Zeitouni, Lectures on Probability Theory and Statistics. Ecole d'Eté de Probabilités de Saint-Flour XXXI-2001. Editor: J. Picard (2004)

Vol. 1838: A.J. Ganesh, N.W. O'Connell, D.J. Wischik, Big Queues. XII, 254 p, 2004.

Vol. 1839: R. Gohm, Noncommutative Stationary Processes. VIII, 170 p, 2004.

Vol. 1840: B. Tsirelson, W. Werner, Lectures on Probability Theory and Statistics. Ecole d'Eté de Probabilités de Saint-Flour XXXII-2002. Editor: J. Picard (2004)

Vol. 1841: W. Reichel, Uniqueness Theorems for Variational Problems by the Method of Transformation Groups (2004)

Vol. 1842: T. Johnsen, A.L. Knutsen, K3 Projective Models in Scrolls (2004)

Vol. 1843: B. Jefferies, Spectral Properties of Noncommuting Operators (2004)

Vol. 1844: K.F. Siburg, The Principle of Least Action in Geometry and Dynamics (2004)

Vol. 1845: Min Ho Lee, Mixed Automorphic Forms, Torus Bundles, and Jacobi Forms (2004)

Vol. 1846: H. Ammari, H. Kang, Reconstruction of Small Inhomogeneities from Boundary Measurements (2004)

Vol. 1847: T.R. Bielecki, T. Björk, M. Jeanblanc, M. Rutkowski, J.A. Scheinkman, W. Xiong, Paris-Princeton Lectures on Mathematical Finance 2003 (2004)

Vol. 1848: M. Abate, J. E. Fornaess, X. Huang, J. P. Rosay, A. Tumanov, Real Methods in Complex and CR Geometry, Martina Franca, Italy 2002. Editors: D. Zaitsev, G. Zampieri (2004)

Vol. 1849: Martin L. Brown, Heegner Modules and Elliptic Curves (2004)

Vol. 1850: V. D. Milman, G. Schechtman (Eds.), Geometric Aspects of Functional Analysis. Israel Seminar 2002-2003 (2004)

Vol. 1851: O. Catoni, Statistical Learning Theory and Stochastic Optimization (2004)

Vol. 1852: A.S. Kechris, B.D. Miller, Topics in Orbit Equivalence (2004)

Vol. 1853: Ch. Favre, M. Jonsson, The Valuative Tree (2004)

Vol. 1854: O. Saeki, Topology of Singular Fibers of Differential Maps (2004)

Vol. 1855: G. Da Prato, P.C. Kunstmann, I. Lasiecka, A. Lunardi, R. Schnaubelt, L. Weis, Functional Analytic Methods for Evolution Equations. Editors: M. Iannelli, R. Nagel, S. Piazzera (2004)

Vol. 1856: K. Back, T.R. Bielecki, C. Hipp, S. Peng, W. Schachermayer, Stochastic Methods in Finance, Bressanone/Brixen, Italy, 2003. Editors: M. Fritelli, W. Runggaldier (2004)

Vol. 1857: M. Émery, M. Ledoux, M. Yor (Eds.), Séminaire de Probabilités XXXVIII (2005)

Vol. 1858: A.S. Cherny, H.-J. Engelbert, Singular Stochastic Differential Equations (2005)

Vol. 1859: E. Letellier, Fourier Transforms of Invariant Functions on Finite Reductive Lie Algebras (2005)

Vol. 1860: A. Borisyuk, G.B. Ermentrout, A. Friedman, D. Terman, Tutorials in Mathematical Biosciences I. Mathematical Neurosciences (2005)

Vol. 1861: G. Benettin, J. Henrard, S. Kuksin, Hamiltonian Dynamics - Theory and Applications, Cetraro, Italy, 1999. Editor: A. Giorgilli (2005)

Vol. 1862: B. Helffer, F. Nier, Hypoelliptic Estimates and Spectral Theory for Fokker-Planck Operators and Witten Laplacians (2005)

Vol. 1863: H. Fürh, Abstract Harmonic Analysis of Continuous Wavelet Transforms (2005)

Vol. 1864: K. Efstathiou, Metamorphoses of Hamiltonian Systems with Symmetries (2005)

Vol. 1865: D. Applebaum, B.V. R. Bhat, J. Kustermans, J. M. Lindsay, Quantum Independent Increment Processes I. From Classical Probability to Quantum Stochastic Calculus. Editors: M. Schürmann, U. Franz (2005)

Vol. 1866: O.E. Barndorff-Nielsen, U. Franz, R. Gohm, B. Kümmerer, S. Thorbjønsen, Quantum Independent Increment Processes II. Structure of Quantum Levy Processes, Classical Probability, and Physics. Editors: M. Schürmann, U. Franz, (2005)

Recent Reprints and New Editions

Vol. 1200: V. D. Milman, G. Schechtman (Eds.), Asymptotic Theory of Finite Dimensional Normed Spaces. 1986. - Corrected Second Printing (2001)

Vol. 1471: M. Courtieu, A.A. Panchishkin, Non-Archimedean L-Functions and Arithmetical Siegel Modular Forms. - Second Edition (2003)

Vol. 1618: G. Pisier, Similarity Problems and Completely Bounded Maps. 1995 - Second, Expanded Edition (2001)

Vol. 1629: J.D. Moore, Lectures on Seiberg-Witten Invariants. 1997 - Second Edition (2001)

Vol. 1638: P. Vanhaecke, Integrable Systems in the realm of Algebraic Geometry. 1996 - Second Edition (2001)

Vol. 1702: J. Ma, J. Yong, Forward-Backward Stochastic Differential Equations and Their Applications. 1999. - Corrected Second Printing (2000)

4. For evaluation purposes, manuscripts may be submitted in print or electronic form (print form is still preferred by most referees), in the latter case preferably as pdf- or zipped ps-files. Lecture Notes volumes are, as a rule, printed digitally from the authors' files. To ensure best results, authors are asked to use the LaTeX2e style files available from Springer's web-pages at:

ftp://ftp.springer.de/pub/tex/latex/mathegl/mono/ (for monographs) and
ftp://ftp.springer.de/pub/tex/latex/mathegl/mult/ (for summer schools/tutorials).

Style files for other TeX-versions, and additional technical instructions, if necessary, are available on request from lnm@springer-sbm.com.

Careful preparation of the manuscripts will help keep production time short besides ensuring satisfactory appearance of the finished book in print and online. After acceptance of the manuscript authors will be asked to prepare the final LaTeX source files (and also the corresponding dvi-, pdf- or zipped ps-files) together with the final printout made from these files. The LaTeX source files are essential for producing the full-text online version of the book

(http://www.springerlink.com/openurl.asp?genre=journal&issn=0075-8434).

The actual production of a Lecture Notes volume takes approximately 8 weeks.

5. Authors receive a total of 50 free copies of their volume, but no royalties. They are entitled to a discount of 33.3 % on the price of Springer books purchased for their personal use, if ordering directly from Springer.

6. Commitment to publish is made by letter of intent rather than by signing a formal contract. Springer-Verlag secures the copyright for each volume. Authors are free to reuse material contained in their LNM volumes in later publications: A brief written (or e-mail) request for formal permission is sufficient.

Addresses:

Professor J.-M. Morel, CMLA,
École Normale Supérieure de Cachan,
61 Avenue du Président Wilson, 94235 Cachan Cedex, France
E-mail: Jean-Michel.Morel@cmla.ens-cachan.fr

Professor F. Takens, Mathematisch Instituut,
Rijksuniversiteit Groningen, Postbus 800,
9700 AV Groningen, The Netherlands
E-mail: F.Takens@math.rug.nl

Professor B. Teissier, Institut Mathématique de Jussieu,
UMR 7586 du CNRS, Équipe "Géométrie et Dynamique",
175 rue du Chevaleret
75013 Paris, France
E-mail: teissier@math.jussieu.fr

Springer-Verlag, Mathematics Editorial I, Tiergartenstr. 17,
69121 Heidelberg, Germany,
Tel.: +49 (6221) 487-8410
Fax: +49 (6221) 487-8355
E-mail: lnm@springer-sbm.com